International Conference on Communication, Image and Signal Processing (CCISP 2016)

Journal of Physics: Conference Series Volume 787

Dubai, United Arab Emirates
18 – 20 November 2016

Editors:

Yucong Duan
Raj Jain

ISBN: 978-1-5108-3685-3
ISSN: 1742-6588

Printed from e-media with permission by:

Curran Associates, Inc.
57 Morehouse Lane
Red Hook, NY 12571

CURRAN ASSOCIATES INC.
proceedings
.com

Some format issues inherent in the e-media version may also appear in this print version.

Printed by Curran Associates, Inc. (2017)

For permission requests, please contact the Institute of Physics
at the address below.

Institute of Physics
Dirac House, Temple Back
Bristol BS1 6BE UK

Phone: 44 1 17 929 7481
Fax: 44 1 17 920 0979

techtracking@iop.org

Additional copies of this publication are available from:

Curran Associates, Inc.
57 Morehouse Lane
Red Hook, NY 12571 USA
Phone: 845-758-0400
Fax: 845-758-2633
Email: curran@proceedings.com
Web: www.proceedings.com

Table of contents

Volume 787

2016 International Conference on Communication, Image and Signal Processing (CCISP 2016)

18–20 November 2016, Dubai, UAE

Accepted papers received: 7 December 2016
Published online: 8 February 2017

Preface

011001
OPEN ACCESS
2016 International Conference on Communication, Image and Signal Processing (CCISP 2016)

011002
OPEN ACCESS
Committees

011003
OPEN ACCESS
Peer review statement

Visualization

012001
OPEN ACCESS
Interactive Visualization of DGA Data Based on Multiple Views

Yujie Geng, Ying Lin, Yan Ma, Zhihong Guo, Chao Gu and Mingtao Wang.....1

012002
OPEN ACCESS
NEB in Analysis of Optical Flow 4 x 4 and 6 x 6-Patches

Shengxiang Xia and Di Liang.....6

012003
OPEN ACCESS
Congestion analysis of pilgrims in Hajj and Umrah congregation using block matching and optical flow

Sumaiyya Farooq, Shoab A Khan and M Usman Akram.....12

012004
OPEN ACCESS
Approaching to Genuine Monochromatic Light Filters Employing A Serially Coupled Triple Ring Resonator System

Phichai Youplao.....17

012005
OPEN ACCESS
A Three-dimensional Topological Model of Ternary Phase Diagram

Yingxue Mu and Hong Bao.....23

012006
OPEN ACCESS
HTTM - Design and Implementation of a Type-2 Hypervisor for MIPS64 Based Systems

Qurrat ul Ain, Usama Anwar, Muhammad Amir Mehmood and Abdul Waheed.....29

012007
OPEN ACCESS
Golden Rectangle Treemap

Liangfu Lv, Shiliang Fan, Maolin Huang, Weidong Huang and Guolan Yang.....34

Image Processing

012008
OPEN ACCESS
Bayesian Framework with Non-local and Low-rank Constraint for Image Reconstruction

Zhonghe Tang, Shengzhe Wang, Jianliang Huo, Hang Guo, Haibo Zhao and Yuan Mei.....40

012009
OPEN ACCESS
Electrical Inspection Oriented Thermal Image Quality Assessment

Ying Lin, Menglin Wang, Xiaojin Gong, Zhihong Guo, Yujie Geng and Demeng Bai.....52

012010
OPEN ACCESS
The segmentation of Thangka damaged regions based on the local distinction

Bi Xuehui, Liu Huaming, Wang Xiuyou, Wang Weilan and Yang Yashuai.....58

012011
OPEN ACCESS
A multiple classifier system based on Ant-Colony Optimization for Hyperspectral image classification

Ke Tang, Li Xie and Guangyao Li.....66

012012
OPEN ACCESS
A Novel Method for Edge Detection in Images Based on Particle Swarm Optimization

C Baby Sherin and L. Mredhula.....75

012013
OPEN ACCESS
Micro-Expression Recognition based on 2D Gabor Filter and Sparse Representation

Hao Zheng.....80

012014
OPEN ACCESS
New Denoising Method Based on Empirical Mode Decomposition and Improved Thresholding Function

Wahiba Mohguen and RaïsEl'hadiBekka.....86

012015
OPEN ACCESS
Identity Recognition Algorithm Using Improved Gabor Feature Selection of Gait Energy Image

LIANG Chao, JIA Ling-yao and SHI Dong-cheng.....92

012016
OPEN ACCESS
Normal Vector Projection Method used for Convex Optimization of Chan-Vese Model for Image Segmentation

W B Wei, L Tan, M Q Jia and Z K Pan.....98

012017
OPEN ACCESS
Video Stabilization Using Feature Point Matching

Shamsundar Kulkarni, D.S Bormane and S.L Nalbalwar.....107

012018
OPEN ACCESS
GPU accelerated Foreign Object Debris Detection on Airfield Pavement with visual saliency algorithm

Jun Qi, Guoping Gong and Xiaoguang Cao.....113

012019
OPEN ACCESS
Adaptive classifier for steel strip surface defects

Mingming Jiang, Guangyao Li, Li Xie, Mang Xiao and Li Yi.....119

012020
OPEN ACCESS
Weighted Discriminative Dictionary Learning based on Low-rank Representation

Heyou Chang and Hao Zheng.....129

012021
OPEN ACCESS
Morphology operators construction by adaptive elliptical structuring elements based on nonlinear structure tensor

Chunming Tang, Xinlei Liu, Yanjie Li and Hongbo Zhao.....135

012022
OPEN ACCESS
Architecture and Key Techniques of Augmented Reality Maintenance Guiding System for Civil Aircrafts

Zhou hong and Lu Wenhua.....143

012023
OPEN ACCESS
Counterfeit Electronics Detection Using Image Processing and Machine Learning

Navid Asadizanjani, Mark Tehranipoor and Domenic Forte.....150

Wireless

012024
OPEN ACCESS
Noncoherent detection of DQPSK in OFDM systems using predictive VA

Vineel K Veludandi and K Vasudevan.....156

012025
OPEN ACCESS
Adaptive Beamforming with Inadequate Snapshots

Jing YU and Yaan LI.....161

012026
OPEN ACCESS
DOA estimation based on multiple beamspace measurements sparse reconstruction for manoeuvring towed array

J Yuan, H Xiao, Z M Cai and C Xi.....166

Computing

012027
OPEN ACCESS
Fast Simulation Method for Ocean Wave Base on Ocean Wave Spectrum and Improved Gerstner Model with GPU

Wenqiao Zhang, Jing Zhang and Tianchi Zhang.....172

012028
OPEN ACCESS
Doppler Feature Based Classification of Wind Profiler Data

Swati Sinha, T.V. Chandrasekhar Sarma and Mary Lourde. R.....182

012029
OPEN ACCESS
Rearrangement and Grouping of Data Bits for Efficient Lossless Encoding

Ajitha Shenoy K B, Meghana Ajith and Vinayak M Mantoor.....189

012030
OPEN ACCESS
Alternative Line Coding Scheme with Fixed Dimming for Visible Light Communication

M T Niaz, F Imdad and H S Kim.....194

012031
OPEN ACCESS
Time Granularity Transformation of Time Series Data for Failure Prediction of Overhead Line

Yan Ma, Wenbing Zhu, Jinxia Yao, Chao Gu, Demeng Bai and Kun Wang.....200

012032
OPEN ACCESS
Optimizing System Call Latency of ARM Virtual Machines

Song-Woo Sok, Young-Woo Jung and Cheol-Hun Lee.....205

Image Processing

012033
OPEN ACCESS
Detection of Pigment Networks in Dermoscopy Images

Khalid Eltayef, Yongmin Li and Xiaohui Liu.....210

012034
OPEN ACCESS
Detection of Melanoma Skin Cancer in Dermoscopy Images

Khalid Eltayef, Yongmin Li and Xiaohui Liu.....217

PREFACE

CCISP 2016 was held in Dubai, UAE during November 18-20, 2016. CCISP 2016 was organized by Asia Pacific Institute of Science and Engineering, technical sponsored by HKSME, supported by other organizations. The conference provided a useful and wide platform both for displaying the latest research and for the exchanging of research results and thoughts in broad Communication, Image and Signal Processing. The participants of the conference were from different parts of the world, with backgrounds of either academia or industry, even well-known enterprise. The success and prosperity of the conference is reflected in the high level of the papers received.

The proceedings are a compilation of the accepted papers and represent an interesting outcome of the conference. This book cover 4 chapters: Visualization; Image Processing; Wireless; Computing.

We would like to acknowledge all of those who supported CCISP 2016. Each individual's and institution's help were very important for the success of this conference. Especially we would like to thank the organizing committee for their valuable advices in the organization and helpful peer review of the papers.

We sincerely hope that CCISP 2016 will be a forum for excellent discussions that will put forward new ideas and promote collaborative researches. We are sure that the proceedings will serve as an important research source of references and the knowledge, which will lead to not only scientific and engineering progress but also other new products and processes.

Prof. Yucong Duan, Hainan University, China
Prof. Raj Jain, Washington University in St.Louis, USA

COMMITTEES

Conference Committee Co-Chairs

Prof. Yucong Duan, Hainan University, China

Prof. David Zhang, Hong Kong Polytechnic University, Hong Kong

Dr. Yaser Qudaih, Kyushu Institute of Technology, Japan

Program Commitee Chairs

Prof. Hussain Al-Ahmad, Khalifa University, The United Arab Emirates

Prof. Albert C. S. Chung, Hong Kong University of Science and Technology, Hong Kong

Publication Chairs

Prof. Majid Mirmehdi, University of Bristol, UK

Dr. Navid Asadi, University of Florida, USA

International Technical Committee

Prof. Kasturi Vasudevan, Indian Institute of Technology Kanpur,India (chair)

Prof. Nizar Bouguila, Concordia University, Canada

Dr. Fatma Taher, khalifa University,The United Arab Emirates

Dr. Ahmed Rashed Kulaib AlTunaiji, khalifa University,The United Arab Emirates

Dr. Taolue Chen, Middlesex University London, United Kingdom

Dr. Fehmida Hussain, Middlesex University Dubai, Dubai

Prof. Kumar Vaibhav Srivastava, Indian Institute of Technology, Kanpur, India

Prof. Pradip Sircar, Indian Institute of Technology, Kanpur, India

Dr. Ruksar Fatima, KBN College of Engineering, India

IOP Conf. Series: Journal of Physics: Conf. Series **787** (2017) 011002 doi:10.1088/1742-6596/787/1/011002

Dr. Haikel Salem Hichri, King Saud University, Saudi Arabia

Dr.Abdessamad Ben Hamza, Concordia University,Canada

Dr. Nursabillilah binti mohd ali, University Teknikal Malasia Melaka, Malasia

Dr. MohammadMehdi Kafashan, Washington University in St. Louis, USA

Dr. Niranjana S., Manipal Institute of Technology (MIT), India

IOP Conf. Series: Journal of Physics: Conf. Series **787** (2017) 011003 doi:10.1088/1742-6596/787/1/011003

Peer review statement

All papers published in this volume of *Journal of Physics: Conference Series* have been peer reviewed through processes administered by the proceedings Editors. Reviews were conducted by expert referees to the professional and scientific standards expected of a proceedings journal published by IOP Publishing.

Interactive Visualization of DGA Data Based on Multiple Views

Yujie Geng[1], Ying Lin[1], Yan Ma[1], Zhihong Guo[1], Chao Gu[1] and Mingtao Wang[2]

[1]State Grid Shandong Electric Power Research Institute, 2000 Wangyue Road, Jinan, Shandong, China

[2]Shandong Zhongshi Yitong Co. Ltd., Xinluo Road, Jinan, Shandong, China

E-mail: gengyujie1225@126.com

Abstract. The commission and operation of dissolved gas analysis (DGA) online monitoring makes up for the weakness of traditional DGA method. However, volume and high-dimensional DGA data brings a huge challenge for monitoring and analysis. In this paper, we present a novel interactive visualization model of DGA data based on multiple views. This model imitates multi-angle analysis by combining parallel coordinates, scatter plot matrix and data table. By offering brush, collaborative filter and focus + context technology, this model provides a convenient and flexible interactive way to analyze and understand the DGA data.

1. Introduction

Traditional dissolved gas analysis (DGA) has a long analysis period and couldn't continuously monitor equipment. Application of on-line monitoring apparatus for dissolved gases in transformer oil makes up for those weaknesses [1]. However, due to the limitation of sensor technology on-line monitoring device is not satisfactory on its reliability and sensitivity [2]. So it is difficult to make a correct judgment for condition of equipment by norms threshold or gas ratio boundaries. With the large range commission and operation of online monitoring device, online monitoring DGA data accumulates continuously. The data analysis capacity of people has lagged far behind the ability to obtain data. Volume and high-dimensional has brought a huge challenge to the Monitoring and analysis of data.

With the development of computer technology, information communication, polymer materials and chromatographic analysis, online dissolved gas analysis monitoring is becoming more and more advanced. But due to the limitation of sensor technology, on-line monitoring device is not satisfactory on its reliability and sensitivity. Besides, there are seven kinds of gas which can contribute to judge the internal fault of the transformer, so more than one gas need to be comprehensively considered in order to judge the status of the transformer correctly.

Visual analysis is an important method used in data analysis. A picture wins a thousand words. For human beings more than 80% outside information are obtained by the visual system. In this paper, we present a novel interactive visualization model of DGA data based on multiple views. This model imitates multi-angles analysis model of users by combining parallel coordinates, scatter plot matrix and data table. We organize massive data into multidimensional data by using multiple views collaborative visual analysis. Users can manage and analysis data directly from different dimension, so that information hidden in the data can be truly understand by users and reflect the status of the transformer correctly.

2. Related Work

The combination of oil and insulation paper is widely used in power transformer insulation. Transformer oil is a kind of mineral oil which is distilled and refined. Insulation oil and paper will be aging and decomposing gradually when affected by the thermal, electrical and mechanical force during transformer operating [3]. Decomposition speed and gas production rate will increase when transformer fault occurs. The decomposed gas will be dissolved in the oil and gas relay after diffusion and convection.

For the diagnosis analysis of dissolved gas, scholars puts forward many visual analysis methods. Triangle method is one of them. In 1971, Davies proposed a diagnosis method based on three elements (H, O, C) according to experiment and experience [4]. In 1974, Duval improved Davies's method and diagnose the fault of transformer by measuring the relative components of CH4, C2H4, C2H2 [5]. As shown in Figure.1, by checking the position of the measured data in the map, people can diagnose the type of fault according to the fault-location look-up table.

Duval triangle method can realize the quick judgment of equipment fault. However, it depends on the accuracy of the data heavily. Since on-line monitoring device is not satisfactory on its reliability and sensitivity, it is difficult to make an effective judgment by using Duval triangle method. At the same time, the characteristics of high dimension and multi element bring a big challenge to the traditional low dimensional visualization technology.

In addition to the challenges brought by the expansion data scale, the problem caused by high dimension is also a focus of research under the background of big data times. Multidimensional data refers to the data variables with multiple dimension attributes. The goal of multidimensional data analysis is to explore the distribution patterns of multidimensional data items, and to reveal the hidden relationships among different dimension attributes. Keim [6] summarized the basic methods of multidimensional visualization including geometric projection techniques, pixel-oriented techniques, icon-based techniques and graph-based techniques. Among them, the multidimensional visualization method based on geometric projection techniques is the main research direction in recent years.

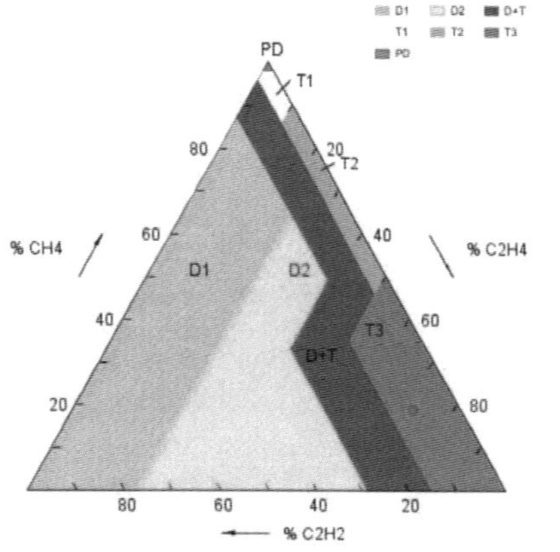

Figure 1. Duval Triangle method

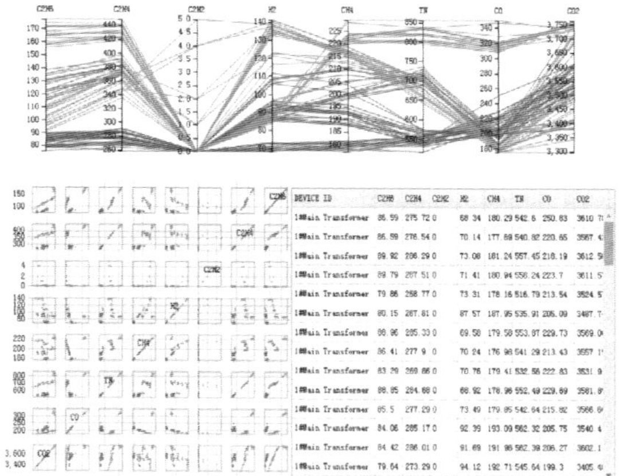

Figure 2. Overall of our visualization model

3. Visualization

There are seven kinds of gas which can contribute to judge the internal fault of the transformer. Single visual analysis technology just analyzes and displays data from a certain aspect and couldn't judge the status of the transformer correctly. So in this paper, we make use of the advantages of various visualization techniques and analyze complex multidimensional data by multiple views. According to the features of online dissolved gas data, we use parallel coordinates, scatter plot matrix and data table to display and analyze information from different aspects.

IOP Conf. Series: Journal of Physics: Conf. Series **787** (2017) 012001 doi:10.1088/1742-6596/787/1/012001

Our Multiple views visualization model contains 10 dimensions. They are device ID, monitoring time, hydrogen (H2) content, methane (CH4) content, ethane (C2H6) content, ethylene (C2H4) content, acetylene (C2H2) content, carbon monoxide (CO) content, carbon dioxide (CO2) content and total hydrocarbon content respectively. We use color to code the device ID. H2 content, CH4 content, C2H6 content, C2H4 content, C2H2 content, CO content, CO2 content and total hydrocarbon content are mapped as the eight properties of parallel coordinates. In scatter plot matrix, we constitute a 8 * 8 scatter matrix using the above 8 kinds of gases. In data table, we list all the monitor data in accordance with the order from near to far. Figure 2 shows the overall of our visualization model.

3.1 Parallel Coordinates

Parallel coordinates is an important visualization technique used to display multidimensional data and analyze its relationship. The main idea of parallel coordinates is to map the N-dimensional data attribute to two-dimensional plane by N equidistant parallel axis. Each axis represents a dimension attribute. Its range is well distributed according to the min and max value of the corresponding attribute. A set of points on a line in n-space is transformed to a polyline in parallel coordinates intersecting at all the n−1 points. Those points represent the data points of the N dimensional data and the polyline can be expressed by N-1 linear independent equations as follows:

$$\frac{x_1-a_1}{k_1}=\frac{x_2-a_2}{k_2}=\frac{x_3-a_3}{k_3}=\cdots=\frac{x_n-a_n}{n} \quad (1)$$

From formula（1）we can know:

$$\mathrm{x}_{\mathrm{i}+1}=\mathrm{m}_{\mathrm{i}}\,\mathrm{x}_{\mathrm{i}}+\mathrm{b}_{\mathrm{i}}\ ,\mathrm{i}=1,2,\ldots,\mathrm{n}-1 \quad (2)$$

Where $m_i = k_{i+1}/k_i$ is the slope and $b_i = a_{i+1} - m_i a_i$ is the intercept of flat x_i/x_{i+1} on axis x_{i+1}.

The sequence of dimension could affect people's analysis and judgment for the data in parallel coordinates visualization. Good dimension order allows people to find relevant features and abnormal data efficiently. The usual method is to calculate the similarity between each dimension attribute and rearrange axis. Arrange the biggest similarity axes together can effectively reduce disorder of the data in parallel coordinates and allow users to view the trends and characteristics clearly. However, the time complexity to calculate similarity is O(N!). With the increase of the data dimension and amount, the time and space cost to calculate similarity will increase exponentially. Besides, the relationship between two dimensions will be weaker when the two axes are far apart.

In order to solve that problem, we arrange the two feature gases which content ratio are useful as the adjacent axis according to the transformer oil dissolved gas analysis and judgment guidelines. By doing that, we can not only ensure a relationship of adjacent data dimension but also reduce the disorder of data between the axes. Our final order for the parallel coordinates is C2H6, C2H4, C2H2, H2, CH4 , TN, CO and CO2.

3.2 Scatter Plot Matrix

Scatter plot matrix is one of the most popular multidimensional visualization methods. Scatter plot maps only two of all the dimensions into two axes and draw the scatter among them. It can effectively reveal the relationship or connection between two variables. Scatter Plot Matrix makes use of the above idea and builds a matrix of all two combinations of the multidimensional data. Variables of multidimensional data are plot in the corresponding panel of the matrix. People can get the implicit message from any two variables in each panel. Scatter plot matrix can explain the relationship between each two dimensions intuitively without being affected by size of data set and the number of dimension. It plays a an irreplaceable role in showing pairwise relations of multidimensional data .We use color to code different devices and use a scatter plot shows the cluster distribution of different devices. By doing this, we can make comparison and judgment for the state of device.

3.3 Data Table

All the information of DGA data is shown by a two dimension data table in our paper. The table contains 10 columns which are device ID, H2 content, CH4 content, C2H6 content, C2H4 content, C2H2 content, CO content, CO2 content ,total hydrocarbon content and monitoring time respectively. The displayed data items are arranged sequentially from top to bottom according to the monitoring data from near to far.

4. Interaction

In order to analyze and display data from multiple views effectively and conveniently, we apply interaction technology including brush, collaborative filter and focus + context to the multiple views visualization model. Those interactions can associate professionals to analyze data consistently and interactively.

4.1. Brush Technology

Brush technology is a useful visual interactive technology which can reflect the multidimensional data subset intuitively. The basic function is to protrude those data users concerned by highlights and hide those unconcerned data. As shown in Figure 3, click the left button of mouse on the C2H2 axis and drag to a certain point to release, then the parallel coordinates applies an overall filter to DGA monitoring data and only shows the monitoring data whose acetylene values in the chosen ranges.

Users can simply double-click the left mouse on the acetylene axis to cancel the filter. The visualization model also supports multiple brush filters which can apply brush functions to more than one axis for multiple condition analysis. The multiple brush technology can effectively reduce the interference of irrelevant data. It is easy for users to analyze the distribution as well as the relation of interest data.

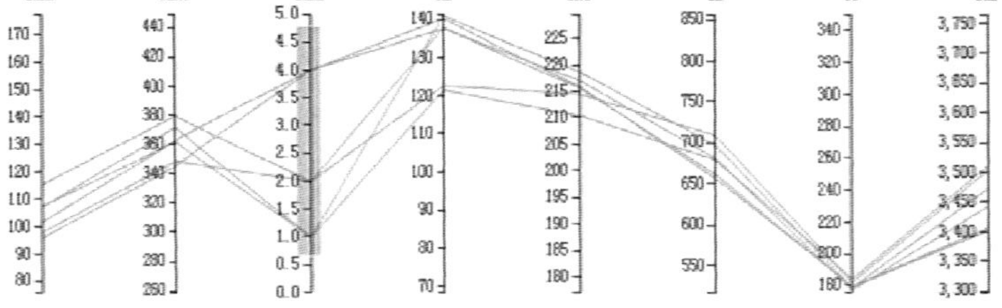

Figure 3. Brush technology

4.2. Collaborative Filter Technology

Collaborative filter technology is the core technology to realize cooperative multi-view analysis. In a single view, filter technology can help to analyze the selected core data from the large amounts of data and reduce the impact of unconcerned information. Collaborative filter technology applies the filter of a view to other related views. For example, when placing the mouse on a row of the data table (the gray item of the data table), visualization model will highlight that row. At the same time, the corresponding data item will also be highlighted in both parallel coordinates and scatter plot matrix (as shown in Figure 4). Since the items in data table are arranged sequentially from top to bottom according to the sampling date monitoring data from near to far, when the mouse move over rows of data table from top to bottom successively, we can view the trends of feature gases from the evolution of data in the visualization model. When using brush technique to select core data from parallel coordinates, the scatter plot matrix and data item table will also filter out non-core data so as to analyze the core data by multiple views (As shown in Figure 5).

4.3. Focus + Context Technology

The focus + context technology (F + C) is originated from fisheye view. This technology shows the

IOP Conf. Series: Journal of Physics: Conf. Series **787** (2017) 012001 doi:10.1088/1742-6596/787/1/012001

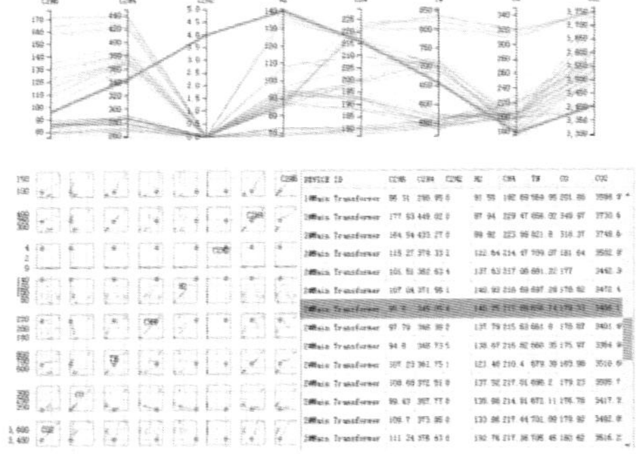

Figure 4. Collaborative filter by data table

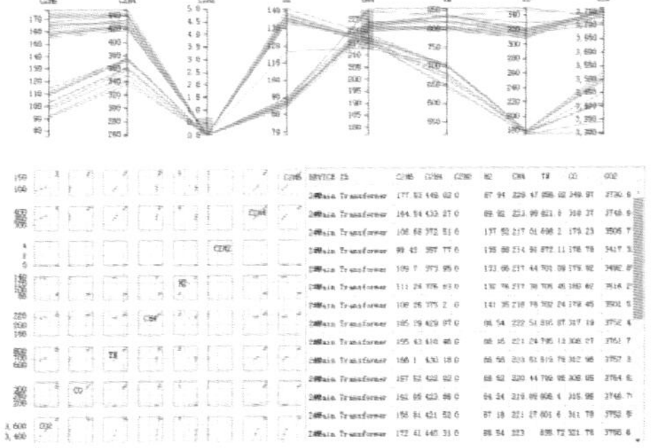

Figure 5. Collaborative filter by parallel coordinates

concerned data (focus) and the overall context (context) in the same views simultaneously. By the focusing function, the technology apply a selective distortion to the object in the views. The focus object will stand out and the surrounding environment in the context will be gradually reduced. The basic cognitive psychology technology is that people often need to maintain the visibility of whole information space when exploring the local information. If the information space is divided into two display areas, people need to constantly switch attention and memory when exploring information and lead to inefficiencies in cognitive behavior. We apply focus + context technology to parallel coordinates and scatter plot matrix. When placing the mouse on any row of the data table, the corresponding data item will become bigger and highlight in parallel coordinates and scatter plot matrix and the surrounding environment data will became light.(As shown in Figure 4)

5. Conclusion

In this paper, we present a novel interactive visualization model of DGA data based on multiple views including parallel coordinates, scatter plot matrix and data table. This model manages and analyzes data directly from different dimension imitating human.to reveal multidimensional data. And by offering brush, collaborative filter and focus + context technology, this model provides a convenient and flexible interactive way to analyze and understand the data. Our model can improve analysis ability of observers and has a great significance for the development of electrical equipment of online monitoring system in substation.

References

[1] Cai-xin, S. 2005 Present Situation and Development of Condition On-line Monitoring and Diagnosis Technology for Power Transmission and Transformation Equipment. *Electric Power*, 2005, 38(2): 1-7.

[2] Jian-zhuang, Z., Chun-lei, N. and Jin-hua. 2007 H. The Status of Monitoring On-line Technology of Transformer in Henan Province. *Henan Electric Power*, 2007, 4:31-33.

[3] Yan-cai, X., Heng-jun, Z., Xiao-yuan, Z., et al. 2006. On-line Monitoring and Diagnosis of Power Transformer Based on Dissolved Gas Analysis. *Electric Power Automation Equipment*, 2006, 26(6): 93-96.

[4] Leibniz, Uniz. 2008 A Software Implementation of the Duval Triangle Method [C]. Electrical Insulation, 2008. ISEI 2008.*Conference Record of the 2008 IEEE International Symposium on*. 2008:124-127

[5] Diego, M., Satish, D. 2010 Gas Analysis of a Thermally Overloaded Oil Immersed Current Transformer[C]. *IEEE/PES Transmission and Distribution Conference and Exposition*. 2010: 645-653

[6] Keim, DA., Kriegel, HP. 1996 Visualization techniques for mining large databases: A comparison. *IEEE Trans. on Knowledge and Data Engineering*, 1996,8(6):923−938.

NEB in Analysis of Optical Flow 4 × 4 and 6 × 6-Patches

Shengxiang Xia[1] and Di Liang[2]

[1, 2] School of Science, Shandong Jianzhu University, Jinan 250101 P.R. China.

E-mail: summer_5069@163.com

Abstract. We apply the nudged elastic band technique to non-lineal high-dimensional datasets, we analyze spaces of 4×4 and 6×6 optical flow patches and detect their topological properties. We experimentally prove that subsets of 4×4 and 6×6 optical flow patches can be modeled a circle, which confirm some results of 4×4 and 6×6 optical flow patches by using a new method-NEB, and expend Adams et al's result to larger patches of optical flow.

1. Introduction

Optical flow is the pattern of clear motion of image objects between two continuous frames caused by the movement of objects [1]. Optical flow has many applications, for example, in object segmentation, motion estimation and video compression. Because the difficulty of collecting data of optical flow statistics, the spatial statistics of optical flow are comparatively unexploited. To analyze non-lineal high-dimensional data is a difficult problem, computational topology is an efficient method for analyzing non-lineal high-dimensional data, recently another tool (NEB) of analyzing nonlinear high-dimensional data appeared. In 2015, Adams, Atanasov, and Carlsson [2] used the nudged elastic band (NEB) technique to analyze data of optical flow, they found a new topological properties of an optical flow patches, that is, optical flow patches have a circle behavior, which initially demonstrate that the NEB method is an effective tool for high dimensional nonlinear data analysis.

In this paper, we use the nudged elastic band method [2] to high-contrast patches of optical flow, analyze , and optical flow patches, and identify topological structures in these data sets, the datasets used here comes from a database built in [3]. We use NEB to prove that there exist subsets of , and optical flow patches which has the homology of a circle, the results have been proven in [4] by using computational topology. For the same data sets, we used a completely different method from paper [4] to analyze them, and obtain the same result as in [4], therefore, these topological properties of the sets are their inherent nature. The NEB method is simpler than the method of computational topology, in NEB, we only use several cell complexes to identify topological properties of spaces, but in the computational topology method, we need several thousands even tens of thousands of complexes to get same results. This proves once again that NEB is an effective tool for analyzing high-dimensional nonlinear data sets and identifying their topological properties.

2. The spaces of optical flow patches

Our data are picked from the Roth and Black optical flow database [3], a sample is shown in Fig.1. For each flow field sequence in the database, we utilize the second optical flow frame of a sequence, and randomly select data sets of high contrast 4×4, and 6×6 patches from the optical flow database. The spaces M_4, and M_6 are sets of 4×4, and 6×6 patches of high contrast as created in [4], which is similar as [5], [6], [7]. And we use the set symbols of the paper [4]. Undefined concepts and symbols in this paper can be found in [2], [4~7].

We should note that one 4×4 patch is arranged as $\begin{pmatrix} (u_1,v_1) & (u_5,v_5) & (u_9,v_9) & (u_{13},v_{13}) \\ (u_2,v_2) & (u_6,v_6) & (u_{10},v_{10}) & (u_{14},v_{14}) \\ (u_3,v_3) & (u_7,v_7) & (u_{11},v_{11}) & (u_{15},v_{15}) \\ (u_4,v_4) & (u_8,v_8) & (u_{12},v_{12}) & (u_{16},v_{16}) \end{pmatrix}$, where u denotes optical flow in the horizontal direction and v denotes the vertical direction. We consider each 4×4 patch as a vector $\mathbf{x}=(u_1,\ldots,u_{16},v_1,\ldots,v_{16})\in\mathbb{R}^{32}$, and each 6×6 patch as a vector $\mathbf{x}=(u_1,\ldots,u_{36},v_1,\ldots,v_{36})\in\mathbb{R}^{72}$. For $y\in X$ and a positive integer k, we let $\rho_k(y)=|\,y-y_k\,|$, here y_k is the k-th nearest neighbor of y. For a fixed k, we sort points of X by descending density, we choose the points indicated by X (k, p) whose densities are in the top p percent. In this paper, we consider subsets $MS^4(200,30)$ and $MS^6(200,30)$.

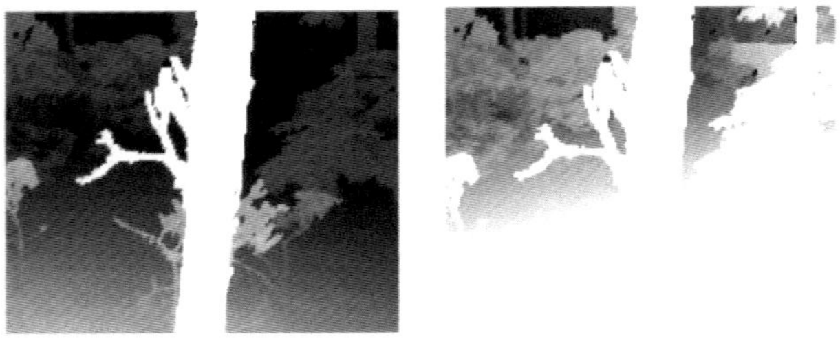

Figure 1. A sample from the optical flow database. Horizontal motion is on the left and vertical motion is on the right.

3. Computing method

The outline of the computing method from the paper [2] is listed as following, for more details, please refer to [2].

3.1. Nudged elastic band

The nudged elastic band (NEB) comes from computational chemistry, that is an efficient method for finding a minimum energy path between a given two states. A NEB is computed by a optimization algorithm from an initial path according to the total force, the initial path is usually built from linear interpolation between the initial and final states. For more details about NEB, please refer to [8], [9], [10].

3.2. CW complexes

A k-dimensional closed ball $\{x\in\mathbb{R}^k \mid \|x\|\le 1\}$ is called a k-cell. A CW complex X is defined by follow inductive steps. The 0-skeleton $X^{(0)}$ of X is a set of 0-cell. The 1-skeleton $X^{(1)}$ is created by gluing the endpoints of 1-cells to the 0-skeleton. Inductively, the k-skeleton $X^{(k)}$ are built by gluing the boundaries of k-cells to the $(k-1)$-skeleton $X^{(k-1)}$.

3.3. Density estimator

Given a set $X \subset \mathbb{R}^n$, let $\Phi_{x,\sigma}: \mathbb{R}^n \to [0, \infty)$ be the probability density of a normal distribution centered at $x \in X$, we apply a differentiable density estimator $g(y) = |X|^{-1} \sum_{x \in X} \Phi_{x,\sigma}(y)$ to approach the unknown density.

3.4. 0-cells

To find 0-cells, we randomly choose an initial point $y_0 \in X$, and iteratively generate a sequence $\{y_0, y_1, \ldots\}$ with $y_{n+1} = m(y_n)$, where $m(y): \mathbb{R}^n \to \mathbb{R}^n$ is the mean shift function given by the formula $m(y) = \frac{\sum_{x \in X} \Phi_{x,\sigma}(y)x}{\sum_{x \in X} \Phi_{x,\sigma}(y)}$. The sequence $\{y_n\}$ converges to a local maxima of g [11]. In order to identify different 0-cells, we use single-linkage clustering to cluster the convergent points, and choose the densest member from each cluster as a 0-cell.

3.5. 1-cells

For two 0-cells, there is a 1-cell between them if we find a convergent band between them by using NEB. For an initial band $[\mathbf{U}_0, \mathbf{U}_1, \ldots, \mathbf{U}_N]$, where $\mathbf{U}_0$ and $\mathbf{U}_N$ are 0-cells. The total force on each midterm node $\mathbf{U}_i$ is computed by $\mathbf{F}_i = (\|\mathbf{U}_{i+1} - \mathbf{U}_i\| - \|\mathbf{U}_i - \mathbf{U}_{i-1}\|)\tau_i - c\nabla g(\mathbf{U}_i)|_{\perp} + \mathbf{F}_{sm}$, where $c = (\sigma\sqrt{2\pi})^n \sqrt{e}$ is the gradient constant. The smoothing force $\mathbf{F}_{sm}$ is given by the similar formula as in [2].

Given a set $X \subset \mathbb{R}^n$ from unknown probability density function $f: \mathbb{R}^n \to [0, \infty)$. For super-level sets $X^{\alpha} = f^{-1}([0, \infty))$, we construct CW complex models Z^{α} to approximate the super-level sets X^{α}.

We construct only one-dimensional cell complexes. First step, we establish a differentiable density estimator to approach the unknown probability density function by using the given set. Then, we find local maxima of the density estimate to get 0-cells. Lastly, we randomly produce initial bands, then find the convergent bands by NEB, thus we get 1-cells. Hence we can detect the topology properties of the data set X by constructing cell complexes Z^{α}.

4. Experimental results

The author of the paper [4] use persistent homology to discuss the topological structure of spaces M_4 and M_6 of 4×4 and 6×6 Optical flow patches, and discover that the homology of the subsets changes from a circle to a 3-circle space. Specially, he experimentally show that there exist core subsets $MS^n(200,30)$ of M_n (n = 4, 6), their homology is that of a circle, and M_4, M_6 have subset $CC_3^4(400)$, $CC_3^6(400)$ respectively with the topology of the three circle space C_3.

In this paper we select three kinds subsets of M_n to analyze (n=4, 6): (1) random subsets $MR^n(10000)$ of M_n with size 10000; (2) core subsets $MS^n(200,30)$; (3) selected subset $CC_3^n(400)$. For a tidy projected graph, we make DCT transformation for our considered sets. For example, we transform u-component and v-component of $MS^4(200,30)$ respectively, then points of the set $MS^4(200,30)$ become 30-dimensional points. In fact, DCT transformation does not change the topology of a space.

4.1. 4 × 4 optical flow patches

Our data set information are displayed in Table 1. For the set $MR^4(10000)$, we take standard deviation $\sigma=$ 0.35, we obtain four 0-cells with densities in [1.285, 3.844], and four 1-cells with densities in [1.133, 2.052], all these cells form a loop. Thus for α =1.133, the Z^{α} is a circle (Fig.2).

Table 1 Data set information

	$MR^4(10000)$	$MS^4(200,30)$	$CC_3^4(400)$
size of data set	10000	15000	1200
dimension n	30	30	30
standard deviation σ	0.35	0.30	0.45

For the set $MS^4(200,30)$, let $\sigma=0.30$, we get four 0-cells with densities in [194.8, 769.4] and four 1-cells having densities in [69.29, 214.6], these cells compose a circle (Fig.3).

For the set $CC_3^4(400)$, we take standard deviation σ = 0.45, we have four 0-cells with densities in [0.003253, 0.00429] and four 1-cells having densities in [0.002607, 0.003009], these cells have the topology of a circle (Fig.4). When we take σ to be values of 0.25, 0.30, 0.35, 0.40, we can't conclude a definite conclusion. We can't find that $CC_3^4(400)$ has the topology of C_3 using the current method.

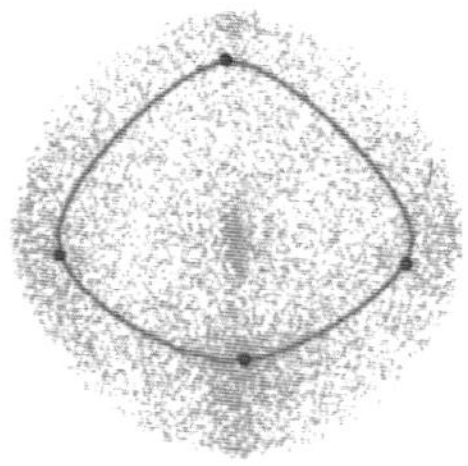

Figure 2. $MR^4(10000)$ and the circle and the $Z^{1.133}$, projected to a plane.

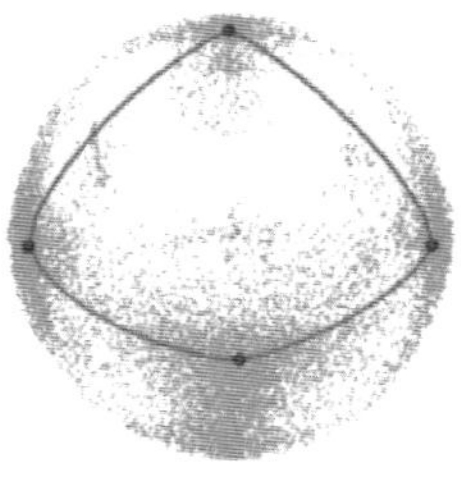

Figure 3. $MS^4(200,30)$ and the circle $Z^{69.29}$, projected to a plane.

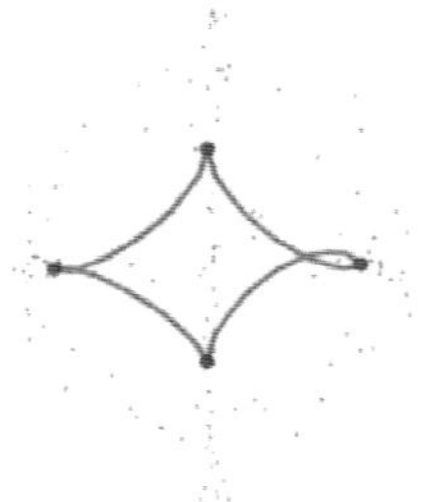

Figure 4. $CC_3^4(400)$ complex $Z^{0.002607}$,projected to plane.

4.2. 6 × 6 optical flow patches

Table 2 gives information of data sets. For the set $MR^6(10000)$, taking standard deviation σ = 0.33, we find four 0-cells, and four 1-cells, which form a loop. The four 0-cells have densities 11400, 17210, 20670, 43200, and four 1-cells have densities 8425, 9866, 17210, 18250 respectively. Hence for $\alpha=8425$, Z^{α} is a circle (Fig.5).

Table 2 Data set information

	$MR^6(10000)$	$MS^6(200,30)$	$CC_3^6(400)$
size of data set	10000	15000	1200
dimension n	70	70	70
standard deviation σ	0.33	0.35	0.45

For the set $MS^6(200,30)$, supposed $\sigma=0.35$, we get three 0-cells with densities 342.4, 623.6, 2315, and three 1-cells with densities 192.7, 209.8, 600.7 respectively, all these cells form a loop (Fig.6). If we take σ =0.30, we get four 0-cells, and four 1-cells, which form a loop.

For the set $CC_3^6(400)$, we take σ =0.45, 0.50, for each case we find four and four 1-cells, these cells form a loop. For example, when σ =0.45,we have four 0-cells with densities in [0.0000279, 0.00003068] and four 1-cells with densities in [0.00002147, 0.00002391] (Fig.7). When we take σ =0.25, 0.30, 0.35 and 0.40, we do not find a definite conclusion, and we can't discover that $CC_3^6(400)$ has the topology of C_3 for other values of σ.

5. Conclusion

In this short note we utilize the nudged elastic band method to discuss topological qualitative analysis of optical flow small patches. We experimentally prove that the spaces of 4×4 and 6×6 high contrast patches have subsets modeled as a circle. The most important parameter of the NEB method is standard deviation σ, we should choose its proper values to get a stable result. The advantage of cell complexes for analyzing high-dimensional dada is its simplicity. For example, we only use several cells to model $MS^6(200,30)$ as a circle, for the same result, we need several tens of thousands witness complexes to model $MS^6(200,30)$ as a circle. If we apply witness complexes to analyze subsets of M_4 and M_6, we detect that the topology of the subsets changes from a circle to a 3-circle s space, but we can't find a 3-circle model for various subsets of M_4 and M_6 by cell complexes. Hence the NEB method may only identify coarse topology of a high-dimensional data set.

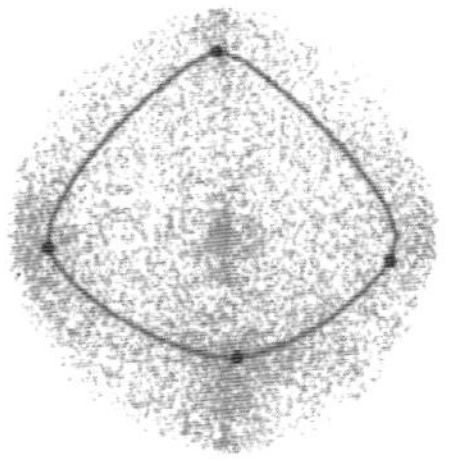

Figure 5. $MR^6(10000)$ and the circle Z^{8425}, projected to a plane.

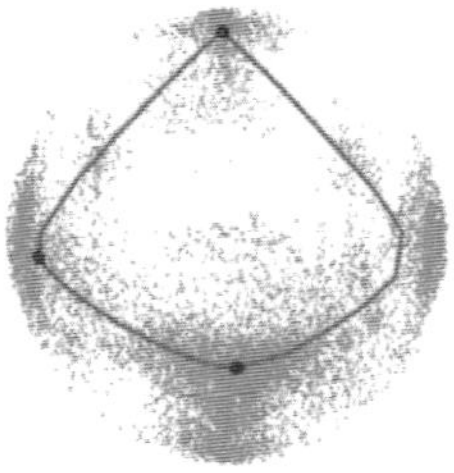

Figure 6. $MS^6(200,30)$ and the circle, $Z^{192.7}$, projected to a plane.

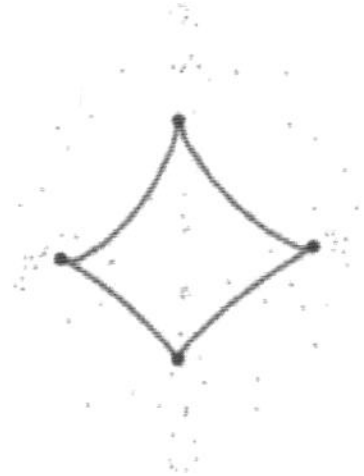

Figure 7. $CC_3^6(400)$ and the complex $Z^{0.00004127}$, projected to a plane.

Acknowledgements

The project is supported by the National Natural Science Foundation of China (No. 61471409).

References

[1] D. H. Warren, and E. R. Strelow, Electronic Spatial Sensing for the Blind: Contributions from Perception. Springer.

[2] H. Adams, A. Atanasov, and G. Carlsson, Nudged Elastic Band in Topological Data Analysis. *Topological Methods in Nonlinear Analysis*, 45(2015), 247-272.

[3] S. Roth, and M. J. Black, On the Spatial Statistics of Optical Flow. *International Journal of Computer Vision*, 74(2007), 33–50.

[4] S. Xia, A topological analysis on patches of optical flow. *J. Nonlinear Sci. Appl.*, 9 (2016), 1609–1618.

[5] H. Adams and G. Carlsson, On the nonlinear statistics of range image patches. *SIAM J. Image Sci.*, 2(2009), 110–117.

[6] G. Carlsson, T. Ishkhanov, V. de Silva, A. Zomorodian, On the local behavior of spaces of natural images. *International Journal Computer Vision*, 76(2008), 1–12.

[7] A. B. Lee, K. S. Pedersen, and D. Mumford, The non-linear statistics of high-contrast patches in natural images. *International J. Computer Vision*, 54(2003), 83–103.

[8] G. Henkelmana and H. Jónsson, Improved tangent estimate in the nudged elastic band method for finding minimum energy paths and saddle points, *Journal Chemical Physics*, 113(2000), 9978–9985.

[9] G. Henkelman, B. Uberuaga, H. Jónsson, A climbing image nudged elastic band method for finding saddle points and minimum energy paths, *Journal Chemical Physics*, 112 (2000), 9901–9904.

[10] H. Jegou, M. Douze, and C. Schmid, Hamming Embedding and Weak geometry consistency for large scale image search, *Proceedings of the 10th European conference on Computer vision*, October (2008), 304–317.

[11] Y. Cheng, Mean shift, mode seeking, and clustering, *IEEE Trans. Pattern Anal.*, 17 (1995), 790–799.

IOP Conf. Series: Journal of Physics: Conf. Series 787 (2017) 012003 doi:10.1088/1742-6596/787/1/0120

Congestion analysis of pilgrims in Hajj and Umrah congregation using block matching and optical flow

Sumaiyya Farooq, Shoab A Khan and M Usman Akram

Department of Computer Engineering, College of E&ME, National University of Sciences and Technology, Islamabad, Pakistan

E-mail: sumaiyya20@hotmail.com

Abstract. A novel method has been proposed to classify the motion of pilgrims with respect to congestion level in the holy mosque of Makkah. Millions of Muslims visit this mosque during Hajj and Umrah every year. A large number of security personnel is required to maintain the smooth flow of pilgrims in order to avert any catastrophe. Therefore, it is inevitable to design a computer aided system to reduce human effort. The proposed system pre-processes input images to segregate the moving shadows and pilgrims in order to nullify the false motion due to moving shadows. A hybrid method consisting of block matching and optical flow techniques has been used for the computation of motion vectors. Decision tree classifier is used on the number of motion vectors having non-zero magnitude. Experiments show that the proposed system has promising results yielding an accuracy of 90.58% for the congestion classification of pilgrims.

1. Introduction

A huge number of Muslims from all over the world gather at the holy mosque of Makkah every year to perform Hajj and Umrah. The number of pilgrims remains very high at both these events. A large number of security personnel is needed to be deployed to manage smooth flow of pilgrims to avoid undesirous situations. A large number of surveillance cameras are installed in the mosque. But manual monitoring of enormous data acquired from these cameras is a highly tedious job. Improvements in the fields of image processing, computer vision and machine learning has sparked a lot of scope to explore this area. More novel, robust and efficient techniques can be developed that can help security staff to monitor the congestion level and suggest some suitable measures in case some emergency situation takes place.

Chen and Huang proposed a motion-based system for detection of abnormal activity in human crowd [1]. Optical flow was computed and the characterization of crowd was performed using unsupervised learning techniques on the features obtained by optical flow. Prediction of crowd behaviour was done using the force field model. H. Idrees et al. [2] incorporated multiple bases of information to make a reliable crowd estimation system in which texture analysis, head count using HOG features and frequency domain analysis was merged. Direct counting of people was not possible because pixel count per person was low and images were highly occluded.

Counting of people was done using head detection method by V.B. Subburaman [3]. Interest points were detected by calculating gradient of the image and head contours were used to train SVM classifier to detect and count human heads. J. Yang et al. estimated the crowd density using the texture features based on gray level co-occurrence matrix (GLCM) in [4]. Texture features such as energy, entropy, contrast, homogeneity, were passed to SVM classifier for density detection and linear regression model was used to count number of people.

To deal with Hajj and Umrah data is much different and challenging as compared to other crowd scenarios. Two reasons for this strenuous situation include: 1) low contrast between the white marble floor and the clothing of pilgrims [5], 2) the motion of pilgrims is highly congested and 3) heavy congestion leads to highly occluded scenes. Figure 1(a) shows the occlusion problem in which moving pilgrims overlap each other and different congestion levels such as high and low are depicted in (b). Many researchers have touched this area in recent years and a brief review of their work is presented below.

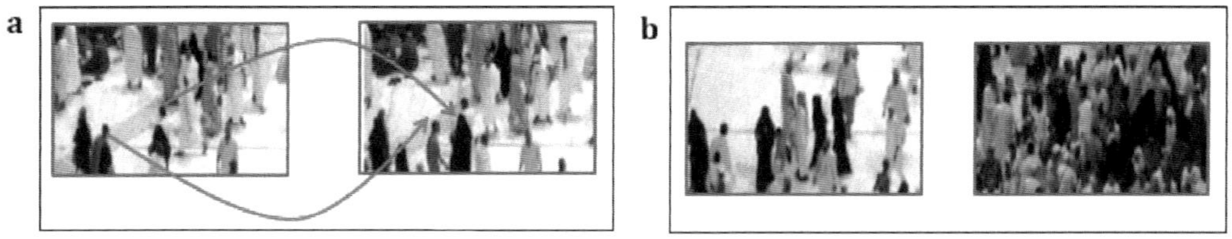

Figure 1. Problems of (a) occlusion (b) low and high congestion scenes.

Hussain et al., in 2011, proposed a crowd density estimation system (CDES) [6] for medium density level specifically for Al-Haram mosque, Makkah. Pixel based features of foreground blobs were extracted and density was classified into five levels by building backpropagation neural network on foreground pixel features. In 2013, M. Arif et al. presented a framework to count the number of pilgrims moving in the Mattaf area with more than 2000 persons per frame. The estimation was done using blob analysis techniques on the segmented image data because the size of the blob was found to be correlated with the actual number of people [7].

Not much work has been done to design a viewpoint dependent framework which is able to characterize the motion of pilgrims in extremely dense and occluded scenarios. This paper addresses the above mentioned scenario by analysing the level of congestion of pilgrims. To achieve the above said task, computation of motion vectors (MVs) is performed and number of motion vectors having non-zero magnitude is used to classify the congestion level.

The paper is organized in the following manner: Section 2 describes the proposed methodology thoroughly. In section 3, results of the proposed framework are presented. Section 4 discusses the results and Section 5 concludes the paper.

2. Proposed methodology

The proposed system aims to estimate and analyse the congestion level of pilgrims in Hajj and Umrah congregation. Top level diagram of the proposed framework is shown in figure 2. The input images are taken in temporal sequence and display the outside view of holy mosque of Kabah near the entrance gate and have a size of 1440x1080 pixels. It is assumed that the camera is held stationary and mounted at an elevated place while capturing the images.

2.1. Preprocessing

Preprocessing includes following steps:

(i) Images are first converted to gray scale and cropped to limit the ROI to 432x1440 pixels.

(ii) An adaptive binary mask is generated to separate the pilgrims from their moving shadows to discard false motion.

(a) Input image is divided into 50x50 sized patches.

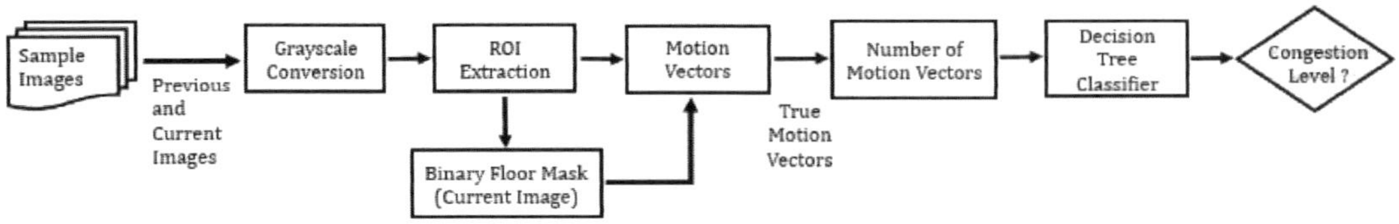

Figure 2. Flow of proposed framework for congestion analysis of pilgrims.

(b) An adaptive threshold 'T' is set equal to 80% of maximum intensity level as shown in equation (1).

$$T = 0.8^* \text{ (Max Intensity Level)} \qquad (1)$$

(c) Binary mask is generated using equation (2). Take the complement of the mask.

$$\text{mask} = \begin{cases} 0; & pixel\ Intensity\ < T \\ 1; & pixel\ Intensity\ \geq T \end{cases} \qquad (2)$$

2.2. Estimation of motion

MVs have been computed using the hybrid technique proposed by S.H. Chan et al. in [8]. At first step, block matching technique based on Sum of Absolute Differences (SAD), block size of 8x8 and search window of 16x16 pixels computes the integer part of displacement. After that, Taylor refinement has been applied on the translated reference image block to calculate the sub-pixel motion.

A significant modification has been proposed to reduce the computation time of calculating motion vectors. Motion vectors have been calculated only for those blocks that hold '1' in the binary mask. The modified algorithm to calculate motion vectors is given below:

Algorithm 2.1: modified algorithm to compute (V_x, V_y)

for i ← 1 **to** rows of target image
  for j ← 1 **to** columns of target image
    if mask (i, j) = 1
      (V_x, V_y) ← x and y components of motion vector for block (i, j) respectively

2.3. Classification of Congestion

Number of MVs having non-zero magnitude "N" is a potential indicator of congestion level in the scene. The MV grid is divided into 3x6 blocks and MVs are calculated for each block separately. In this way, congestion level is analysed in local blocks. Decision tree classifier is used to divide the congestion level of each block into five classes (figure 3). The upper branch of the tree is followed if the condition of relevant node is true and vice-versa.

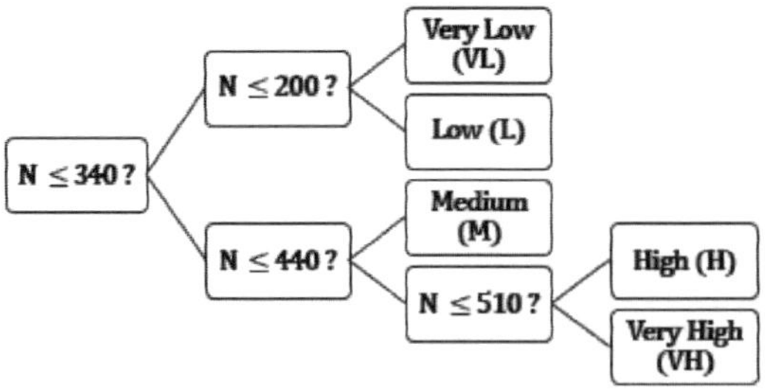

Figure 3. Decision tree for congestion classification.

3. Experimental Results

The formulation of congestion analysis system is tested on images captured at the entrance gate of Al-Haram mosque. Since this data is not publicly available, it is manually annotated to set the ground truth for congestion analysis.

Figure 4(a) shows the original image with red box indicating the ROI. Floor mask is generated for the extracted ROI as shown in figure (b) and (c) to segment the pilgrims and their moving shadows on the floor.

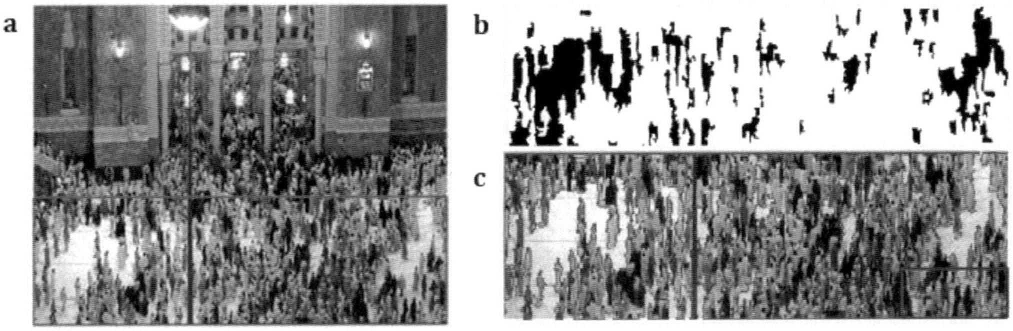

Figure 4. Preprocessing results (a)original frame (b)floor mask (c)result of floor segmentation.

MVs are calculated using binary mask as a prior check to eliminate the false motion. Results of it are presented in figure 5. It is quite clear that redundant MVs are cleaned up to a greater extent.

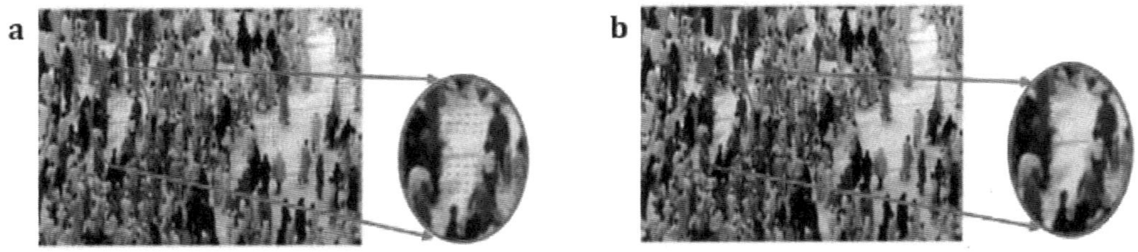

Figure 5. Motion vectors (a)without floor mask condition (b)with floor mask condition.

To make the system more accurate, number of non-zero MVs are calculated for all 3x6 sub-images separately. The classification results for one sub-image from each class of congestion are shown in figure 6. The title of each sub-image includes number of MVs having non-zero magnitude 'N', ground truth (GT) class of congestion and predicted class respectively.

Figure 6. Results of congestion classification for each class.

A total of 1050 images are tested and number of non-zero MVs are averaged out for a group of 10 consecutive images to discard any irregularities. Each image further consists of 18 sub-images. Hence we have a total of (1050/10) x 18 = 1890 test subjects. The classification accuracy is found to be 90.58%.

4. Discussion

Illumination variation and occlusion are the two main reasons for wrong classification. Illumination does not remain constant across the image. This poses a challenge on the floor extraction method. The reason is primarily the inclusion of static floor along with moving pilgrims because shadow makes the floor appear a bit darker than the rest of the floor. The blue box in the figure 4(c) region explains the scenario. Images are more prone to wrong classification if actual number of motion vectors lie at the periphery of two classes. In such a case, a slight increment in their number throw them in next class of congestion.

Second reason of anomalous results in some cases is the occlusion of pilgrims in highly dense areas of images. Despite the fact that camera is mounted at an elevated place, it still occludes the persons and remains problematic for the correct classification of congestion level of pilgrims. Because of large density of persons, no algorithm of image processing can overcome the problem of occlusion.

To my best knowledge, a significant work has not been done yet to build a crowd analysis system for Hajj and Umrah events where number of pilgrims are in millions. Therefore, a true comparison cannot be performed between our proposed methodology and the existing state-of-the-art techniques.

5. Conclusion

We have proposed a novel framework to classify the congestion level of moving pilgrims at the entrance gate of Al-Haram mosque. Floor mask is generated to segment out the static floor and shadow of moving pilgrims. A hybrid technique comprising of block matching and optical flow methods is used to calculate the motion of pilgrims. Block matching computes the integer part of displacement while sub-pixel motion has been computed using optical flow. Further, a modification is proposed to calculate the motion vectors. Floor mask is incorporated as a prior check to see whether to compute motion vectors for a given block of input image or not. To classify the congestion level of pilgrims, number of motion vectors having non-zero magnitude is considered. In future, there is a great scope to work in this area. For example, decision support systems can be built on the basis of such statistics of motion of pilgrims to avoid any unfortunate situation.

References

[1] Chen D Y and Huang P C 2011 *J. Visual Communication and Image Representation* **22** pp 178-86
[2] Idrees H, Saleemi I, Seibert C and Shah M 2013 *Proc. IEEE Conf. on Computer Vision and Pattern Recognition* pp 2547-54
[3] Subburaman V B, Descamps A and Carincotte C 2012 *Advanced Video and Signal-Based Surveillance (AVSS), 2012 IEEE 9th Int. Conf. on (IEEE)* pp 470-75
[4] Yang J, Li J and He Y 2014 *J. of Multimedia* **9** 1152-59
[5] Al-Khaffaf H S, Haron F, Sarmady S, Talib A Z and Abu-Sulyman I M 2012 *Frontiers in Computer Education (Springer)* pp 727-36
[6] Hussain N, Yatim H S M, Hussain N L, Yan J L S and Haron F 2011 *Safety Science* **49** 824-33
[7] Arif M, Daud S and Basalamah S 2013 *IAES Int. J. of Artificial Intelligence* **2** 51
[8] Chan S H, Nguyen T Q et al. 2010 *2010 IEEE Int. Conf. on Acoustics, Speech and Signal Processing (IEEE)* pp 722-25

IOP Conf. Series: Journal of Physics: Conf. Series **787** (2017) 012004 doi:10.1088/1742-6596/787/1/012004

Approaching to Genuine Monochromatic Light Filters Employing A Serially Coupled Triple Ring Resonator System

Phichai Youplao

Electrical Engineering Department, Faculty of Industry and Technology, Rajamangala University of Technology Isan Sakon Nakhon Campus, 199, Phungkon, Sakon Nakhon 47160, Thailand

E-mail: phichai3112@yahoo.com

Abstract. This paper presents the design and analysis of an optical channel drop filters configuration based on a serially coupled triple ring resonator system. The filtering characteristics of such the proposed ring configuration have been investigated by using the mathematical model simulation. In this work, the optical signal as a Gaussian pulse with a specified wavelength ranges from 1545 nm to 1555 nm is used to be the input signal that launches into and propagates through the ring resonators. Whereupon, the required optical channels are obtained at the drop port. The center wavelength of these filtering optical channels can be specified and controlled by selectively applying the suitable parameters to the system, such as each the ring radius and the coupling coefficients. From the simulation results, an ultra-narrow band optical channel with a spectral width at FWHM of approximately 12 pm can be achieved. Furthermore, an attempt to tune the filtering responses of the proposed system, by customizing the refractive index of the waveguide material, has been illustrated. By din of the tuning technique, the free spectral range shift between adjacent channels of approximately 46 pm, which corresponds to be less than 6.25 GHz of a channel spacing in DWDM system, can be actualized. The potential of using such the triple ring device for filtering application is performed and discussed.

1. Introduction

Present day, optical filters are one of the most important components in modern fiber-optic communication systems. Especially, for a wavelength division multiplexing (WDM) method, the optical filters are used to separate required optical channels from a combined optical signal without any required electronic systems. Various kinds of the optical filters have been investigated, such as arrayed waveguide gratings [1], Fabry–Perot filters [2], and photonic crystals [3]. In addition, although the photonic crystals are expected to be a promising platform for photonic integrated circuits due to its significant properties, such as the compactness, and spectral tunability. The filtering channel that can be obtained from the photonic crystal systems has remained as a large spectral width, e.g., about 10 nm [4].

Ring resonators have also been being received extensive attention as a useful component in optical communication systems. Many of its potential applications are demonstrated both theoretically and experimentally in many investigations, such as optical logic operators, optical switch, signal processing, and filters [5-8]. However, it still needs to be further developed in order to improve the filtering responses.

In this paper, the primary aim is to achieve the analytical transfer functions of the serially coupled triple ring resonator system to be useful as an ultra-narrow band optical filter. Furthermore, the filtering characteristics that can be obtained by the proposed system and the attempt to tune its filtering response, by customizing the refractive index of the waveguide material, are demonstrated and discussed.

2. The Triple Ring Resonator System

The proposed ring resonator system of which the schematic diagram is shown in figure 1 is employed as an optical channel drop filter, which has an effective in filtering characteristics. In operation, the Gaussian pulse with a specified wavelength from the light source is used to be the optical input signal. This input light pulse is considered as a function that consists of a constant light field amplitude, E_0, and a random phase modulation, which is the combination of an attenuation term, α, and a phase constant term, ϕ_0, results temporal coherence degradation. The time-dependent input optical field, E_{in}, can be expressed as equation (1), where L is a propagation distance (waveguide length).

$$E_{in}(t) = E_0 \exp[-\alpha L + j\phi_0(t)] \tag{1}$$

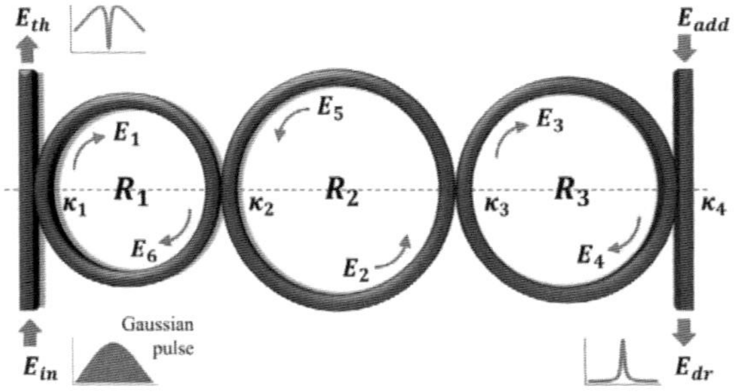

Figure 1. A schematic diagram of the serially coupled triple ring resonator system

When a Gaussian pulse is propagating through the ring resonator system, in the direction as shown in the figure, the separate resonant outputs are formed. These output optical fields each at the throughput port and the drop port are represented by E_{th} and E_{dr}, respectively. The relations between each of the output, E_{th} or E_{dr}, and the input optical field, E_{in}, in each round trip calculation can be expressed as:

$$\left|\frac{E_{th}}{E_{in}}\right|^2 = \frac{(1-\kappa_1)(x_1-x_2+x_3)^2-2(x_1-x_2+x_3)(x_4-x_5+x_6)\sqrt{1-\kappa_1}e^{-\frac{\alpha}{2}L_1}\cos(knL_1)+(x_4-x_5+x_6)^2e^{-\alpha L_1}}{(x_1-x_2+x_3)^2-(x_1-x_2+x_3)(x_4-x_5+x_6)\sqrt{1-\kappa_1}e^{-\frac{\alpha}{2}L_1}\cos(knL_1)+(1-\kappa_1)(x_4-x_5+x_6)^2e^{-\alpha L_1}} \tag{2}$$

$$\left|\frac{E_{dr}}{E_{in}}\right|^2 = \frac{\kappa_1\kappa_2\kappa_3\kappa_4 e^{-\frac{\alpha}{2}(L_1+L_2+L_3)}}{(x_1-x_2+x_3)^2-(x_1-x_2+x_3)(x_4-x_5+x_6)\sqrt{1-\kappa_1}e^{-\frac{\alpha}{2}L_1}\cos(knL_1)+(1-\kappa_1)(x_4-x_5+x_6)^2e^{-\alpha L_1}} \tag{3}$$

Where the optical field that propagates within each the part i of the ring resonator system is E_i and the specified constant quantities of C_n and x_n are expressed as in Table 1.

Table 1. The expression of each the optical field E_i and the constant quantity of C_n and x_n

The optical fields: E_i	The specified constant quantities: C_n and x_n
$E_1 = E_{in} j\sqrt{\kappa_1} + E_6\sqrt{1-\kappa_1}e^{-\frac{\alpha L_1}{2\,2}-jkn\frac{L_1}{2}}$	$C_1 = 1-\sqrt{1-\kappa_2}\sqrt{1-\kappa_3}e^{-\frac{\alpha}{2}L_2-jknL_2}$
$E_2 = E_1 j\sqrt{\kappa_2}e^{-\frac{\alpha L_1}{2\,2}-jkn\frac{L_1}{2}} + E_5\sqrt{1-\kappa_2}e^{-\frac{\alpha L_2}{2\,2}-jkn\frac{L_2}{2}}$	$C_2 = \sqrt{1-\kappa_3}\sqrt{1-\kappa_4}e^{-\frac{\alpha}{2}L_3-jknL_3}$
$E_3 = E_2 j\sqrt{\kappa_3}e^{-\frac{\alpha L_2}{2\,2}-jkn\frac{L_2}{2}} + E_4\sqrt{1-\kappa_3}e^{-\frac{\alpha L_3}{2\,2}-jkn\frac{L_3}{2}}$	$C_3 = \sqrt{1-\kappa_2}\sqrt{1-\kappa_4}e^{-\frac{\alpha}{2}L_2-jknL_2}e^{-\frac{\alpha}{2}L_3-jknL_3}$
$E_4 = E_3\sqrt{1-\kappa_4}e^{-\frac{\alpha L_3}{2\,2}-jkn\frac{L_3}{2}}$ (where $E_{add} = 0$)	$C_4 = \sqrt{1-\kappa_2}-\sqrt{1-\kappa_3}e^{-\frac{\alpha}{2}L_2-jknL_2}$
$E_5 = E_4 j\sqrt{\kappa_3}e^{-\frac{\alpha L_3}{2\,2}-jkn\frac{L_3}{2}} + E_2\sqrt{1-\kappa_3}e^{-\frac{\alpha L_2}{2\,2}-jkn\frac{L_2}{2}}$	$C_5 = \sqrt{1-\kappa_2}\sqrt{1-\kappa_3}\sqrt{1-\kappa_4}e^{-\frac{\alpha}{2}L_3-jknL_3}$
$E_6 = E_5 j\sqrt{\kappa_2}e^{-\frac{\alpha L_2}{2\,2}-jkn\frac{L_2}{2}} + E_1\sqrt{1-\kappa_2}e^{-\frac{\alpha L_1}{2\,2}-jkn\frac{L_1}{2}}$	$C_6 = \sqrt{1-\kappa_4}e^{-\frac{\alpha}{2}L_2-jknL_2}e^{-\frac{\alpha}{2}L_3-jknL_3}$ and $x_n = \lvert C_n\rvert$

Equation (2) and (3) indicate that the ring resonator system in this particular case is very similar to a Fabry–Perot cavity, which has an input and output mirror with a field reflectivity, $(1-\kappa)$, and a fully reflecting mirror, κ, where κ is the coupling coefficient. The details of the specific variables are as follows: the linear absorption coefficient is α, a roundtrip loss coefficient is $e^{-\alpha L_i/2}$, the linear phase shift is $\phi_i = knL_i$, where $kn = 2\pi n/\lambda$, and the refractive index of the ring waveguide is n. The waveguide length of the ring radius R_i is $L_i = 2\pi R_i$.

The input optical field, E_{in}, i.e. a Gaussian pulse from a laser source with the specified wavelength, is launched into the proposed ring resonator system with the appropriate parameters, such as the radius of the ring resonators, and the coupling coefficients, so that the transmitted output signals can be controlled.

From figure 1, in principle, the input light pulse is divided and sliced as the discrete signal propagates within the first ring, R_1, through the first coupling location of κ_1. Thereafter, it is spread to the other rings, R_2 and R_3, through the coupling locations of κ_2 and κ_3, respectively, in the direction as shown in the figure. Finally, after propagation in the proposed ring resonator, the required signals are obtained via the throughput port and the drop port of the system.

3. The Simulation Results

The optical Vernier effects that can be used for high-resolution distance measurement are commonly generated by using a resonant structure such as a micro-ring resonator [9]. For instance, a Gaussian input with a specified wavelength ranges from 1.545–1.555 μm (C-band), 10 nm bandwidth, and a peak power of 0.2 W, as shown in figure 2(a), is launch and propagating through the proposed system. Thereafter, the resonant outputs each at the throughput port and the drop port are obtained with the optical Vernier effects. In addition, the Vernier effects are valuable and can be applied to advantage for filtering the optical ultra-narrow band under appropriate conditions.

In this work, the triple ring resonator system is proposed to be useful as an effective optical filter by din of the optical Vernier effects that resulting in its resonant outputs. In order to associate the system with a practical device [10, 11], the system parameters are specified as follows: the ring radii R_1 = 15 μm, R_2 = 450 μm and R_3 = 60 μm. The refractive index of the waveguide material is fixed to be n = 3.47 (Si–Crystalline silicon). The coupling coefficients, κ, of the system are ranged from 0.2 to 0.5. The waveguide loss is considered to be α = 0.5 dB/mm. For simplification, the optical field relation does not take into account the coupling losses [12].

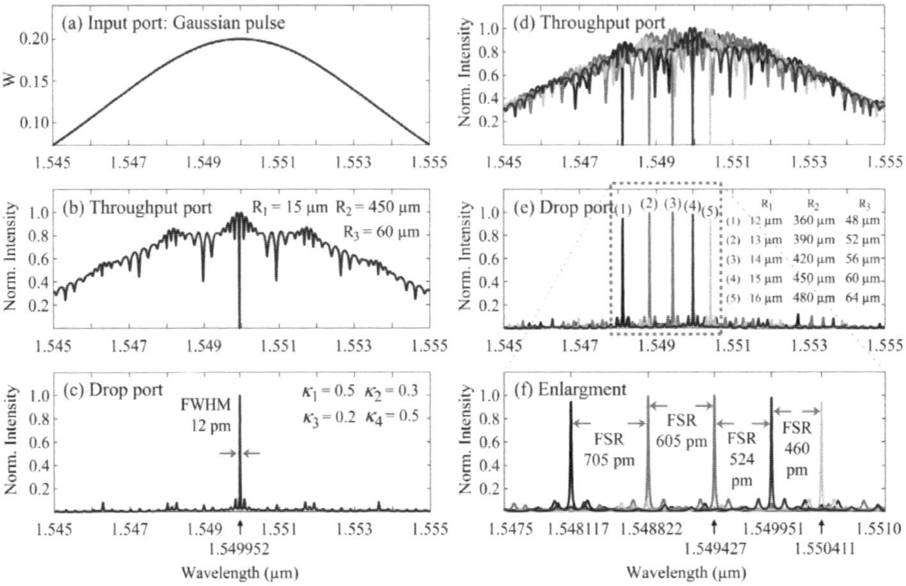

Figure 2. The simulation results, (a) the Gaussian input, (b) and (c) the filtering characteristics that are obtained at the throughput port and drop port, respectively, (d) and (e) the filtering characteristics for changing the appropriate ring parameters as the five sets of the ring radius obtained at the throughput port and drop port, respectively, and (f) the enlargement of the figure (e).

Figure 2(b) and 2(c) show the filtering characteristics that can be obtained via the throughput port and the drop port, respectively, where the filtering channel with a center wavelength of 1.549952 μm and a spectral width of 12 pm, at FWHM, can be achieved.

The simulation results of the same input signal with the appropriate ring radius change for five sets as given in the figure are shown in figure 2(d) and 2(e). The output signals from the throughput port and the drop port of the system are superimposed in the figure, respectively. The enlargement of figure 2(e) is also demonstrated in figure 2(f). The filtering channels obtained from the drop port for each ring parameter set are at the center wavelengths of 1.548117 μm, 1.548822 μm, 1.549427 μm, 1.549951 μm, and 1.550411 μm. The free spectral range shift between the adjacent channels is approximately of 705 pm, 605 pm, 524 pm and 460 pm, respectively.

4. Discussion

In practice, the appropriated ring parameters that are selectively applied to the system, such as the ring radius, and the coupling coefficients, might have unavoidable errors in the fabrication process and result in the filtering characteristics of the actual device. However, the resonant characteristic of a micro-ring resonator can be tuned by dint of several methods. The most typical and straightforward approach for tuning the resonant response is to change the refractive index of the waveguide material, such as the thermo-optic effect method, which applies the heat to the material [13], and the electro-optic effect method, which applies the electrical field to the waveguide [14]. Therefore, the waveguide tuning techniques must be considered in the fabrication work for the actual device in order to function properly. Hence, in this study, an attempt to tuning the filtering characteristics of the proposed ring system is demonstrated. It can be performed by customizing the refractive index of the waveguide material, e.g., to be within ±0.015% of the base refractive index value of the waveguide material (Si–Crystalline silicon; n = 3.47). The simulation results for customizing the refractive index, with the same of the other ring parameters as shown in figure 2(b) and 2(c), are demonstrated in figure 3(a). By the well-known linear regression, the linearity relation of the free spectral range shift as a function of the change in the refractive index is also illustrated in figure 3(b). As shown in the figure, the center wavelength of a filtering channel obtained from the drop port can be varied by customizing the refractive index of the waveguide material. Herein, the free spectral range shift between the adjacent channels of approximately 46 pm, which corresponds to be less than 6.25 GHz of a channel spacing in DWDM system (recommendation of ITU-T G.694.1) [15], can be actualized.

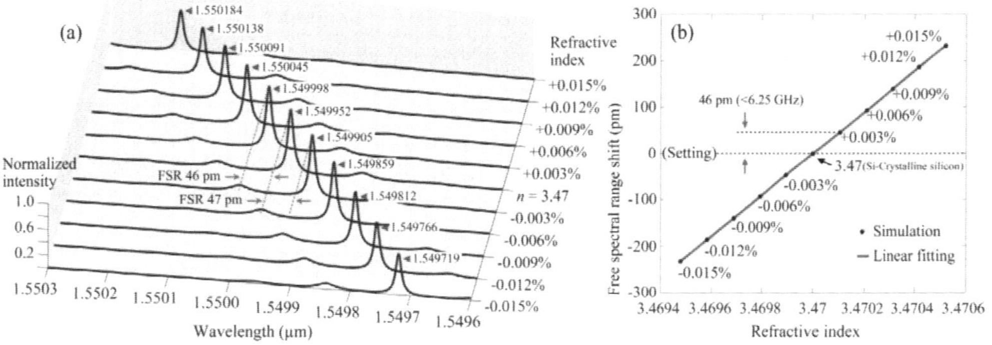

Figure 3. The results for customizing the refractive index of the waveguide material.

5. Conclusion

This article presents an optical waveguide configuration based on a serially coupled triple ring resonator system of which the design and analytical model are demonstrated to be useful as an optical channel drop filter. The resonant characteristics of the proposed system can be investigated by dint of its mathematical model simulation. These resonant responses can be controlled by selectively applying the suitable parameters to the ring system. From the simulation results, an ultra-narrow band optical channel with a spectral width of approximately 12 pm can be achieved. Moreover, by customizing the refractive index of the waveguide material, the filtering responses of the ring system can be tuned as approximately 46 pm of free spectral range shift between the adjacent channels. Therefore, due to such significant properties as the compactness, the spectral tunability, and especially in the filtering ability, the proposed ring system will be valuable for improving the performance of modern fiber-optic systems.

References

[1] W. Liu, G. Hu, H. Zhu, Y. Chen, Waveguide-metal–dielectric bi-layer gratings for reflective filteringand color security, International Journal for Light and Electron Optics. 126 (2015) 1245-1248.

[2] Y. Li, B. Zhou, L. Zhang, S. He, Tunable Fabry–Perot filter in cobalt doped fiber formed by optically heated fiber Bragg gratings pair, Optics Communications. 344 (2015) 156-160.

[3] H.A. Banaei, F. Mehdizadeh, M.H. Kashtiban, A new proposal for PCRR-based channel drop filter using elliptical rings, Physica E. 56 (2014) 211-215.

[4] A. Taalbi, G. Bassou, M.Y. Mahmoud, New design of channel drop filters based on photonic crystal ring resonators, International Journal for Light and Electron Optics. 124 (2013) 824-827.

[5] J.K. Rakshit, J.N. Roy, Micro-ring resonator based all-optical reconfigurable logic operations, Optics Communications. 321 (2014) 38-46.

[6] X. Yan, C.S. Ma, C.T. Zheng, X.Y. Wang, D.M. Zhang, Analysis Of Polymer Electro-optic microring resonator switches, Optics & Laser Technology. 42 (2010) 526-530.

[7] Y. Dumeige, C. Arnaud, P. Fe´ron, Combining FDTD with coupled mode theories for bistability in micro-ring resonators, Optics Communications. 250 (2005) 376-383.

[8] J.S. Li, Fast-tunable terahertz wave filter based on Kerr medium, Optics & Laser Technology. 56 (2014) 263-268.
[9] R. Boeck, N.A.F. Jaeger, N. Rouger, L. Chrostowski, Series-coupled silicon racetrack resonators and the Vernier effect: theory and measurement, Optics Express. 18(24) (2010) 25151-25157.
[10] H. Takara, T. Ohara, T. Yamamoto, H. Masuda, M. Abe, H. Takahashi, T. Morioka, Field demonstration of over 1000-channel DWDM transmission with supercontinuum multi-carrier source, Electronics Letters. 41(5) (2002) 270-271.
[11] Q. Xu, V.R. Almeida, R.R Panepucci, M. Lipson, Experimental demonstration of guiding and confining light in nanometer-size low-refractive-index material, Optics Letters. 29(14) (2004) 1626-1628.
[12] D.G. Rabus, M. Hamacher, U. Troppenz, H. Heidrich, Optical Filters Based on Ring Resonators With Integrated Semiconductor Optical Amplifiers in GaInAsP-InP, IEEE Journal of Selected Topics in Quantum Electronics. 8(6) (2002) 1405-1411.
[13] M.B.J. Diemeer, Polymeric thermo-optic space switches for optical communications, Optical Materials. 9(1-4) (1998) 192-200.
[14] I.L. Gheorma, R.M. Osgood, Fundamental limitations of optical resonator based high-speed EO modulators, IEEE Photonics Technology Letters. 14(6) (2002) 795-797.
[15] Information on https://www.itu.int/rec/T-REC-G.694.1-201202-I/en

IOP Conf. Series: Journal of Physics: Conf. Series **787** (2017) 012005 doi:10.1088/1742-6596/787/1/012005

A Three-dimensional Topological Model of Ternary Phase Diagram

Yingxue Mu and Hong Bao

University of Science and Technology Beijing,China

muyingxue2013@163.com, baohong@ies.ustb.edu.cn

Abstract. In order to obtain a visualization of the complex internal structure of ternary phase diagram, the paper realized a three-dimensional topology model of ternary phase diagram with the designed data structure and improved algorithm, under the guidance of relevant theories of computer graphics. The purpose of the model is mainly to analyze the relationship between each phase region of a ternary phase diagram. The model not only obtain isothermal section graph at any temperature, but also extract a particular phase region in which users are interested.

1. Introduction

The phase diagram can reflect many important content with our industrial production and scientific research. Therefore, the study of phase diagram is of important practical significance. The traditional ternary phase diagram is mainly expressed by plane projection, isothermal section, vertical section and other plane figures. The ternary phase diagram figured by the thermodynamic calculation software not only needs to go through a series of complex computations, but also stays on the plane figure [1]. Thus it is necessary to establish a three-dimensional model of ternary phase diagram. The visualization of three-dimensional phase diagram can help people understand the complex internal structure of ternary phase diagram intuitively.

The spatial model of ternary phase diagram is described as a special kind of three prisms, which is shown in Figure 1. The equilateral triangle ΔABC of the bottom represents the concentration information of the three constituent elements. Each point in the equilateral triangle can be represented by the relative content of the three elements. The sides of the prism are the relevant three binary phase diagrams A-B, B-C, A-C, the height which is vertical to the bottom of ΔABC represents the temperature.

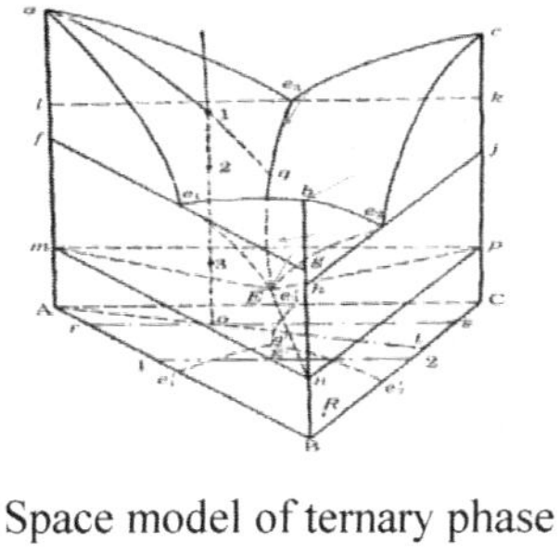

Figure 1. Space model of ternary phase diagram

2. Modeling Approch

The modeling methods of ternary phase diagram mainly use multivariate linear regression methods [2,3] to obtain a three-dimensional phase diagram, the spatial model made from which is barely satisfactory. As the rapid development of computer graphics, the Bezier method for phase diagram expression is proposed [4]. However the method is confined to the liquidus surface of ternary phase for three-dimensional reconstruction. According to the designed data structure and improved algorithm, the line frame model was adopted to realize a three-dimensional topology model of ternary phase diagram in this paper. Not only the model improves the drawing efficiency of isothermal section of phase diagrams, but also can be established easily.

2.1. Phase Diagram Selection

As a new type of alloy with permanent magnetic properties the ternary phase diagram B-Fe-Nd can be processed simply and has low price. It has captured great attention both at home and abroad [5]. So the ternary phase diagram B-Fe-Nd is selected as the research object.

2.2. Data Source

FactSage is one of the largest fully integrated database computing systems in chemical thermodynamics in the world. With FactSage users can calculate the conditions for multiphase, multicomponent equilibria, with a wide variety of tabular and graphical output modes, which is under a large range of constraints. For example, general N-component phase diagram sections can be easily generated with a wide choice of axis variables. FactSage has several hundred industrial, governmental and academic users in materials science, pyrometallurgy, hydrometallurgy and other different fields. It is used internationally in graduate and undergraduate teaching and research. It can provide users with more authorized accurate data.

As shown in Figure 2(a), the vertical section calculated by Factsage was shaped when the ratio of element Fe and element B is equal to 1:1. Vertical sections reflect the topological relations between different phase regions. Since each phase region is a closed set, so we treat the single phase region as the body element to collect, then select corresponding phase boundary face, phase boundary edge and vertices at the single phase region. The phase region Nd(s)+B5Nd2(s)+Nd5Fe17(s) numbered as 1 will be an example to introduce.

A line separating the two phase regions on the vertical section is extending into a plane or a curved surface in space. As shown in Figure 2(a), the phase boundary temperature of phase NO.1 are 924K and 300K. In order to intercept the phase boundary faces of the phase, FactSage is used to calculate the isothermal section of 300K and 924K, as shown in Figure 2(b)&(c). Through isothermal section, the phase boundary faces and vertices 1,2,3,4,5,6 component concentration information of phase NO.1 could be obtained.

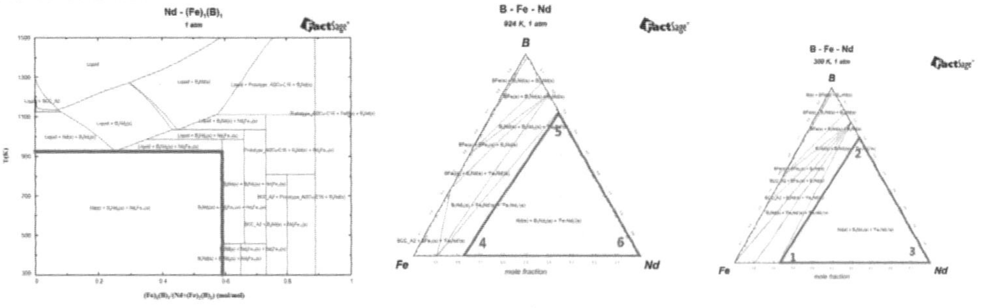

(a) (b) (c)

Figure 2. (a) Vertical section; (b) Isothermal section at temperature 924K; (c) Isothermal section at temperature 300K

2.3. 3D Coordinate Transformation

The composition of ternary phase diagram is usually expressed as an equilateral triangle, which is called the concentration triangle. Any point within a concentration triangle can be expressed as a percentage of the concentration of three components. After the steps mentioned above, the component concentration information of six vertices of phase NO.1 can be figured out. However, to describe the ternary phase diagram in three-dimensional rectangular coordinate system, it's necessary to convert the data in concentration triangle into Cartesian Coordinate System (CCS). Because the meaning of temperature in the coordinate system is the same as that of CCS, we only need to convert the coordinates of the component concentration into CCS. As shown in Figure 3, point A is the origin of CCS. If the length of the equilateral triangle is l, and the composition of the point Q is (B, C), the relationship between the point Q(x,y) in CCS and the composition of the point Q (B, C) satisfies equation 1.

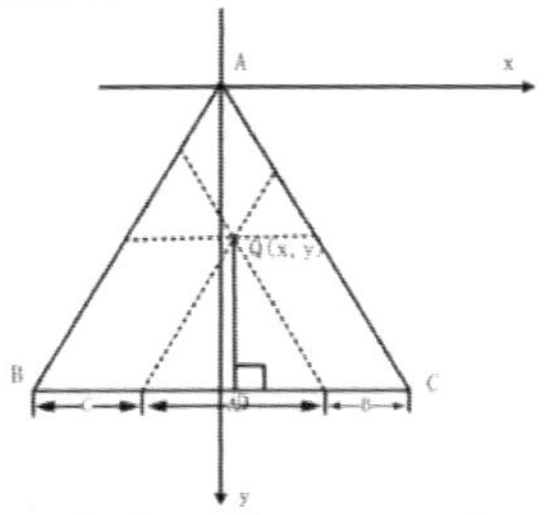

$$\begin{cases} A = l - B - C \\ x = C + A/2 - l/2 \\ y = \sqrt{3}/2(l - A) \end{cases} \quad (1)$$

Figure 3. Coordinate transformation.

2.4. Data Structure Constructing

A series of discrete space points are obtained by means of three-dimensional coordinate transformation. It's necessary to build up a high efficient data structure to organize these discrete points in space. Thus users can quickly access the phase boundary faces, phase boundary edges and vertices of each phase region. The winged-edge data structure is a typical data structure. The structure is well adapted for the representation of a convex polyhedron. In the ternary phase diagram system B-Fe-Nd, the shape of solid phase regions are most convex polyhedron. Secondly, the winged-edge structure clearly describes the geometric and topological features of the faces, edges and vertices of three or more intersecting surfaces. Based on and starting from the edges, it can retrieve the two adjacent surfaces and four adjacent edges of the edge's starting point, as well as the edge's own attributes. In order to simplify the winged-edge structure, an improvement is made to draw the whole body with the right adjacent surface of the winged-edge structure [6,7].

DATA structure of the four tables {

Data object: S, F, E, P;

S represents the voxel, F represents the face structure, E represents the edge structure, P represents the point set;

Data relationship: R={SFR,FER,EPR};

SFR={<s, f>|s∈ S, f∈ F; represents face f in s};

FER={<f, e>|f∈ F, e∈ E; represents edge e in f};

EPR={<e, p>|e∈ E, p∈ P; represents point p in e};

}

Storage mode of winged-edge data structure:

(1)Voxel structure

```
class solid
{
Id solid; /*the number of phase region*/
Face *sface; /*the face number contained in the phase region*/
}
```

The voxel structure is the upper layer of the whole data structure. The phase boundary faces, phase boundary edges, vertices and other elements that constitute the polyhedron can be found by traversing the voxel structure.

(2)Face structure

```
class face
{
Id facenum; /*the number of the face */
 Edge edgeid; /*the edge number contained in the face*/
Int flag; /*a flag of whether to be traversed*/
}
```

Face structure contains the face number, edge number contained in the face and the flag of whether to be traversed. If the flag is 1, it has been traversed.

(3)Edge structure

```
class edge
{
Id edgeid; /*the number of edge*/
Int PS; /* Initial vertex*/
Int PE; /* End vertex*/
Id facenum; /* Right adjacent surface*/
}
```

Edge structure includes the edge number edgeid, starting point PS, ending point PE and adjacent right face facenum.

(4)Vertex set

```
class point
{
id point; /*the number of vertex*/
Int x, y, z; /* vertex coordinates*/
}
```

Vertex set includes the number of vertex and the coordinate of the vertex(x, y, z).

According to the above data structure, the phase region NO.1 is stored in the computer as shown in Table 1.

Table 1. Four tables structure of phase region NO.1

Voxel structure	
Solid	Face num
1	1,2,3,4,5

Face structure		
facenum	Flag	edgeid
1	0	1,2,3
2	0	4,5,6,7
3	0	8,9,10,11
4	0	12,13,14,15
5	0	16,17,18

Vertex set			
point	X	Y	Z
1	-27	50	300

Edge structure			
edgeid	PS	PE	facenum
1	1	2	2
2	2	3	3
3	3	1	4
4	2	1	1
5	1	4	4
6	4	5	5
7	5	2	3
8	3	2	1
9	2	5	2
10	5	6	3
11	6	3	4
12	1	3	1
13	3	6	3

2	14.5	14.5	300		14	6	4	5
3	50	86.6	300		15	4	1	2
4	-27	50	924		16	5	4	2
5	14.5	14.5	924		17	4	6	4
6	50	86.6	924		18	6	5	3

2.5. Transformation from the World Coordinate System (WCS)to the Screen Coordinate System (SCS)
Currently, we have obtained the three-dimensional coordinates of phase region in WCS and organized the discrete coordinate points using the designed data structure. However in order to output the three-dimensional phase diagrams on computer screen, the conversion from WCS to SCS will be involved in. Detailed conversion processes are shown in Figure 4.

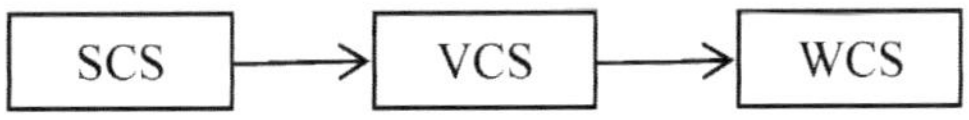

Figure 4. Conversion processes

The coordinate system where three-dimensional phase diagram locate is WCS, which is defined using right-handed coordinate system [8]. While Viewing Coordinate System (VCS) is a coordinate system, which a view point and projection plane can be defined. And by standing on such view point, the three-dimensional objects can be projected to projection plane. SCS is a coordinate system where the diagram on projection plane is displayed on screen. WCS, VCS and SCS are correlated with each other. VCS is evolved on the basis of WCS. The object displayed in SCS is the projection of three-dimensional object in projection plane of VCS.

2.6. Display a single phase region
After the steps mentioned above, the visual expression of the phase region could be displayed on the computer screen intuitively. Currently, the existing written phase diagram data or some phase diagram calculation software such as FactSage and Pandat can't present the specific form of a single phase region. However based on the data structure defined above, it is easy to realize the reconstruction of a single phase region. The voxel structure as the root node store the number of phase regions, which is equal to the index data and guide us to extract patch set displayed in the phase regions. It isn't hard to obtain the information of edge structure of all the patch sets in the phase region by searching the number of patch set in face structure. Accordingly, all the information of vertices can be get according to the edge structure. Then the result of phase region NO.1 Nd(s)+B5Nd2(s)+Nd5Fe17(s) and region NO.5 B4Nd(s)+Nd2Fe17(s)+Nd5Fe17(s) was listed in Figure 5(a)&(b).

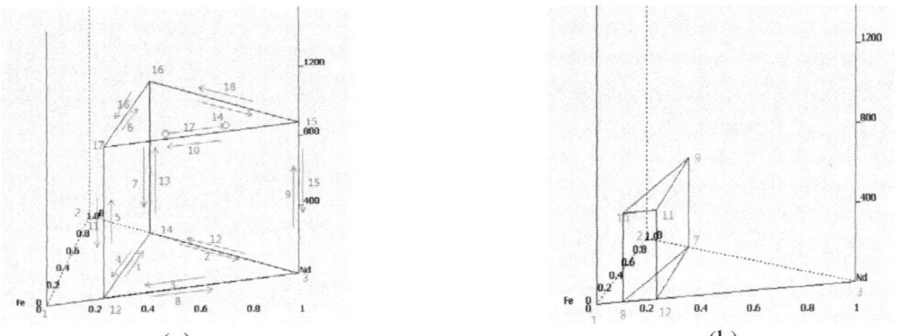

(a) (b)

Figure 5. (a) Phase region Nd(s)+B5Nd2(s)+Nd5Fe17(s); (b) Phase region B4Nd(s)+Nd2Fe17(s)+Nd5Fe17(s)

3. Discussion

According to the method proposed in this paper, the three-dimensional topological model of any ternary phase diagram can be realized through a small amount of data and the isothermal section of model is drawn. And the writer extracts the single-phase region in which users interested. The visualization of ternary phase diagram help people to understand the complex internal structure of ternary phase diagram.

4. Conclusion

Based on the concept from whole to part, the paper firstly divided the whole system into different phase regions, and then restored all phase regions with well-designed data structure and algorithm, finally combined all phase regions into a whole body. Thus the topological model of ternary phase diagram B-Fe-Nd was realized, shown in Figure 6. Regarding this system, it can also obtain the isothermal section diagram of a random temperature. See the isothermal section diagram at 900K in Figure 7.

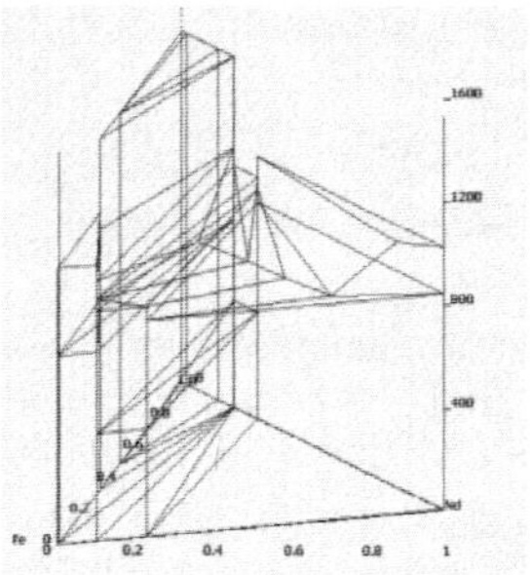

Figure 6. Topological model of the ternary phase diagram B-Fe-Nd.

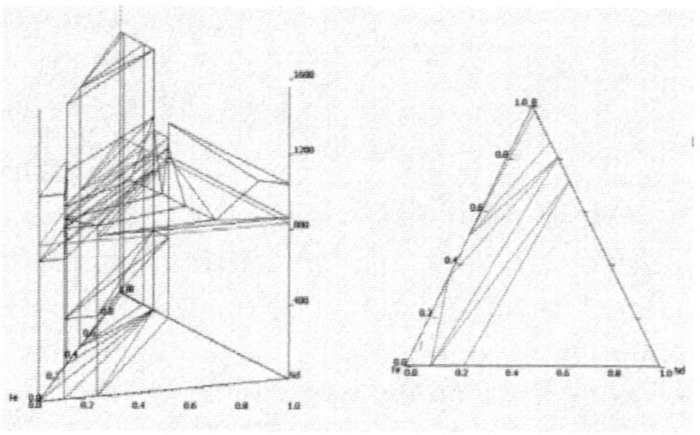

Figure 7. Isothermal section diagram at temperature 924K.

Acknowledgments

The research work was supported by National Natural Science Foundation of China under Grant No.51090381 and National Program on Key Basic Research Project (973 Program) under Grant No.2013CB632601.

References

[1] A reference Ming Zheng, Chenguang Bai, Lingyan Dong, etc 2005 The research progress of digital phase diagram system, J. *Journal of China's rare earths*. 23.206-210.

[2] Ming Zheng 2005 *Digital research for ternary oxide phase diagram*, D. Chongqing University.

[3] Lingyan Dong, Dengfu Chen, Chuanjun Wang, etc. 2004 Digital research for Al2O3-SiO2-TiO2 ternary phase diagram, *J. Journal of China's rare earths*. 22. 101-104.

[4] Tian Hang 2013 *The Bezier method expression for phase diagram*, D. Beijing University of Science and Technology.

[5] Pingguo Xu, Guoquan Liu, Weijing Zhang 1995 The computer expression of the topological structure of the three-dimensional phase diagram,*J.Journal of iron and steel research*, (5):55-60.

[6] Hong Bao, Wenling Dong 2015 *The calculation and the drawing of equivalent sections of model-based 3D graphics*[C]// International Congress on Image and Signal Processing. IEEE.

[7] Ying Xu 2009 A 3D Facial Model Editing System Based on Modified Winged-edge Data Structure. *Journal of Mianyang Normal University*,28(2): 89-93.

[8] Mingtian Ni, Liangzhi Wu 1999 *Computer Graphics[M]*. Peking University Press.

HTTM - Design and Implementation of a Type-2 Hypervisor for MIPS64 Based Systems

Qurrat ul Ain[a], Usama Anwar, Muhammad Amir Mehmood and Abdul Waheed

Al-Khwarizmi Institute of Computer Science, University of Engineering and Technology, Lahore, Pakistan

[a] qurratulain@kics.edu.pk

Abstract. Virtualization has emerged as an attractive software solution for many problems in server domain. Recently, it has started to enrich embedded systems domain by offering features such as hardware consolidation, security, and isolation. Our objective is to bring virtualization to high-end MIPS64 based systems, such as network routers, switches, wireless base station, etc. For this purpose a Type-2 hypervisor is a viable software solution which is easy to deploy and requires no changes in host system. In this paper we present the internal design HTTM –A Type-2 hypervisor for MIPS64 based systems and demonstrate its functional correctness by using Linux Testing Project (LTP) tests. Finally, we performed LMbench tests for performance evaluation.

1. Introduction

Recent years have seen manifold increase in the processing power of computing devices. However, these hardware advancements become ineffective if the software running on top of such hardware is not able to fully utilize its processing capabilities. Virtual environment enables us to safely use a single hardware for multiple purposes. Virtualization is an effective software solution for multiple fields of computing. For instance, in the form of Type-1/ bare-metal hypervisors (e.g., Xen [1]), it has revolutionized the sever domain while Type-2/ hosted hypervisors (e.g., VMWare workstation [2]) have proved their benefits for client domain.

Virtualization has attracted developers of embedded systems to consider it as a viable solution for their problems such as security, consolidated infrastructure, isolated debugging environment, etc., [3-5]. Companies like Motorola and VMware have used virtualization as a possible solution by creating Evoke [6] and Mobile Virtualization Platform (MVP) [7] respectively. Gu et al. [8] have implemented a real-time application on integrated automotive gateway platform. They used KVM (Type-2 hypervisor) as their virtualization layer and created two guest systems. One guest executes security critical tasks e.g., parking assistance while the other run applications that can compromise security due to connection to external network e.g., GPS navigation or mobile office.

There are number of MIPS based devices already installed and functional in the current network infrastructure including security gateways, network routers, switches, base station, etc. To benefit from the valuable aspects of virtualization, these devices can use a Type-2 hypervisor which can be easily installed as a user process without altering existing infrastructure.

In this paper, we have designed HTTM -- **H**ypervisor **T**ype-**2** **M**IPS-On-MIPS which support full-virtualization i.e., it does not need the host or guest Linux OS to be patched or changed. In Section 2,

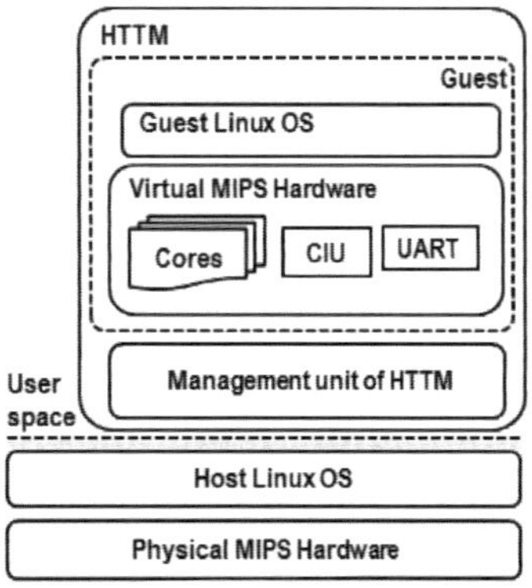

Figure 1. Overview of HTTM infrastructure on Linux systems.

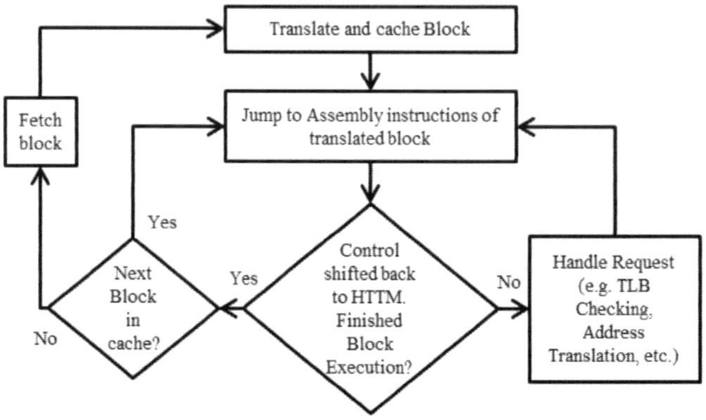

Figure 2. Execution flow of each core thread.

we present the internal design of hypervisor while in Section 3 we describe the results of its functional testing. In Section 4 we conclude this presented work.

2. Design of HTTM

HTTM is a user space process for Linux based MIPS64 host system. It creates an abstract layer of virtual resources for the MIPS64 guest Linux system running on it. For development and debugging purposes we used Cavium Octeon MIPS64 evaluation boards. Octeon II CN6860 board with 12 cores, 1.2 GHz, and 4GB RAM was used as a host system, which runs their customized 2.6.32 Linux kernel. Octeon CN5750 was the guest/ target system with 12 cores, 800 MHz, and 1GB RAM.

The virtual resources consist of CPU cores, memory management Unit (MMU) and some I/O devices. Figure 1 shows an overview of HTTM which is running on host Linux system along with the other user space processes. HTTM can broadly be categorized in two sections: i) Virtual resource and ii) Management. The virtual resource section is responsible for correct representation of guest/ target system and its on-board units that includes cores, Translation Look-aside Buffer (TLB), General Purpose (GP) and Co-Processor (CP) registers, etc. While the management section helps that virtual resource to complete its tasks correctly and efficiently. It includes Dynamic Binary Translation (DBT) unit, caches, guest-to-host translation unit, and signal handler.

Multithreaded approach was implemented in HTTM with cores and devices executing in separate threads. The basic execution cycle of a core thread is simple *fetch-translate-execute*. A set of instructions with control shifting instruction at the end (forming a block) is fetched from the guest code and translated into set of host-safe instructions which is then directly executed on the host. If interrupts or exceptions are generated during the execution, control is transferred to the management section which handles it accordingly. Figure 2 shows a simplified version of execution cycle in a core thread. Next we give a brief description of different units of HTTM and virtual guest system.

2.1. DBT for MIPS64 instruction set

As HTTM executes in the user space rather than directly on hardware, particular constraints of user space hinders in direct execution of guest MIPS instruction over host system. For instance host Linux system inhibits the user space process to write over its *ra* (return address) and *sp* (stack pointer) registers, therefore we cannot directly map the guest GP register to host's GP registers. Hence, we need DBT to translate the guest assembly instructions into host-safe assembly instructions and then execute them directly on hardware. Other than basic translation mechanism for ISA virtualization, correct execution of guest code requires addition of some set of instructions in the translated block e.g., privilege check for privileged instructions, Program Counter (PC) increment at every instruction, etc. Due to this fact the translated block becomes larger than the guest code block. We have provided translation for 184 MIPS instructions out of which 159 were standard while 25 were Cavium specific instructions. Cache is maintained for translated blocks to avoid translation overhead.

2.2. *Memory management unit*

Standard MIPS64 memory map is divided into different regions depending on i) whether the target address requires TLB for translation, ii) the current mode of execution (kernel or user), and iii) whether the address is cached or not. Also there was an additional Cavium specific memory segment called *CVMSEG*, whose implementation was inevitable for the correct execution of guest system. The guest uses it as a scratch pad memory during context switching. In Cavium Linux system the memory management is done through TLB which provides virtual to physical address translation. TLB is managed by the kernel with four basic assembly instructions. i) *tlbp* probes the TLB for a entry, ii) *tlbr* reads the found entry, iii) *tlbwr* writes the TLB at random index, and iv) *tlbwi* write the entry on a particular index. HTTM implements the required functionality by making the intended changes in guest's TLB against these instructions. As this is a virtualized environment, guest TLB provides a mapping from guest virtual address (GVA) to host virtual address (HVA) so that guest can load or store to host actual memory. For doing so, we have to provide a two level of translation i.e., GVA-to-GPA and GPA-to-HVA. In case of invalid address access or any other error, relevant exceptions are generated by the hypervisor and dispatched to the guest.

2.3. *Timing infrastructure*

Linux kernel keeps track of the time by receiving a periodic timer interrupt that provides the basic framework for the time synchronization in the OS. Correct implementation is crucial for a stable guest system. In native system, simple *Count-Compare* timing mechanism is implemented in which *count* register is incremented every clock cycle. Due to the fact that our host system is also Cavium Octeon we use host *count* register to calculate the current time for guest. When the guest reads its *count* register, HTTM intercept the call and in action reads the host *count* register and write to guest *count* register after converting the value according to the guest clock. In case if the guest system needs a timer interrupt, HTTM registers a timer of required duration with the host. Upon timer expiration, host sends a SIGALRM signal and HTTM notifies the guest by sending an interrupt.

2.4. *Interrupt and exceptions*

Interrupts originates from an external source and occur asynchronously. Cavium Octeon system has Central Interrupt Unit (CIU) which dispatches the external and inter-core interrupts to the destined core. Implementation of CIU was critical due to the fact that it is involved in interrupt handling and proper working of multiple cores. Exception is any event that can cause a change in the execution flow that can be programmed or un-programmed too. In case of programmed exception, control shifting instructions of guest are replaced with instructions that shift the control to HTTM during DBT. During execution control is shifted back to HTTM which handles the case accordingly. While during un-programmed exceptions, specific guest registers are set with the current status information and control is shifted towards guest's exception table. At the end of exception service routine *eret* instruction occurs, on which HTTM directs the guest back to its normal execution by restoring its PC value.

2.5. *I/O devices*

Implementation of I/O devices is one the most important feature in the virtual guest. We have added the support for UART, disk, and networking in HTTM. Following is brief description of these devices.

2.5.1. UART. Console is a basic necessity of any virtual system and for that purpose UART implementation is required at a very early development stage. UART provides serial communication of characters to and from different cores. UART implemented in HTTM consists of two threads: i) Receiver thread that continuously checks for the availability of input characters from the keyboard and whenever input is available, it generates an interrupt for a particular core through CIU. ii) Transmitter thread, which helps processor in printing all the messages on the console. After transmitting data, it notifies the processor through CIU that UART is free for transmission.

2.5.2. Network. Communication with the external world is a important part of MIPS based devices. Instead of following the typical inefficient approach of device emulation, HTTM employs Virtio API [9] and Vhost_net module to implement an efficient network virtualization solution. Virtio_net driver act as the front-end driver in the guest and can easily be configured in the guest Linux kernel as MMIO (Memory Mapped Input Output). Being MMIO device it doesn't require PCI bus for communication. Vhost_net provide the emulation of network card in the host kernel. Although, Vhost_net does not emulate a complete Virtio PCI adapter, instead it restricts itself to Virt-queues operations only. These two drivers are connected through network device in HTTM. It manages bi-directional control signals between Virtio_net and Vhost_net using device specific registers, e.g., VIRTIO_MMIO_STATUS, which are used by the guest driver for Virt-queues sharing and device controlling.

2.5.3. Disk. The disk is one of the key requirements of any virtual system to enable data persistence. In general, Virtio and Vhost Linux drivers can be used for both character and block devices, (e.g., console and disk respectively). Following the footsteps from network implementation, we have used Virtio_Blk driver for disk. A simple raw file is created on host system through which data can be transferred from host to guest and act as a disk for guest. Virtio_Blk is also a MMIO device similar to Virtio_net and thus do not rely on PCI bus. For making Virtio_Blk front-end driver for virtual disk in the guest, it needs to be explicitly configured. Virtual disk file is passed to Virtio_Blk by the backend driver Vhost_Blk. Note that Vhost_Blk is not the part of the standard Linux kernel. For our purpose, we choose the basic implementation of the Vhost_Blk from [10], and modified it for the Cavium Octeon MIPS64 kernel. Once this module is specifically built, it is loaded in the host kernel. To enable the communication between front-end and backend drivers, Virtual Block Device was implemented which passes the control signals in both directions. It notifies the host kernel when data is provided by the guest or sending interrupt to the guest when Vhost_Blk completes its assigned task.

3. Performance Evaluation

Further we present functional correctness of the single-core guest system followed by the performance evaluation using standard benchmarking tool. Here the native system refers to the real target system and guest is the virtualized system.

3.1. Linux testing project

To verify that the virtual guest system created by HTTM is functionally correct, we rely on Linux Testing Project (LTP) [11]. We successfully tested various subsystems including memory, file system, scheduling, Interprocess communication (IPC), timer, system calls, and network. Some of the tests include *ftest(01, 02), mem(01, 02), trace_schedule, shmem_test_(01, 07), signal_test_(03,04), alarm(01-07), mmap09, vfork(01-02)* and *pingpong.*

3.2. LMbench

Next, we rely on LMbench [12] which is a benchmarking tool for performance evaluation that contains different types of tests. These tests were performed on native hardware and virtual guest system. We define "overhead" as the ratio of native and guest system data transfer rates. It highlights the performance penalty due to the virtualized environment of HTTM. Table 1 shows the test results of different LMbench tests. *lat_ops* measures the latency of basic CPU operations, such as add, mul, bit (XOR) etc. For memory bandwidth measurements, we executed *bw_mem* tests for basic operations including read, write, and copy. These tests were performed with 4MB memory size. For network testing, we performed *bw_tcp* test that measures TCP bandwidth between a server on the guest and a client on an external system. In general, we found that virtualization overhead varies from 5.5 to 24.11

in different tests. We note that the overhead in different design units of HTTM is varying, e.g., ISA translation for guest instructions takes a considerable amount of time.

Table 1. Performance results of LMbench.

	int64 bit (nsec)	**int64 add (nsec)**	**int64 mul (nsec)**	**bw_mem rd (Mb/sec)**	**bw_mem wr (Mb/sec)**	**bw_mem cp (Mb/sec)**	**bw_tcp (Mb/sec)**
Native	1.25	1.31	0.50	821.45	2918.56	718.97	11.60
Guest	12.24	31.59	5.33	116.00	128.81	65.44	2.10
Overhead	9.79	24.11	10.66	7.08	22.66	10.99	5.50

For detailed information we refer to working demonstration of HTTM and documentation available at [13]. This research work is funded by National ICT R&D Fund, Pakistan under the project "Development of Type-2 Hypervisor for MIPS64 Based Systems".

4. Conclusion

In this paper, we present a design and implementation of full-virtualized Type-2 hypervisor for MIPS64 based high-end devices. We present a brief description of its software architecture and brief detail of each unit. Finally we tested the guest system created by hypervisor using LTP. In future we plan to analyse performance of HTTM in detail and optimize it for real-time applications to demonstrate real-time efficacy.

References

[1] Barham P Dragovic B Fraser K Hand S Harris T Ho A Neugebauer R Pratt I and Warfield A 2003 Xen and the art of virtualization *ACM SIGOPS Operating Systems Review* **37** 164
[2] Vmware *VMWare Workstation* 2013 [Online] Available: www.vmware.com/products/player/
[3] Cox L P and Chen P M 2007 Pocket Hypervisors: Opportunities and Challenges *Proc. of the 8th IEEE Workshop on Mobile Computing Systems and Applications* (Tucson, Arizona: IEEE) p46
[4] Aguiar A and Hessel F 2010 Embedded systems virtualization: The next challenge? *21st IEEE Int'l Symp. on Rapid System prototyping (RSP)* (Virgina, USA: IEEE) p1
[5] Heiser G 2011 Virtualizing embedded systems: why bother? *Proc. of the 48th Design Automation Conference* (NewYork, USA: ACM) p901
[6] Heiser G 2009 The Motorola Evoke QA4-A Case Study in Mobile Virtualization
[7] Barr K Bungale P Deasy S Gyuris V Hung P Newell C Tuch H and Zoppis B 2010 The VMware mobile virtualization platform: is that a hypervisor in your pocket? *ACM SIGOPS Operating Systems Review* **44** 124
[8] Gu Z Wang Z Li S and Cai H 2012 Design and implementation of an automotive telematics gateway based on virtualization *15th IEEE Int'l Symp. on Object/Component/Service-Oriented Real-Time Distributed Computing Workshops* (China: IEEE) p53
[9] Jones M T, 2010 *Virtio: An I/O virtualization framework for Linux*. IBM developerWorks.
[10] Pulavarty B 2010 *vhost-blk implementation* [Online] Available: http://lwn.net/Articles/379864/
[11] Larson P 2002 Testing Linux® with the Linux Test Project *Ottawa Linux Symp.*(Ottawa, Canada) p265
[12] Mcvoy L W and Staelin C 1996 lmbench: Portable Tools for Performance Analysis *USENIX Annual technical Conf.* (San Diego, USA: usenix) p279
[13] *Type-2 Hypervisor for MIPS64 based Systems* [Online] Available: http://www.kics.edu.pk/project/downloads/53

Golden Rectangle Treemap

Liangfu Lv[1,2], Shiliang Fan[1], Maolin Huang[3], Weidong Huang[4] and Guolan Yang[1]

[1] School of Software, Tianjin University, Tianjin, P. R. China.
[2] Mathematics Department, Tianjin University, Tianjin, P. R. China.
[3] Faculty of Engineering and IT, University of Technology, Sydney, Australia.
[4] School of Engineering and ICT, University of Tasmania, Tasmania, Australia.

Mao.Huang@uts.edu.cn

Abstract. Treemaps, a visualization method of representing hierarchical data sets, are becoming more and more popular for its efficient and compact displays. Several algorithms have been proposed to create more useful display by controlling the aspect ratios of the rectangles that make up a treemap. In this paper, we introduce a new treemap algorithm, generating layout in which the rectangles are easier to select and hierarchy information is easier to obtain. This algorithm generates rectangles which approximate golden rectangles. To prove the effectiveness of our algorithm, at the end of this paper several analyses on golden rectangle treemap have been done on disk file system

1. Introduction

Hierarchical structures of information are widely used nowadays: computer program, family trees, directory structure and so on. It's easy to locate information in small hierarchical structures, but large hierarchical structures are hard to handle. Several algorithms have been presented to solve this problem. Treemap[1] provide a method to generate rectangles recursively according to the given relations of nodes. This method enjoys the advantages such as high efficiency in generating layouts and great effectiveness in showing the size of final elements. But shortcomings are also significantly, emergence of thin and elongated rectangles and unclearness of hierarchical structures are the mutual problems of the treemap methods.

Among all the methods, Squarified treemaps algorithm generates the lowest aspect ratio on static data. Squarified treemaps[2] are introduced to solve the problem of thin and elongated rectangles, generating sub rectangles which approximate squares. In addition, this method makes full use of the display area.[3]

However, all of the traditional methods are focusing on improving space utilization and layout formation. Generated layout still has thin and elongated rectangles. In this paper, we combine traditional treemaps algorithms and golden rectangle to introduce the advantages of golden rectangle into traditional treemap algorithms. The golden rectangle has been known since ancient times as one having a pleasing shape. Many artists and architects have been fascinated by the presumption that the golden rectangle is considered aesthetically pleasing and balance. So we propose this method to generate a more aesthetically pleasing and rationally layout.

In geometry, a golden rectangle[4] is a rectangle whose side lengths are in the golden ratio. The golden ratio is a special number found by dividing a line into two parts so that the longer part divide by the smaller part is also equal to the whole length divided by the longer part. It is often symbolized using phi. In mathematical description, that is $\varphi = a/b = (a+b)/a = \frac{1+\sqrt{5}}{2} = 1.6180339887 \approx 1.618$. [5][6][7]

2. Related work

Many methods are introduced to go through and display hierarchical structures which also known as tree. File browsers are the best known example[2][8], in which a listing of the file and directories are displayed, where the levels in the tree is showed by indentation. Only limited number of files and directory can be showed. Another well-known example is tree diagrams[1], it contains node and link. Elements are shown as node and relations are shown as links. However, these methods are not capable of giving overview of the whole structure[8]. Moreover, only limited information can be displayed via these methods.

In 1990s, treemaps[1] were firstly proposed to deal with the above problems, which use full use of the displaying area to show the tree content. Each node has a name and size. The size of leaves represents the size of related file. The size of non-leaf node is the summation of all its children nodes' size. Treemaps are generated via recursive separation of rectangles.

Squarified treemaps were introduced to avoid the emergence of thin and elongated rectangle[2]. Generated square-like items are easier to detect and to compare in size. Besides, efficiency of space usage is improved. However, this algorithm does not work well with dynamic data.

In this paper, golden rectangle treemaps is proposed, which combines golden rectangle with Treemaps. In geometry, a golden rectangle[4] is a rectangle whose side lengths are in the golden ratio. The golden ratio is a special number found by dividing a line into two parts so that the longer part divide by the smaller part is also equal to the whole length divided by the longer part. And golden ratio is calculated as $\varphi = \frac{1+\sqrt{5}}{2} = 1.6180339887 \approx 1.618$. People have a long time using golden rectangles, for example the Parthenon Temple. Figure 1 shows golden rectangle in Parthenon temple[9].

Figure 1. Parthenon Temple[9]

Kong's experiment shows that not only extreme aspect ratio but also squares should be avoided in treemaps.[10] So using golden rectangle can also solve the problem of Squarified treemaps.

In this paper, we introduce aesthetic pleasing factors into treemaps methods. We generate layout in which rectangle approximate golden rectangle. And the presentation of nodes and leaves as golden rectangle has several advantages:

-efficient in display space using.

-easier to grasp relation between different nodes.

-golden rectangles are easier to detect and point at

-size comparison is easier between rectangles which approximate golden rectangles.

3. Golden rectangle treemap

During segmentation, we use heuristic algorithm to subdivide initial rectangle. The key idea is based on two thoughts. First, we don't take all levels of a tree into consideration at a time. We treat siblings at same level as a whole and fit these nodes in the rectangle which represent their parents' node. Then we apply the same method recursively to next level of the tree. Thus the rectangles that the next level nodes divide into are golden-rectangle-like, which offers good chances for good separation. Second, we sort siblings in descending order according to their weight. When computing position for nodes to fit in, we choose the option that making golden rectangle parameter $GRP = (MAX\left(\frac{w}{h},\frac{h}{w}\right) - \varphi)^2$ closer to 0.

We calculate the weight of as nodes follows: define tree T in composed of N nodes. The root of T has M children.

$$w(v) = \begin{cases} w(v), node\ v\ is\ a\ tree\ node \\ \sum_{j=0}^{k-1} w(v_j), node\ v\ has\ k\ children \end{cases} \quad (1)$$

And the tree T have v_0 as root are defined as $T(v_0) = \{v_0, v_1, \cdots, v_{n-1}\}$ $v_i(i \in \{0,1,\cdots,n-1\})$ are the whole siblings of tree T. $W(v)$ represent the weight of node v.

3.1 Algorithm

Here we present our algorithm for golden rectangle treemaps. this algorithm is not capable of providing optimal result, because optimal result is time consuming. We decide to generate layout in reasonable time. So we use heuristic function during the whole algorithm,

$$GRP = (MAX\left(\frac{w}{h},\frac{h}{w}\right) - \varphi)^2 \quad (2)$$

This parameter is used to judge optimal situation during dividing. we think optimum is reached when adding next rectangle

$$GRP_{after} - GRP_{before} > 0 \quad (3)$$

GRP_{after} refers to GRP after adding next rectangle, GRP_{before} stands for GRP before adding next rectangle. In this situation, we think optimum is reached before this insertion. So we will stop adding this rectangle into the section which have already reached optimum. This rectangle is going to be added into R_{remain} (unprocessed domain of current processing rectangle). That means, when processing a rectangle, a decision is made between two alternatives. Either rectangle is added into current section, or current section is fixed or reaching optimum, new insertion should be started in the remaining area of current processing rectangles.

According to the shape of rectangle being processed, division is different. If parameter

$$vert = w_p - h_p > 0 \quad (4)$$

w_p represents the width of processing rectangle, h_p stands for the height of processing rectangle, subdivision starts from the left part. And new rectangle will be added above original rectangle until reaching optimum. To subdivide a horizontal rectangle. We will add rectangles from bottom, and new rectangles will be added on the right of previous rectangles until reaching optimum in this section.

The whole algorithm can be summed up as following,

Step1: Deciding processing rectangle shape.

$$shape = \begin{cases} vertical, & vert > 0 \\ horizontal, & vert \le 0 \end{cases} \quad (5)$$

And adding next rectangle into current section according to their shape. After adding in, compute parameter GRP.

Step2: Add next rectangle into current section, and parameter GRP will be computed. Then, decide whether current section has reached optimum before adding in this rectangle.

Step3: if current section has reached its optimum, then addition in step2 will be abandoned. And the rectangle to be added will be placed into remaining section. Then repeat step1 and step2.

Step4: optimum has not been reached in current section, keep repeating step2 until reaching optimum.

Stop condition: all the rectangles have been added into initial rectangle. This algorithm stops.

4. Experiments and results

To evaluate the performance of golden rectangle treemaps layout algorithms, we performed two experiments. The first experiment is visualizing disk file system, the original goal of designing treemaps. Our goal was to compare out algorithm with standard treemaps algorithm and Squarified treemaps algorithm. Besides, we also did a user study to compare layout readability between 6 algorithms.

4.1 Disk file system

Visualizing file system is the original goal of treemaps. In this experiment, we read in file system information from folder A in local computer. Root folder A is composed of 12 folders and 13 files. One of its child folders D contains another 3 folders, so the data structure of folder A has 4 levels. Every file is represented with a rectangle and rectangle's area shows the size of file. Structure of folder A is shown in Table 1.

Figure 2 shows result of standard treemaps. In Figure 3 (a) (A) is the output of Squarified treemaps and (B) is the enlarged view of red region in (A).

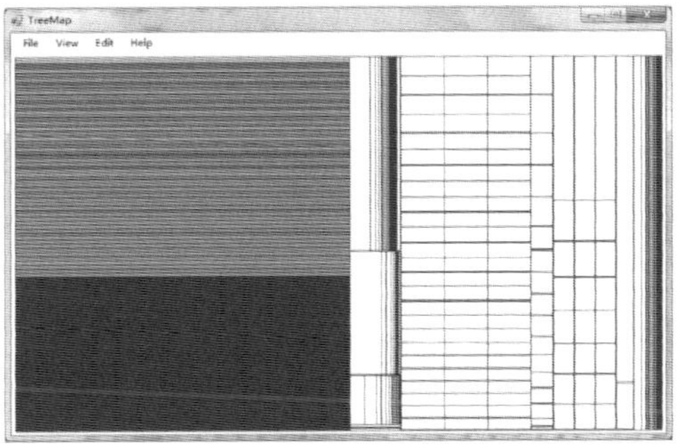

Figure 2. Standard Treemap Result

Table 1. Structure of folder A

Folder Name	Parent Folder	Children Folder Number	Contain File Number
A	— —	12	13
B	A	0	229
C	A	0	6
D	A	3	10
E	A	0	19
F	A	0	1
G	A	0	2
H1	A	0	8
H2	A	0	8
H3	A	0	8
I1	A	0	25
I2	A	0	25
I3	A	0	25
D1	D	0	10
D2	D	0	10
D3	D	0	20

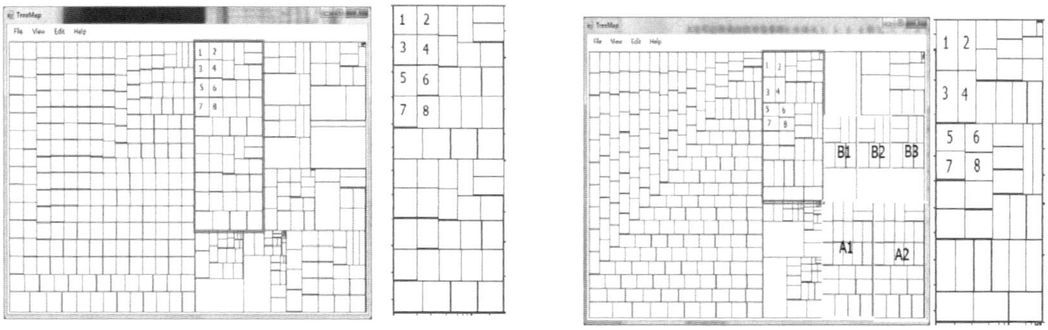

(A) (B)
(a) Squarified Treemap Result

(A) (B)
(b) Golden Rectangle Treemap Result

Figure 3. Comparison Between Squarified Treemap and Golden Rectangle Treemap

In Figure 3 (a)(B), it is hard to identify files 1, 2, ..., 8 belongs to which parent folder. So it loses structural hierarchical information. However, in Figure 3 (b)(B), we can tell easily that 1,2,3,4 and 5,6,7,8 belongs to different folders. (B) is enlarged view of (A). Besides, from Figure 3 (b)(A), we can see region A1 and A2 has the same structure and size. So is B1, B2, B3. Although Squarified treemap has this advantages, but it appears far less often than our algorithm.

Figure 4 shows the result of combining Cushion treemap[11] and our algorithm, using 3D effect highlights hierarchical structure.

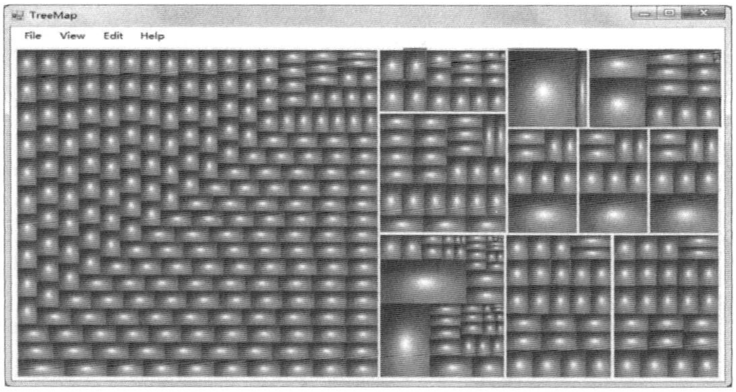

Figure 4. Combine Cushion Treemap with Golden Rectangle Treemap

5. Conclusions

In this paper, we proposed a new method to generate layout for hierarchical structures of information. We forced rectangles in the layout to be more golden rectangle like. This algorithm not only takes in advantages of squarified treemaps but also have two more advantages. First, our algorithm shows hierarchical structure more clearly. Second, comparison in size is more friendly. Size comparison between golden rectangles is easier than that between squares.

Acknowledgments

The authors wish to thank anonymous reviewers for spending valuable time in reviewing our paper and providing us with important comments to improve the paper. This work was supported in part by Science Foundation of Tianjin grant 15JCQNJC00200 and the FEIT Ph. D post thesis publication scholarship of University of Technology, Sydney, Australia.

References

[1] Shneiderman B 1992 *ACM Transactions on graphics (TOG)* **11(1)** 92-99.
[2] Bruls M, Huizing K and Van Wijk J J 2000 Data Visualization 2000. *Springer Vienna* **2000** 33-42.
[3] Marson F and Musse S R 2010 *International Journal of Computer Games Technology* **2010** 7.
[4] Benjafield J 1976 *The American journal of psychology*
[5] Livio and Mario 2002 *The Golden Ratio: The Story of Phi, The World's Most Astonishing Number* (New York: Broadway Books)
[6] Piotr Sadowski 1996 *The knight on his quest: symbolic patterns of transition in Sir Gawain and the Green Knight* (University of Delaware Press) p 124
[7] Richard A Dunlap 1997 *The Golden Ratio and Fibonacci Numbers* (World Scientific Publishing)
[8] R. Chimera, K. Wolman and B. Shneiderman 1991 *Evaluation of three interfaces for browsing hierarchical tables of contents* (Technical Report Technical Report CAR-TR-539 CS-TR-2620 University of Maryland)
[9] The Parthenon and Phi, the Golden Ratio: *http://www.goldennumber.net/parthenon-phi-golden-ratio/*
[10] Kong N, Heer J and Agrawala M 2010 *IEEE Trsnscations on Visualization and computer Grgapics* **16(6)** 990-8.
[11] Van Wijk J J and Van de Wetering H 1999 *Information Visualization* **1999** 73-78.

Bayesian Framework with Non-local and Low-rank Constraint for Image Reconstruction

Zhonghe Tang, Shengzhe Wang, Jianliang Huo, Hang Guo, Haibo Zhao and Yuan Mei

Department of Guided and Information Engineering, Southwest Institute of Technology and Physics, 630811, Sichuan, China

E-mail: sinawang21@126.com, huoliaang@163.com, 408306714@qq.com, haibozhao125@sina.com, meiyuan1984@foxmail.com

Abstract. Built upon the similar methodology of 'grouping and collaboratively filtering', the proposed algorithm recovers image patches from the array of similar noisy patches based on the assumption that their noise-free versions or approximation lie in a low dimensional subspace and has a low rank. Based on the analysis of the effect of noise and perturbation on the singular value, a weighted nuclear norm is defined to replace the conventional nuclear norm. Corresponding low-rank decomposition model and singular value shrinkage operator are derived. Taking into account the difference between the distribution of the signal and the noise, the weight depends not only on the standard deviation of noise, but also on the rank of the noise-free matrix and the singular value itself. Experimental results in image reconstruction tasks show that at relatively low computational cost the performance of proposed method is very close to state-of-the-art reconstruction methods BM3D and LSSC even outperforms them in restoring and preserving structure.

1. Introduction

The search for efficient image denoising methods still is a valid challenge. The local smoothing methods and the frequency domain filter aim at a noise reduction and at a reconstruction of the main geometrical configurations but not at the preservation of the fine local structure, details and texture. Early local denoising method relied on various smoothness assumptions, such as anisotropic filtering, total variation or image decomposition on fixed basis. In some degree, non local methods all explicitly exploit the self-similarities of natural images and represent each clean image patch as a linear combination of a few elements from a basis set called a dictionary. In essence, they are all data driven representation methods. However, the difference among these denoising algorithms is the source of atoms in the dictionary and the constraints to the representation coefficients. In NLM [1], the atoms are from the noisy image and the coefficients depend on the similarly between two noisy image patches. Based on a nonlocal regularity assumption, the NLM [1] estimates an unknown pixel by a weighted average of local and non-local pixels throughout the entire image. To BM3D [2], it is based on classical fixed orthogonal dictionaries and requires the corresponding representation coefficient to be sparse. It restores noisy patches by finding more similar patches in the noisy image using patch matching, stacking fragments together into a 3D signal group, and denoising the group by using collaborative filtering. Inspired by BM3D[2] , K-SVD[3] scheme, employs a joint sparse decomposition based on a learned, possibly over-complete, dictionaries adapted to specific image patches and assumes if two image patches are similar[4], the coefficients should be similar in structure or the distribution of nonzero entries.

Compared with conventional NLM [1] and BM3D [2], LSSC [4] takes both the structure and the sparsity into consideration. However, there still exist some shortcomings. To improve universality of dictionary, its size should be huge, but the learning of a huge dictionary and the determining of a high-dimensional but sparse coefficient matrix are very computationally expensive. In essence, the structured similarity between the columns of coefficient matrix is equivalent to the constraint that the image matrix or coefficient matrix [2, 4, 5, 6, 7] has a low rank. Data matrix stacked by similar image patches is born with low rank characteristic [4, 7], but this property is rarely exploited directly in spatial domain.

In this paper, an image denoising method is proposed which is built upon the methodology of 'grouping and collaboratively filtering'. It combines two now classical ideas into a single framework: The low rank joint decomposition and the non-local means idea. The fundamental idea is to convert the problem of removing noise to a low rank matrix approximation problem and a weighted nuclear norm is defined to replace the classical nuclear norm. we focused primarily on low rank constraint in the spatial domain instead of the transform domain. For each image patch centered at the current pixel a similar patch series is found in a search window. The numerical rank of corresponding noise-free matrix is predicted based on the estimation of noise variances and singular values of noisy data matrix, then some significant singular values are retained according the numerical rank. Then, for each retained singular value, an adaptive thresholds or weight are calculated. Then shrinkage is carried out on the retained singular values. Lastly, the recovered image patch can be easily reconstructed based on the shrunk singular values.

2. Related work

Essentially there are three types of image denoising method: local, non-local and the mixture of them. Our approach is a non-local approach. Thus, we will only discuss the most related non-local techniques.

In the non-local image denoising, we always are required to search n-1 similar image patches $s_{i,j}$, $j=1,\ldots,n-1$ for a given reference patch s_i from a search window $\mathbf{W}_i$ sharing the same centre with given reference patch, then group them into an image patch set $\mathbf{S}_i=\{s_i,s_{i,1},\ldots,s_{i,n-1}\}$. Measuring the similarity within patches set $\mathbf{S}_i$ is one of fundamental problems in image denoising. Another key issue is to how to explore the similarity and gather all the helpful information about the noise-free version of s_i distributing among the n image patches to restore the noisy patch s_i. The key prior is these image patches S_i are similar and correlated[8].

In the conventional NLM [1], the weight $w_{i,j}$ is computed based on the similarity between the current reference patch s_i and candidate patch $s_{i,j}$. Then, the noise-free version of the estimated s_i is computed as

$$\hat{s}_i=\sum_{j=1}^{n}\hat{w}_{i,j}s_{i,j},\hat{w}_{i,j}=w_{i,j}/\sum_{j=1}^{n}w_{i,j} \tag{1}$$

To find enough image patches whose noise-free versions are almost same as that of given s_i, the size of search window $\mathbf{W}_i$ should be very large enough, but it will cause the search very computationally expensive. At the same time, only the similarity between s_i and $s_{i,j}$ is considered and exploited instead of any two patches in the n patches set.

In the BM3D [2] method, A shrinkage in 3D transform domain such as wavelet shrinkage and Wiener filter are applied on the 3D coefficient array of patch set $\mathbf{S}_i$ to suppress noise. The result is obtained by synthesizing a clean image from the de-noised patches. However, essentially fixed orthogonal transform is a kind of local processing, and only local interactions of neighboring pixels and patches are considered or the similarity among a local range is exploited. The definition of local range depend on the grouping order how the 2D patches are grouped into the 3D array. In a word, it is not completely non local de-noising method.

To solve the mentioned issues, some sparse models for de-noising are proposed. In the K-SVD [3], an over-complete dictionary $\mathbf{D}$ adapted to specific patches will be learned. It is assumed the each image patches s_i can be sparsely represented by the over-complete dictionary $\mathbf{D}$. To determine the sparse coefficient, we need solve the following optimization problem

$$\min_{\alpha} |\alpha|_{l^p},\ 0 \le p \le 1 \text{ and } | s_i - \mathbf{D}\alpha | < \delta \tag{2}$$

To exploit the structure self-similarity of natural image, in the LSSC, some similar image patches $s_{i,j}$, $i = 1,\ldots,n-1$ are employed to restrict the sparse coefficient of s_i and its structure. An important assumption in LSSC [4] is that the sparse coefficients of some similar image patches $s_{i,j}$, should possess similar structure and decomposition, especially the distribution or locations of nonzero entries. To determine the sparse coefficient matrix $\mathbf{A}$, we need solve the problem

$$\min_{\mathbf{A}} |\mathbf{A}|_{p,q}, | \mathbf{S}_i - \mathbf{DA} | < \delta \tag{3}$$

In fact, the similarities of the structures of the image patches can be represented from many aspects, not only limited in the distribution of nonzero entries. At same time, all the denoising methods based on overcomplete dictionary suffer from a drawback: the learning of overcomplete dictionary and the determination of structured coefficient vector or matrix are very computationally expensive [9,10].

3. Image denoising model based on non-local and low-rank

3.1. Image denoising and low-rank decomposition

Concretely, firstly let us define a similar image patch set $\mathbf{S}_i$ for the given patch s_i, namely, $\mathbf{S}_i = \{s_{i,j} \mid \mathrm{sim}(s_i, s_{i,j}) > \varepsilon, s_{i,j} \subset \mathbf{W}_i\}$, where $\mathrm{sim}(s_i, s_{i,j})$ is used to measure the similarity between reference patch s_i and candidate block $s_{i,j}$. Two more similar patches lead a higher $\mathrm{sim}(s_i, s_{i,j})$ value. And the higher the similarity threshold ε is, the less the image patches with higher similarity are grouped into the set S_i. The effectiveness of de-noising degraded image largely depends on how to measure the similarity between two patches and how to utilize the similarity in the de-noising methods. In this paper, in order to simplify the problem the simple Euclidean distance is used to measure the similarity. The focus is to exploit structure self-similarity of natural image by low-rank approximation or decomposition to restore the noisy patches.

In this paper, the image patch $s_{i,j}$ is stacked as the column vector, and then image patch array $\mathbf{S}_i$ can be considered as a 2D data matrix $\mathbf{S}_i \in R^{m\times n}$, where n is the number of patches in the patch set $\mathbf{S}_i$ and m is the number of the pixels in each image patch. In general case, n is bigger than m. According the analysis above, the patch group $\mathbf{S}_i$ can be decomposed into three parts: commonness information $\mathbf{A}_i \in R^{m\times n}$, the difference $\mathbf{E}_i \in R^{m\times n}$, and the noise $\mathbf{N}_i \in R^{m\times n}$, i.e., $\mathbf{S}_i = \mathbf{A}_i + \mathbf{E}_i + \mathbf{N}_i, \mathrm{rank}(\mathbf{A}_i) \le \min(m,n),.\ \|\mathbf{E}_i\|_{l^0} < m \times n$. In fact the commonness and difference are relative. Generally, the bigger the rank of commonness $\mathrm{rank}(\mathbf{A}_i)$ is, the less the energy $\|\mathbf{E}_i\|_{l^2}$ of difference $\mathbf{E}_i$ is and the sparser the difference is. That means by adjusting the rank of the commonness, we can assign partial difference to the commonness $\mathbf{A}_i$ and other unimportant difference to the uncertainty information $\mathbf{N}_i$. At same time, to image de-noising problem both $\mathbf{A}_i$ and $\mathbf{E}_i$ are signals we need recover[9,11,12]. It is not necessary to recover the commonness and difference separately. Therefore, the low-rank sparse decomposition can be simplified as a low-rank decomposition or approximation problem mathematically

$$\mathbf{S}_i = \mathbf{A}_i + \mathbf{N}_i, \text{rank}(\mathbf{A}_i) < \min(m,n) \tag{4}$$

Therefore, our focus is to solve the following optimization problem

$$\mathbf{A}_i^* = \min_{\mathbf{A}} \text{rank (A),s.t.} \|\text{S}_i\text{-A}\|_{l^2} < \varsigma \tag{5}$$

The parameter ς depends on the variance of noise. Applying the relaxation to the problem (5) yields a new optimization problem as follows

$$\mathbf{A}_i^* = \arg\min_{\mathbf{A}}\{\frac{1}{2}\|\mathbf{A}-\mathbf{S}_i\|_F^2 + \tau \times \text{rank}(\mathbf{A})\} \tag{6}$$

Unfortunately, (6) is a highly no convex optimization problem, and corresponding minimization problem is extremely difficult (it always is NP-hard and hard to approximate). At present, no efficient numerical solution is known[13].

To solve the problem (6), Candes ect. [8] replaced the $\text{rank}(\mathbf{A})$ with the nuclear norm $\|\mathbf{A}\|_* = \sum_i \sigma_i$ where σ_i is the singular value of data matrix $\mathbf{A}$. Simple replacement yields a tractable optimization problem:

$$\mathbf{A}_i^* = \arg\min_{\mathbf{A}}\{\frac{1}{2}\|\mathbf{A}-\mathbf{S}_i\|_F^2 + \tau\|\mathbf{A}\|_*\} \tag{7}$$

Fortunately, this optimization problem (7) is convex and can be solved efficiently. If we denote the singular value decomposition of data matrix S_i as $S_i = \mathbf{U}_i\Sigma_i\mathbf{V}_i^T$, where $\Sigma_i = diag(\lambda_{i,1},\ldots,\lambda_{i,r})$, $r = \min(m,n)$ and $\lambda_{i,j}$ is singular value of data matrix S_i (including zero). Then, the optimal solution of (8) is computed as [14, 15]

$$T_\tau(\mathbf{S}_i) = \arg\min_{A}\{\frac{1}{2}\|\mathbf{A}-\mathbf{S}_i\|_F^2 + \tau\|\mathbf{A}\|_*\} \tag{8}$$

Here, $T_\tau(\mathbf{S}_i)$ is the singular value shrinkage operation and can be defined as

$$T_\tau(\mathbf{S}_i) := \mathbf{U}_i T_\tau(\Sigma_i)\mathbf{V}_i^T, T_\tau(\Sigma_i) = diag\{(\lambda_{i,j}-\tau)_+\} \tag{9}$$

Where $(t)_+$ denote the positive part, defined as $\max(0,t)$. The remarkable difference between $\text{rank}(\mathbf{A})$ and $\|\mathbf{A}\|_*$ is that the former only depends on the number of nonzero singular values and not the specific value, but $\|\mathbf{A}\|_*$ is equal to their sum and depends directly on their values. This is just like the difference between l_0-norm and l_1-norm.

Essentially, the singular value shrinkage operation simply applies a soft-thresholding rule to the singular values of $\mathbf{S}_i$. Just like the common soft thresholding denoising in wavelet domain [16]. The determination of proper parameter or threshold τ is a key issue. It is necessary to utilize an adaptive threshold determined by the similarity of patch set, the variance of noise and the distribution of signal information and noise[17].

3.2. Effect of perturbation on the rank and the truncated SVD

Suppose that $\mathbf{S}_i$ is a matrix containing many column signals that are contaminated with a certain level of noise. The SVD resolves the data in noisy $\mathbf{S}_i$ into r mutually orthogonal components [18] as following

$$\mathbf{S}_i = \mathbf{U}_i \Sigma_i \mathbf{V}_i^T = \sum_{j=0}^{r} \lambda_{i,j} u_{i,j} v_{i,j}^T = \sum_{j=0}^{r} \lambda_{i,j} z_{i,j} \tag{10}$$

Where $z_{i,j} = u_{i,j} v_{i,j}^T$. $u_{i,j}$ and $v_{i,j}$ are column vectors of orthogonal matrix $\mathbf{U}_i$ and $\mathbf{V}_i$. In other words, the SVD decomposition (10) can be regarded as Fourier expansion or wavelet decomposition, and consequently, $\lambda_{i,j}$ can be interpreted as the proportion of signal $\mathbf{S}_i$ lying in the "direction" of $z_{i,j}$ whose rank is low and equal to one. Therefore, the SVD can be a useful tool in applications involving the need to sort through noisy data and lift out relevant information.

The noise-free version $\mathbf{A}_i$ always is assumed to be low-rank. But in most case, the matrix $\mathbf{A}_i$ and its rank is unknown. Whether the rank of noise-free matrix can be estimated accurately based on the noisy matrix is a key problem to low- rank decomposition. The effect of noise on singular values is important for the estimation the rank of $\mathbf{A}_i$. The following theory will show the perturbation of singular values due to a certain level of noise.

Theorem.1 Perturbations and Rank [18]. For $\mathbf{A} \in \Re^{m\times n}$, where m the size of image is patch and n is the number of image patches. Denote the noisy version of $\mathbf{A}$ as $\mathbf{S} = \mathbf{A} + \mathbf{N} \in R^{m\times n}$. where matrix $\mathbf{N}$ contains perturbation or noise. If we let $\{\lambda_1, \lambda_2, \cdots, \lambda_r\}$ and $\{\beta_1, \beta_2, \cdots, \beta_r\}$ denote the singular value (nonzero as well as any zero ones) of matrix $\mathbf{S}$ and $\mathbf{A}$ as, $r = \min\{n, m\}$, then

$$|\lambda_t - \beta_t| \le \|\mathbf{N}\|_2^2, \text{ for each } t = 1, \ldots, r \tag{11}$$

The proof can be found in the literature [18].

This theorem explains why singular values are the primary mechanism for numerically determining the rank of a matrix. If the rank of noise-free signal matrix $\mathbf{A}$ is equal to l, then the singular value $\beta_p, p = l+1, \ldots, r$ should be exactly zero. Therefore, Theory above can guarantee that $r - l$ singular value of the computed $\lambda_p, p = l+1, \ldots, r$ cannot be larger than $|\mathbf{N}\|_2^2$. Therefore, if the singular values λ_p obeys

$$\lambda_1 \ge \lambda_2 \ge \ldots \ge \lambda_q > \|\mathbf{N}\|_2^2 = \xi \ge \lambda_{q+1} \ge \ldots \ge \lambda_r \tag{12}$$

Then, it's reasonable to consider q to be the numerical rank of nosiy free signal $\mathbf{A}_i$. ξ is a threshold depending on the standard deviation of noIn the general case, the perturbation $\|\mathbf{N}\|_2^2$ is unknown. Fortunately, if a perturbation is Gaussian white noise, relatively accurate estimation of $\|\mathbf{N}\|_2^2$ can be obtained. Based on the estimation, we can predict the difference between the singular values for $\mathbf{A}$ and $\mathbf{S}$ is just related to $\|\mathbf{N}\|_2^2$.

To verify the theory above, zero mean white Gaussian noise with standard deviation σ =30 is added to the 512×512 Barbara image. From the noisy image and the noise-free image, four 7×7 image patch pairs a_i and s_i are selected randomly. For each noisy patch s_i , the top most similar 60 image patches are choose from a 25×25 search window. Then, the r singular values λ_t, β_t, t =1,...,r (r=49) of $\mathbf{S}_i$ and $\mathbf{A}_i$ are computed. Fig.1 show the difference $e_t = \lambda_t - \beta_t$ between λ_t and β_t.

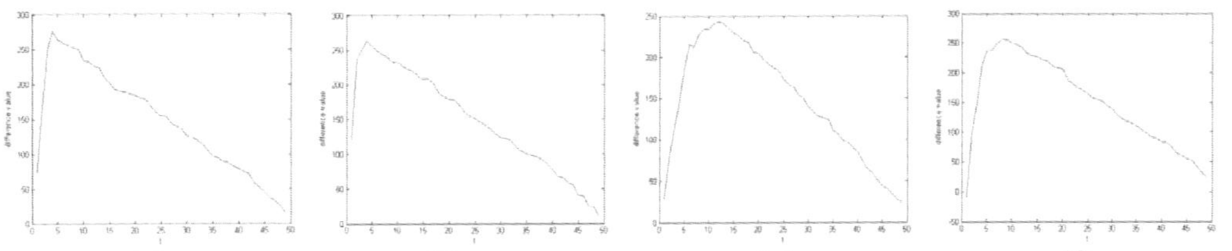

(a) (b) (c) (d)

Figure 1. the difference of the singular values between the noisy data matrix S_i and the noise-free data matrix A_i. (a) the first example (b) the second example (c)the third example (d) the fourth example

It can be easily observed that there exist many similarities[19]. All the singular values β_t can be divided into two parts. The first parts is $\beta_t, t = 1, \ldots, t_s$, and corresponding z_t represent the commonness and the similarity among the $\mathbf{S}_i$ and noise has little effect on the singular values when β_t is bigger or t is small. As the singular values decrease, the difference increases. The second part $\beta_t, t = t_{s+1}, \ldots, r$ is small and always close to zero. And corresponding z_t always represents the randomness and uncertainty information, such as noise and the weak difference of contents. As t increase, the difference decrease. When t is close to rank of noise-free data matrix, the difference e_t reaches its peak point. In general case, the most information of signal is concentrated on the first part, and the second part is not important for the signal recovery.

3.3. Threshold determination of the nonlocal and low-rank denoising

For the singular value shrinkage operation, the key issue is to select a reasonable threshold. Obviously, applying a fixed threshold $\lambda_{i,k}$ (k is the truncated location and is equal to $\text{rank}(\mathbf{A}_i)$) to all singular values λ_t is not reasonable. Although, the truncation can remove most noise, much residue still appear in the retained singular values. At same time, the noise is always distributed uniformly across the $z_{i,j}$, but signal is not. Therefore, it is necessary to design an adaptive threshold for each retained singular values $\lambda_t, i = 1, \ldots, k$ to remove the noise contained in $\hat{\mathbf{S}}_i$ and preserve the information of signal.

According the analysis and experiment above, a fact can be found that the bigger the singular value $\lambda_t, i = 1, \ldots k$ is, the less the difference is between the noisy matrix and the noise-free one. Namely, the bigger singular value is, the smaller the threshold should be. In a word, to adaptively adjust the parameter τ , the parameter τ is assumed to be proportional to noise variance and in inverse proportion to singular value. Therefore, we adopt the following strategy for determining the parameter $\tau_{i,j}$

$$\tau_{i,j} = \begin{cases} c\dfrac{\sigma^2}{\lambda_{i,j}} & if \;\; j < k_i \\ \lambda_{i,k} & else \end{cases} \tag{13}$$

where k_i is the estimation of the rank of $\mathbf{A}_i$ and σ^2 is noise variance, $\lambda_{i,j}$ is the $j-\text{th}$ singular value of $\mathbf{S}_i$, and c is predefined constants. For this, the noise standard deviation σ is estimated using a simple gradient-based estimator as [20,21],

$$\sigma = 1.4826 \times \text{median}\left(\left|\nabla Y - \text{median}(\nabla Y)\right|\right) \tag{14}$$

where ∇Y is the vectorized form of the gradient of the input nosiy image Y , which is calculated as

$$\nabla Y = 0.4082 \times \text{vec}\left(Y * \begin{bmatrix} 2 & -1 \\ -1 & 0 \end{bmatrix} \right) \quad (15)$$

Here, vec(•) and $*$ denote the vectorization operation and the convolution operation, respectively.

3.4. The main frame of proposed denoising scheme

To conclude this section, we will summarize the main step of the proposed image denoising method. The concrete process is shown as follows .

Grouping the similar patches. For the given reference patch s_i, n top most similar patches $s_{i,j}$ are selected based the similarity between s_i and $s_{i,j}$. Then, these vectorized patches are grouped into data matrix $\mathbf{S}_i = [\text{vec}(s_{i,1}),\ldots,\text{vec}(s_{i,n})]$.

Estimating the rank of underlying matrix $\mathbf{A}_i$. Estimate the noise standard deviation σ and parameter $\xi = \kappa n \sigma$, and then compute the singular values of matrix $\mathbf{S}_i$. Lastly, estimate the numerical rank k_i based on theory 1.

Applying the adaptive shrinkage to each singular values. Based the rank estimated rank k_i , truncate the singular values and compute the adaptive parameter $\tau_{i,t}$. Then, apply the shrinkage to the retained singular values $\lambda_{i,t}$.

In our implementation, the image patches are selected with overlapping regions. Thus, each pixel is covered by several de-noised patches. In this paper, the value of each pixel in images is determined by taking the average of all the estimations for same location. In fact, we observe a fact the lower the estimated rank is, the better the quality of recovery is. Intuitively, an aggregation can be performed by a weighted averaging at those pixel positions where there are overlapping block-wise estimates just like BM3D. This is one direction of our future research.

4. Experimental results

In this section, the performance of the proposed denoising scheme is tested. The main innovation in the frame is the combination of non local idea and low rank.

Throughout this study, the patch size in our implementation is set as 7x7 for $\sigma \le 30$ and 9x9 for $\sigma > 30$ .The size of search window is fixed as 25x25. The parameter ξ is set as $\kappa n \sigma$ where n is size of patch and κ is a constant and chosen according to an empirical rule. The parameter ε [4] is set as $32\sigma / n$ for images scaled between 0 and 255 to all test images, which has shown to be appropriate in all of our non local denoising experiments. Iterations times are set as 5. In the experiment, δ and c is fixed as 0.035 and 115.

Table 1. The PSNR(dB) values of the denoised images with respect to different noise levels of Gaussian noise.

Image	σ	NLM	K-SVD	BM3D	LSSC	K-LLD	Proposed
Barbara	5	37.06	38.11	38.33	38.48	36.61	**38.51**
	10	33.16	34.42	34.97	**34.97**	33.03	34.93
	20	30.26	30.81	31.74	**31.57**	29.05	31.59
	25	29.08	29.58	30.72	**30.47**	27.27	30.51
	50	25.66	25.58	27.23	27.06	23.20	27.18
Fingerprint	5	35.15	36.66	36.59	36.70	35.85	**36.71**
	10	30.99	32.42	32.43	32.57	31.55	**32.62**
	20	27.25	28.45	**28.87**	28.78	28.14	**28.87**
	25	26.14	27.25	27.70	27.62	26.99	**27.78**
	50	22.96	23.25	24.53	24.25	22.34	**24.55**
Lena	5	37.94	38.62	**38.71**	38.69	37.61	38.61

	10	34.36	35.51	35.94	35.83	35.12	35.58
	20	31.61	32.38	33.07	32.90	32.37	32.77
	25	30.36	31.36	32.08	31.87	31.34	31.63
	50	27.32	27.75	29.05	28.87	25.15	28.81
Man	5	37.01	37.50	37.80	**37.89**	36.24	37.86
	10	33.11	33.55	33.94	**34.06**	33.09	33.97
	20	29.72	30.05	30.54	**30.64**	30.12	30.60
	25	28.38	29.08	29.62	**29.63**	29.22	29.59
	50	25.33	26.08	**26.81**	26.69	24.38	26.71
Couple	5	36.75	37.30	**37.49**	37.45	35.89	37.39
	10	32.90	33.48	**34.02**	33.98	33.08	33.91
	20	29.01	30.02	30.76	30.69	29.92	30.77
	25	27.99	28.88	**29.72**	29.61	29.02	29.71
	50	24.58	25.29	**26.46**	26.30	23.20	26.37

For performance evaluation, several common test images are considered (including Barbara, Lena, Figureprint, Man and Couple) which are corrupted by additive white Gaussian noise with standard deviation $\sigma \in \{5,10,20,25,50\}$. To illustrate the proposed denoising method more clearly and fully demonstrate the performance gains from the combination of non-local and low-rank, the proposed method is compared with five state-of-the-art methods which include NLM[1], BM3D[2], K-SVD[3], LSSC[4], and K-LLD[6]. Except LSSC algorithm, all the results are based on the source codes or executables released by the original authors. The default parameters are employed in the comparison algorithms. We try to realize LSSC. Unfortunately, our results are inconsistent with the original literature. To make a fair comparison, all the evaluation indexes of LSSC are from the literature [4].

4.1. Comparison of quantitative evaluation

The PSNR results on five test images are reported in Table 1 for different noise variance σ^2. The best performance, the second best and the third best is highlighted by red, green, and blue color in each cell, respectively. From table 1, we can conclude that the proposed non-local and low-rank denoising method significantly outperforms NLM, K-SVD and K-LLD. In term of PSNR, BM3D produce state-of-the-art result, however the results of the three schemes are nearly the same. Even on a certain test image containing similar structure pattern (edge and texture), such as Figureprint and Barbara, proposed scheme outperforms the BM3D and LSSC. That demonstrates low-rank is very suitable for capturing and restoring the patches with similar structure. For Barbara, when variance of noise is low, performance of proposed method outperforms other schemes. But as noise level increase, our method is not better than BM3D, but still better than LSSC. The reason is that when noise is weak, it is easy to capture the common pattern contained in data matrix. As the noise increase, the extraction of patterns becomes more difficult. On contrary, for the figureprint image, their local patterns are single and play a main role. No matter how strong the noise is, it is easy to capture the corresponding local pattern. Therefore, for Figureprint our scheme always outperforms the BM3D and LSSC.

It is not surprising that our method shows outstanding performances for the test image containing some textures or structure patterns. For Barbara image, the proposed method outperforms the K-SVD and NLM, and it is superior to K-SVD by 0.4~1.6dB and to LSSC by 0.2 dB. For fingerprint image, proposed method outperforms the BM3D, LSSC, K-SVD and NLM, and on average it is superior to BM3D by 0.1~0.2dB, to LSSC by 0.1~0. 3dB and to K-SVD by 0.1~1.3dB.

Table 2. The SSIM results for different standard images and several denoising methods.

Image	σ	NLM	K-SVD	BM3D	K-LLD	Proposed
Barbara	5	0.944	0.964	0.965	0.951	0.966
	10	0.928	0.934	0.942	0.920	0.941
	20	0.867	0.881	0.923	0.851	0.919
	25	0.825	0.850	0.887	0.804	0.892
	50	0.671	0.713	0.794	0.572	0.805

Fingerprint	5	0.965	0.987	0.987	0.985	0.988
	10	0.948	0.968	0.968	0.964	0.969
	20	0.889	0.922	0.929	0.924	0.933
	25	0.859	0.898	0.911	0.903	0.913
	50	0.712	0.775	0.830	0.778	0.838
Lena	5	0.914	0.946	0.944	0.928	0.951
	10	0.901	0.910	0.916	0.902	0.916
	20	0.837	0.863	0.877	0.864	0.872
	25	0.790	0.843	0.861	0.840	0.863
	50	0.674	0.761	0.799	0.552	0.801
Man	5	0.895	0.951	0.954	0.937	0.954
	10	0.873	0.900	0.907	0.887	0.908
	20	0.784	0.815	0.833	0.817	0.833
	25	0.724	0.779	0.805	0.789	0.804
	50	0.597	0.663	0.705	0.561	0.707
Couple	5	0.895	0.950	0.951	0.923	0.952
	10	0.877	0.897	0.909	0.890	0.909
	20	0.787	0.815	0.848	0.819	0.848
	25	0.745	0.779	0.820	0.793	0.818
	50	0.566	0.632	0.706	0.562	0.695

Compared with PSNR, SSIM is a popular evaluation for measuring the content and structure similarity between two images, which have been proved to be inconsistent with human eye perception. To understand how the combination of low-rank decomposition and non-local idea restores the structure, we also compare the SSIM of the proposed algorithm and other leading denoising techniques in the literature at different noise levels for 5 test images. The SSIM results on five test images are shown in table 2. The best performance and the second best are highlighted by red and green font in each cell, respectively. We conclude the proposed denoising method has achieved competitive SSIM performance to BM3D, K-SVD, KLLD and NLM for the most test images. It is not surprising that BM3D achieved some competitive PSNR, but in term of SSIM the proposed method is best. The success owe to the fact the proposed method aims to restore and preserve the local structure not minimize the numerical error.

Note that the parameters have not been optimized for speed and for quality in this experiment. They are set just based on experience. Optimized parameters may further improve the performances. This is another direction of future research. Because all the source codes are developed by different program language, it is not fair to measure the speed only depending on running time. Intuitively, all the algorithms (K-SVD, BM3D and LSSC) need search some similar image patches, the time spent on the search should be nearly same. But for time spent for the computation of the coefficient matrix, LSSC are computationally expensive. At same time, K-SVD and LSSC must learn a overcomplete dictionary based on a large amount of samples. It is natural conclusion that the proposed method is faster than LSSC and has the nearly same speed as BM3D algorithm.

4.2. Visual comparisons

Even though BM3D and LSSC perform a little better in terms of PSNR, they suffer from classical artifacts, as shown by the example of Fig.2 and Fig.3. Fig.2 shows the denoising results of Barbara image ($\sigma = 30$) obtained only by BM3D, K-SVD, NLM and proposed method due to space limit. The visual comparisons of the detail and local region, especially the regions containing certain structure and texture, show that proposed method provides better visual quality. Some edges and textures are blurred even lost in BM3D, K-SVD and NLM, but they are better preserved in the proposed method.

In the first row of Fig.3, a certain texture structure is contained in the original image patch (as shown in Fig.3 (a1)). Compared with the results of the K-SVD, NLM and BM3D ($\sigma = 30$), our result is more clear and complete. On the contrary, in other result there exist some obvious blur and loss. In the second row of Fig.3, in the original image block (as shown in Fig.3 (a2)) contains a straight edge.

Compared with the results of the K-SVD, NLM and BM3D, the edge in our result is clear and straight. However, the edge of compared algorithms is seriously deformed and bent. In the smooth region with a very weak structure (as shown in Fig.3 (a3)), noticeable reconstruction artefacts appear in the results obtained with BM3D, because some noise is considered as detail or structure. Unfortunately, in some degree our algorithm only aims to restore the main structure and loses some weak structure (shown in last row of Fig.3). The main reason is the truncation of SVD due to the inaccurate estimation of the rank of the corresponding noise-free data matrix. The loss of weak structure originates from the inaccurate truncation of singular values. Increasing the size of patch can avoid the weak structure loss. But it will increase the computational complexity of patch matching.

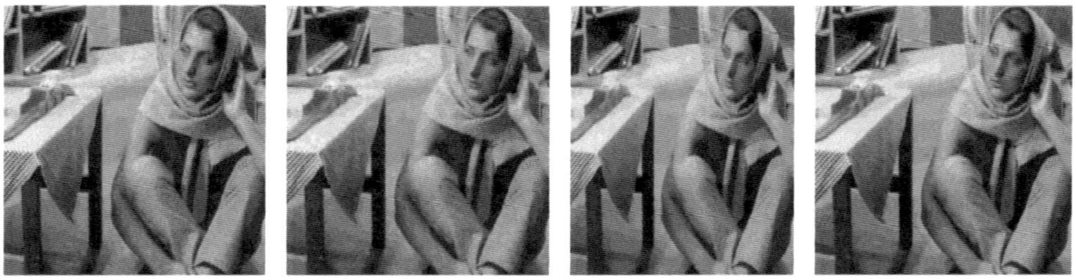

(a) (b) (c) (d)

Figure 2. Visual comparisons of Barbara (a) BM3D (b) K-SVD (c) NLM (d) Proposed method.

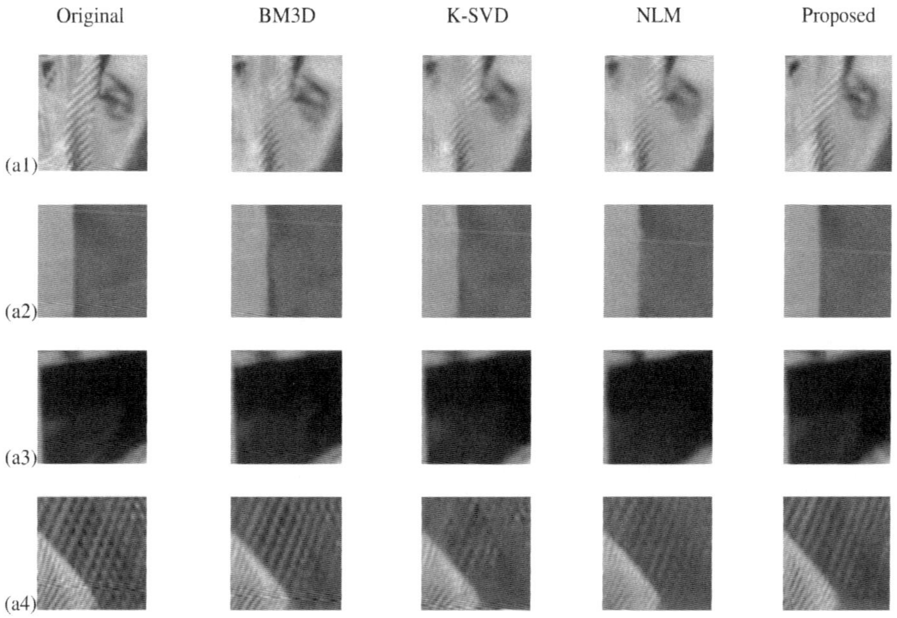

Figure 3. The local detail and structure comparison

5. Conclusion

In this paper, a simple but effective framework for image denoising is presented which works by exploiting the merits of self-similarity and similarity matrices low rank approximation. The weighted nuclear norm is defined and used to replace the classical nuclear norm to improve adaptivity of the threshold used in the singular value shrinkage. The corresponding problems are effectively tackled. We need to take several practical steps: finding similar patches for each reference patch, computing the singular values of noisy data matrix and estimating the rank of corresponding noise-free data matrix, executing low rank approximation for each data matrix based on the weight matrix, reconstructing low rank matrices and obtaining denoised image. Experimental results demonstrate that at relatively low computational cost the proposed method is very close to state-of-the-art denoising technique BM3D and LSSC even outperforms them in some images containing many structure and texture in terms of both PSNR and SSIM.

Acknowledgements
This work was supported by the Southwest Institute of Technology and Physics. The authors thank Chatterjee and Dong, for providing their results.

References

[1]. Buades, B. Coll, and J. M. Morel, A review of image denoising methods, with a new one, Multiscale Model Simulation, 2(4), pp. 490–530, 2005.
[2]. K. Dabov, A. Foi, V. Katkovnik, and K. O. Egiazarian, Image denoising by sparse 3-D transform-domain collaborative filtering, IEEE Transactions on Image Processing, 8(16), pp. 2080–2095, 2007.
[3]. M. Elad and M. Aharon, Image denoising via sparse and redundant representations over learned dictionaries, IEEE Transactions on Image Processing, 12(15), pp. 3736–3745, 2006.
[4]. J. Mairal, F. Bach, J. Ponce, G. Sapiro, and A. Zisserman, Non-local sparse models for image restoration, IEEE 12th International Conference on Computer Vision, pp. 2272–2279, 2009.
[5]. W. Dong, X. Li, L. Zhang, and G. Shi, Sparsity-based image denoising via dictionary learning and structural clustering, in 2011 IEEE Conference on Computer Vision and Pattern Recognition (CVPR), pp. 457–464, 2011.
[6]. P. Chatterjee and P. Milanfar, clustering-based denoising with locally learned dictionaries, IEEE Transactions on Image Processing, 7 (18), pp. 1438–1451, 2009.
[7]. L. Zhang, W. Dong, D. Zhang, and G. Shi, Two-stage image denoising by principal component analysis with local pixel grouping, Pattern Recognition, 4(43), pp. 1531–1549, 2010.
[8]. E. J. Candes and B. Recht, Exact matrix completion via convex optimization, Foundations of Computational Mathematics, 9(6), pp. 717–772, 2009.
[9]. K. Konstantinides and K. Yao, Statistical analysis of effective singular values in matrix rank determination, IEEE Transactions on Speech Signal Processing, 36(5), pp. 757–763, 1988.
[10]. Hui Jiy, Chaoqiang Liuz, Zuowei Sheny and Yuhong Xu, Robust video denoising using low rank matrix completion, 2010 IEEE Conference on Computer Vision and Pattern Recognition (CVPR), pp.1791–1798, 2010.
[11]. Yigang Peng, Ganesh. A., Wright J., Wenli Xu and Yi Ma, RASL: Robust alignment by sparse and low-rank decomposition for linearly correlated images, 2010 IEEE Conference on Computer Vision and Pattern Recognition (CVPR), pp.763-770, 2010.
[12]. E. J. Candès, X. Li, Y. Ma, and J. Wright. Robust Principal Component Analysis? Journal of ACM 58(1), pp.1-37, 2009.
[13]. Tianyi Zhou and Dacheng Tao, GoDec: Randomized Lowrank & Sparse Matrix Decomposition in Noisy Case, In the 28th International Conference on Machine Learning (ICML), pp. 33–40, 2011.
[14]. E. J. Candès and Y. Plan. Matrix completion with noise. Proceedings of the IEEE, 98(6), pp. 925-936, 2009.
[15]. J .F. Cai, E. J. Candès, and Z. W. Shen, A singular value thresholding algorithm for matrix completion, SIAM Journal on Optimization. 20(4), pp. 1956–1982, 2010.

[16]. D. L. Donoho and I. M. Johnstone, Ideal spatial adaptation via wavelet shrinkage, Biometrika, 3(81), pp. 425–455, 1994.
[17]. S. G. Chang, B. Yu, and M. Vetterli, Adaptive wavelet thresholding for image denoising and compression, IEEE Transactions on Image Processing, 9(9), pp. 1532–1546, 2000.
[18]. Carl D.Meyer, Matrix Analysis and Applied Linear Algebra, Society for industrial and Applied Mathematics (SIAM), 2000.
[19]. D. D. Muresan and T.W. Parks, Adaptive principal components and image denoising, In Proceedings of the 2003 International Conference on Image Processing, pp. 1101–1104, 2003.
[20]. P. Chatterjee and P. Milanfar, patch-based near-optimal image denoising patch-based near-optimal image denoising, IEEE Transactions on Image Processing, 21(4), pp. 1635–1649, 2012.
[21]. Jianfeng Cai, Stanley Osher and Zuowei Shen, Split Bregman methods and frame based image restoration, Multiscale Modeling and Simulation: A SIAM Interdisciplinary Journal, 8(2), pp. 337-369, 2009

Electrical Inspection Oriented Thermal Image Quality Assessment

Ying Lin[1], Menglin Wang[2], Xiaojin Gong[2], Zhihong Guo[1] , Yujie Geng[1] and Demeng Bai[1]

[1] State Grid Shandong Electric Power Research Institute, Jinan, China
[2] Zhejiang University, Hangzhou, China

E-mail: lysgwork@163.com

Abstract. This paper presents an approach to access the quality of thermal images that are specially used in electrical inspection. In this application, no reference images are given for quality assessment. Therefore, we first analyze the characteristics for these thermal images. Then, four quantitative measurements, which are one-dimensional (1D) entropy, two-dimensional (2D) entropy, centrality, and No-Reference Structural Sharpness (NRSS), are investigated to measure the information content, the centrality for objects of interest, and the sharpness of images. Moreover, in order to provide a more intuitive measure for human operators, we assign each image with a discrete rate based on these quantitative measurements via the k-nearest neighbor (KNN) method. The proposed approach has been validated in a dataset composed of 2,336 images. Experiments show that our quality assessment results are consistent with subjective assessment.

1. Introduction

Thermal imaging systems are able to capture the surface temperature without contact, so that they are widely used in electric power industry for equipment inspection. However, the quality of captured thermal images highly depends on operating conditions, such as background radiation, the range of operation of thermal imaging sensors [1][2], and so on. Poor images may bring much more challenges for succeeding automatic analysis including equipment detection and fault diagnosis. It is therefore important to assess the quality of thermal images before conducting analysis, but also helpful to improve thermal imaging for human operators.

Image quality assessment (IQA) is a problem that has been studied for decades. Most existing works focus on visible light images [3][4], in which various quantitative measurements have been developed for objective assessment. For instance, the mean squared error (MSE), the peak signal-to-noise ratio (PSNR), as well as some models based on human vision system (HVS) and structural similarity image measurement (SSIM) were proposed to measure the quality of images when reference images are given. In practical applications, reference images are unavailable. Therefore, no-reference IQA approaches [5] have also been developed to evaluate the degree of noise, blur, or block effects. In contrast to the studies on visible light images, the quality assessment for inferred and thermal images are relatively few. Although different modal images should share something in common, inferred and thermal images present different characteristics due to their distinct optical and electrical performance. Therefore, Xu et al. [6] and Yue et al. [7] proposed a couple of quantitative measurements specific to inferred images, for instance, pesudo SNR, histogram based image information capacity, and energy spectrum entropy. Agaian et al. [1] and Amon and Lock [8] designed measurements, such as cross entropy brightness-darkness and brightness-darkness intensity, for thermal images.

This paper aims to assess the quality of thermal images which are used in electrical equipment inspection, and no reference images are available. In contrast to the related works mentioned above, the thermal images that we evaluate have the following characteristics:

(1) In electric power industry, thermal images are commonly displayed with iron palette in order for human to visually interpret them better. Typical examples are shown in Figure 1. Therefore, the images that we evaluate are the iron colored thermal images.
(2) Although thermal cameras can capture the temperature distribution of imaging environments in a great range, the contrast of displayed thermal images highly depend on the range of temperature of interest that are set by human operators.
(3) Electric equipment is often of higher temperature than environment. Therefore, different from thermal images studied in [1], the images in electric power industry are of high contrast between object and background if appropriate temperature range is set.
(4) In order to perform equipment inspection, it is expected to have equipment of interest located at the center of thermal images.

By considering the above phenomena and existing image quality measures, we investigate four quantitative measurements, which are, respectively, one-dimensional (1D) entropy, two-dimensional (2D) entropy, centrality, and No-Reference Structural Sharpness (NRSS) [9]. Moreover, instead of integrating them together to get a final continuous value, we take these measurements as features and classify each image into one of five rating category. Each category is associated with a score following the rating scheme used in mean opinion score (MOS) [10]. The discrete rating score is more helpful for guiding human operators to capture better images. Our approach is validated in a dataset composed of 2,336 images. Experiments show that our quality assessment results are consistent with subjective assessment.

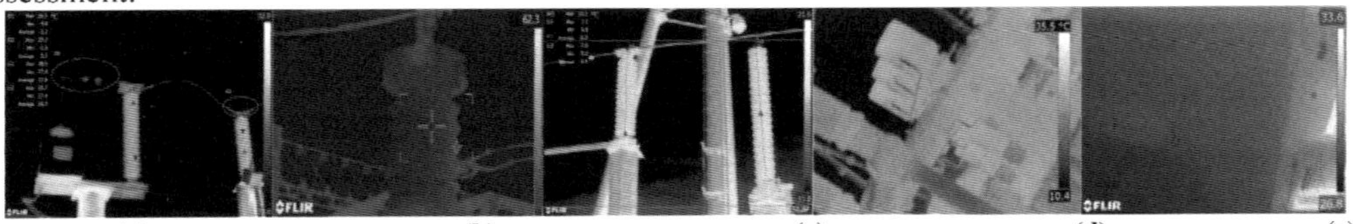

(a) **(b)** **(c)** **(d)** **(e)**

Figure 1. Thermal images with iron palette. From (a) to (e), the image quality degenerates.

2. Proposed approach

Given a thermal image, we assess its quality by considering the amount of information it contains, whether the object is located at the image center, and how clear the image is. With respect to these considerations, four measurements are investigated, which are 1D entropy, 2D entropy, centrality, and NRSS. In this section, we first explain these measurements, and then introduce the way of using them to rate each image.

2.1 Quantitative measurements

2.1.1. 1D Entropy. Entropy is a measure of information content. 1D entropy is to measure the information content based on the histogram of pixels' color. Given a thermal images with iron palette, the 1D entropy is computed by

$$H_{1D} = -\sum_{i=1}^{B} p_i \log(p_i) \quad (1)$$

Here, p_i represents the probability for a pixel whose color falls into the i-th bin, and B is the total number of bins. It is reasonable to expect a small 1D entropy for an image with high quality.

2.1.2. 2D Entropy. The above 1D entropy measures the information content of pixels while simply ignoring spatial information. In order to take spatial information into account, 2D entropy is also adopted. Let i be a pixel's intensity value and j be the mean intensity value of the pixel's neighbors. Then, 2D entropy is computed by

$$H_{2D} = -\sum_{i=0}^{255}\sum_{j=0}^{255} p_{ij} \log p_{ij} \tag{2}$$

Here, p_{ij} is the probability of the spatial pattern in which the center pixel is i intensity value and its neighbors are of j intensity value in average. p_{ij} is estimated from a 2D histogram. Analogously, a high quality image should have a small 2D entropy.

2.1.3. Centrality. The centrality is to measure the degree that objects of interest occupy an image's central part. In our case, the central part refers to the half size region located at the center of the image. Considering that electric equipment has higher temperature than background, we binarize the image with certain thresholds to get foreground and background regions. The percentage that the foreground region occupies within the central region is taken as the centrality. It is

$$C = \frac{N_F}{N_T} \tag{3}$$

in which N_F is the number of foreground pixels with the central region, N_T is the total number of foreground pixels.

2.1.4 NRSS. The no-reference structural sharpness (NRSS) [9] is proposed by Xie et al. to measure the quality of visible light images. It is adopted in this work to evaluate the sharpness of thermal images as well. We briefly introduce it here for the purpose of self-containedness.

Given a thermal image I, the first step is to smooth it using a Gaussian filter to get a reference image I_r. Then, by dividing both images into M blocks, NRSS is computed by

$$NRSS = 1 - \frac{1}{M}\sum_{i=1}^{M} SSIM(x_i, y_i) \tag{4}$$

Here, $SSIM$ is the structural similarity [3] defined as

$$SSIM(x,y) = \frac{2\mu_x\mu_y + C_1}{\mu_x^2 + \mu_y^2 + C_1} \cdot \frac{2\sigma_x\sigma_y + C_2}{\sigma_x^2 + \sigma_y^2 + C_2} \cdot \frac{\sigma_{xy} + C_3}{\sigma_x\sigma_y + C_3} \tag{5}$$

in which x and y are blocks in I and I_r, μ_x and μ_y are the mean intensity values of x and y, and σ_x and σ_y are the corresponding standard deviations. C_1, C_2, and C_3 are three constants empirically set.

2.2 Discrete rating

One way of assessing image quality is to integrate different quantitative measurements by multiplying or summing up them together. Instead of taking this way, we prefer to assign each image with a discrete rate based on the above measurements. The rating scheme follows the one used in mean opinion score (MOS) [10], which is listed in Table 1. In contrast to continuous scores, the 5-level rating score is not only more intuitive for human operators, but also appropriate for succeeding automatic analysis.

Table 1. The discrete rating scores

1	2	3	4	5
Bad	Poor	Fair	Good	Excellent

We treat the discrete rating task as a classification problem. The four quantitative measurements are taken as features associated with each image. Therefore, rating an image is essentially classifying it into one of the five categories. A training set composed of 1,400 images are given, in which each image is manually labeled as one of the 5-level score. The centroids of each category is computed. When a test image is given, the nearest neighbor approach is used to determine its rating score.

3. Experiments

3.1 Datasets

In order to validate our approach, we first construct a dataset that consists of 2,336 images in total. Most of the images are taken for the purpose of inspecting current transformer, lightning protector, and other electric equipment. High quality images are often with an equipment located at the center and with high contrast. While poor images may be blurred, with no obvious equipment, or with low contrast. Besides Figure 1, more typical examples are presented in Figure 2.

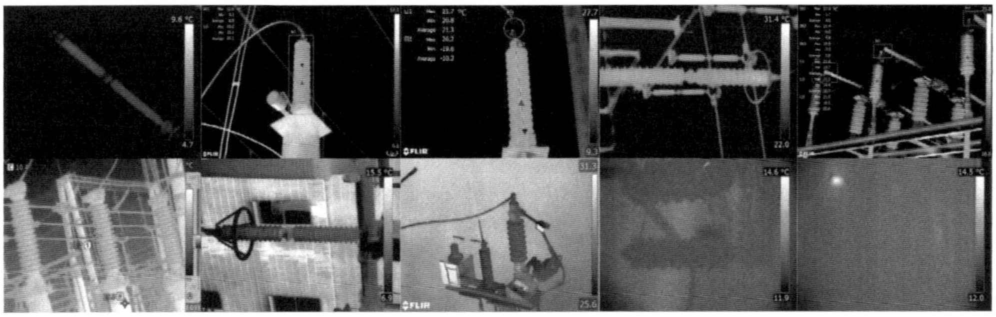

Figure 2. Typical thermal images for electric inspection.

In the dataset, each image is manually labeled with a 5-level score for training and evaluation purpose. Among all the images, 60% images are randomly selected as training samples and the remaining is for testing. Table 2 presents the number of images for each rating level. Each image is colored with iron palette.

Table 2. The number of samples for each score

1	2	3	4	5
517	359	419	369	672

3.2 Experiments

We first investigate the four quantitative measurements. Table 3 lists the results for the images in Figure 1, in which low entropy values and high centrality and NRSS values indicate high quality. From which we can observe that these measurements can access the corresponding considerations well. They are also discriminative and consistent with subjective assessments.

Table 3. Quantitative measurements for example images

	Fig1.a	Fig1.b	Fig1.c	Fig1.d	Fig1.e
1D Entropy	2.657	4.261	5.305	6.528	6.566
2D Entropy	2.446	3.813	4.315	5.022	4.771
Centrality	0.543	0.830	0.690	0.476	0.219
NRSS	0.678	0.717	0.620	0.749	0.432

Figure 3 plots the distributions of the four measurements for all images in the training set. From which we observe that 1D and 2D entropy can separate Bad and Poor images from the others well, centrality alone mostly separates Fair and Excellent images from the others, and NRSS alone can separate Bad images from the others. When integrating the four measurements together, better performance should be expected.

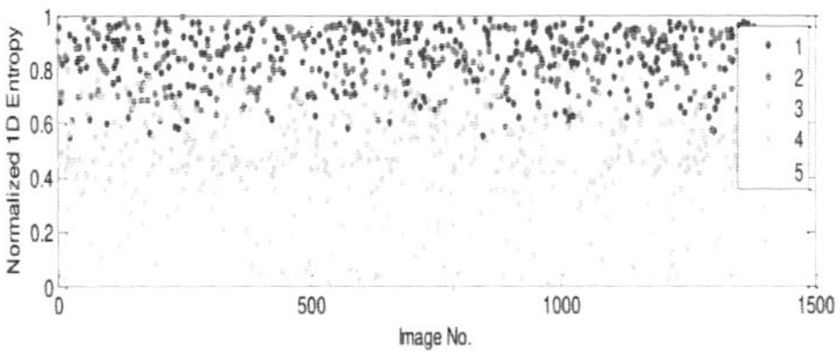

(a)1D entropy

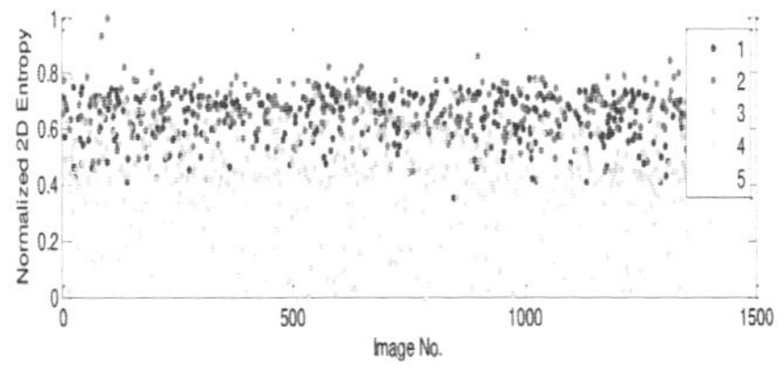

(b) 2D entropy

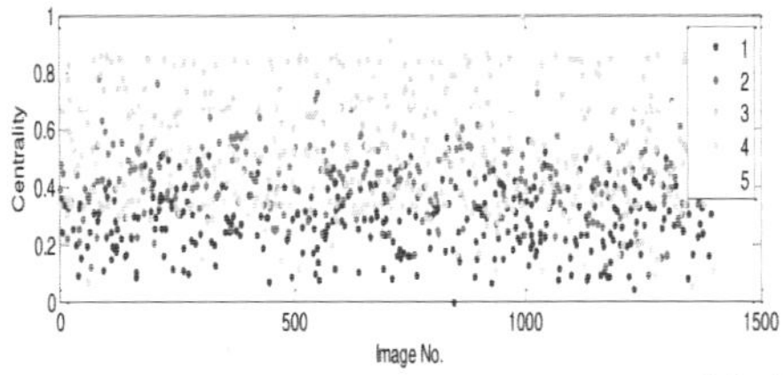

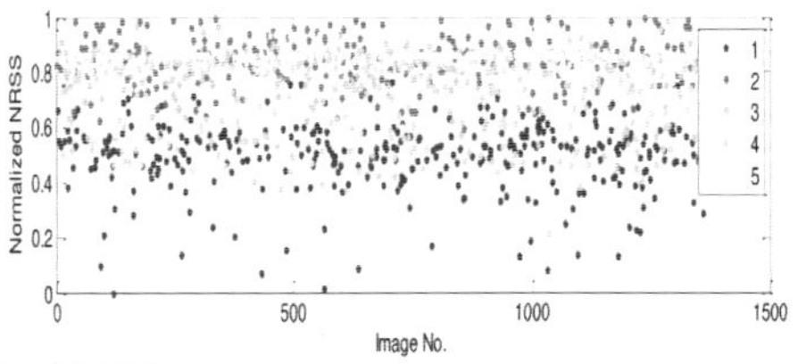

(c) Centrality NRSS

Figure 3. The distributions of four measurements for training images.

For rating test images, we first compute the centroids of each rating category. Each image is then assigned the rating score with the one have the nearest centroid. The accuracy of each rating score images are presented in Table 4. It implies that our approach is highly consistent with subjective assessment.

Table 4. The number of test samples for each score

	1	**2**	**3**	**4**	**5**
No. of images	207	144	168	148	269
Accuracy	96.6%	97.2%	95.2%	95.9%	96.3%

4. Conclusion

In this paper, we have presented an approach to discretely rating thermal images for better imaging and inspecting electric equipment. Four quantitative measurements are investigated with respect to the considerations of information content, location of object of interest, as well as the sharpness of images. Then, the discrete rating task is transformed into a classification problem based on the four measurements and the nearest neighbour method is applied. Experiments have validated the effectiveness of our approach.

References

[1] Agaian, S., Roopaei, M., and Akopian, D. Thermal-Image Quality Measurements, *Proc. of ICASSP*, (2014), 2779-2783,.

[2] Taib, S., Jadin, M., and Kabir S. Thermal Imaging for Enhancing Inspection Reliability: Detection and Characterization. Infrared Thermography, *Nippon Seikeigeka Gakkai Zasshi*, Vol. **45**, No.10, (1971), 834-6.

[3] Wang, Z., Bovik, A., Sheikh, H., and Simoncelli, E. Image Quality Assessment: From Error Visibility to Structural Similarity, *IEEE Transactions on Image Processing*, Vol. **13,** No. 4, (April 2004), 600-612.

[4] Jiang, G., Huang, D., Wang, X., and Yu, M. 2009. Overview on Image Quality Assessment Methods. *Journal of Electronics & Information Technology*, Vol. **32**, No. 1, (Jan. 2010), 219-226.

[5] Wang, Z. Review of No-reference Image Quality Assessment. *Acta Automatica Sinica*, Vol. **41**, No. 6, (June 2015), 1062-1079.

[6] Xu, C., Li, G., Hu, W., and Zhang W. 2004. Image Quality Evaluation of Infrared Image. *Infrared Technology*, Vol. **26**, No. 6, (Nov. 2004), 72-75.

[7] Yue, S., Ren, T., Wang, C., Lei, B., and Zhang, Z. Infrared Image Quality Evaluation Method Without Reference Image, Proc. SPIE 8907, *International Symposium on Photoelectronic Detection and Imaging 2013: Infrared Imaging and Applications*, 89074X (September 11, 2013); doi:10.1117/12.2034887.

[8] Amon, F. and Lock A. Evaluation of Image Quality of Thermal Imagers used by the Fire Service, *NIST Technical Note 1630.*

[9] Xie, X., Zhou, J., and Wu, Q. No-reference Quality Index for Image Blur. Journal of Computer Applications, *Journal of Computer Applications* , Vol. 30, No. 4, (Apr. 2010), 921-924.

[10] https://en.wikipedia.org/wiki/Mean_opinion_score

The segmentation of Thangka damaged regions based on the local distinction

Bi Xuehui[1], Liu Huaming[1], Wang Xiuyou[1], Wang Weilan[2] and Yang Yashuai[1]

[1]Department of Computer and Information Engineering, Fuyang Normal College, 236000, Fuyang, China.
[2]Mathematics and Computer Science, Northwest University for Nationalities, 730030, Lanzhou, China

E-mail: bixuehui888@163.com

Abstract. Damaged regions must be segmented before digital repairing Thangka cultural relics. A new segmentation algorithm based on local distinction is proposed for segmenting damaged regions, taking into account some of the damaged area with a transition zone feature, as well as the difference between the damaged regions and their surrounding regions, combining local gray value, local complexity and local definition-complexity (LDC). Firstly, calculate the local complexity and normalized; secondly, calculate the local definition-complexity and normalized; thirdly, calculate the local distinction; finally, set the threshold to segment local distinction image, remove the over segmentation, and get the final segmentation result. The experimental results show that our algorithm is effective, and it can segment the damaged frescoes and natural image etc.

1. Introduction

In the process of preservation, Thangka which has rich cultural connotation and research values will be damaged differently because of natural and human factors, the positions and shape of damaged region are uncertain. Accurate segmentation of damaged regions is the premise of digital inpainting and the segmentation result will directly affect the quality of inpainting. At present, the proposed algorithms which aim directly at damaged regions are less, researchers usually use the manual annotation [1-4] and semi-automatic segmentation algorithms [5-8] to segment damaged regions when digital inpainting. Those method can segment the parts of damaged regions, they remain to be further improved.

Transition region is refers to the intermediary between target and background region. It's a special region, having the characteristic of the boundary which can separate different regions (foreground and background), and having the characteristic of region which has width and whose area is not zero [9]. Segmentation based on transition region is a new rise of algorithm in the recent years, which has certain advantage in the practical application [10-12]. Therefore, it is an effective way to solve damaged region segmentation from transition region and itself characteristics of damaged regions. In order to segment damaged region of Thangka effectively, a new segmentation algorithm based on local distinction is proposed considering the gray mean, local complexity and local difference of damaged region. Theoretical analysis and experimental results show that the proposed algorithm can obtain better segmentation effect.

2. Related work

2.1. Local gray mean

Ω_k denotes the neighborhood which has k as the center, the neighborhood average gray value is shown in formula(1).

$$\overline{f}(\Omega_k) = \frac{1}{M_k \times N_k} \sum_{u=-M_k/2}^{M_k/2} \sum_{v=-N_k/2}^{N_k/2} f(i+u, j+v) \quad (1)$$

Where $f(i, j)$ is gray value in the region of Ω_k, $M_k \times N_k$ is the local block size.

2.2. Local complexity

The gray distribution of the Thangka damaged regions is fluctuation, its feature is similar with feature of transition region. It has better segmentation effect in practice when using the feature of transition region, the segmentation method based on transition region can make up the deficiency of the traditional algorithm. Among them, the segmentation algorithm [13] based on the local complexity is a typical transition region segmentation algorithm.

$f(i, j)$ denotes the distribution function of gray-scale images, $(i, j) \in S$, S denotes the pixel space coordinates set of integers. Image has 256 gray levels, the size is $M \times N$, and the image histogram is shown in formula (2).

$$h(t) = \sum_{i=1}^{M} \sum_{j=1}^{N} \delta(t - f(i, j)) \quad (2)$$

Where $t \in \{0,1,2,\cdots,255\}$ is the gray level, $\delta(\cdot)$ is the unit impulse function.

Defining a flag function as (3).

$$S_t(h(t)) = \begin{cases} 1, h(t) \neq 0 \\ 0, h(t) = 0 \end{cases} \quad (3)$$

The definition of local complexity is shown in formula (4).

$$C(\Omega_k) = \sum_{t=0}^{255} S_t \quad (4)$$

Within a local window, the local complexity is equal to the number of different gray levels in the window. If the colour is the same, the local complexity is 1, if all colors are not the same, then the local complexity is the number of pixels in the window.

2.3. Local difference

The damaged degree in the damage regions are difference, gray level of damaged regions is large, so the homogeneity in the damaged regions is poor, local complexity value is higher. But the complexity value reflects the difference of the gray level, and it has nothing to do with the gray value. In addition, although the isolated noise has little effect on the local complexity, it also weakens the difference between the pixels, so that the contrast of the transition region is reduced.

In order to better extract the image transition region and improve the difference between the pixels, Chen et al [14] proposed a segmentation algorithm based on the image transition region, which increases the gray level and detail information of the transition region.

The size of the local window Ω_k is $N \times N$, $f(x, y)$ is the gray distribution function, the definition of B_k as formula (5)

$$\left\{B_k \mid B_k = (f'(x, y) - f'(i, j))^2, x > i \; or \; y > j; (x, y, i, j) \in [1, N], k \in [1, C_{N^2}^2]\right\} \quad (5)$$

Where the definition of the function $f'(x, y)$ is formula (6)

$$f'(x,y) = f(x,y) + C(f(x,y) - \overline{f}(x,y)) \tag{6}$$

$f'(x,y)$ is the enhancement function in the local window, $\overline{f}(x,y)$ is the mean value of gray level distribution in the local window(Ω_k), C is the enhancement factor, and reflects the local gray level distribution, the value of C is the local complexity which is normalized. The range of B_k can be stretched by $f(x,y) - \overline{f}(x,y)$, when $f(x,y) - \overline{f}(x,y) > 0$ is true, the value of B_k increase, otherwise, the value is B_k decrease. The definition of the local difference is formula (7).

$$LDC(\Omega_k) = \sum_{k=1}^{C_{N^2}^2} h(B_k) \tag{7}$$

Here, the definition of $h(B_k)$ as shown in the formula (8).

$$h(B_k)\begin{cases} 1\ \delta(B_k) \neq 0 \\ 0\ \delta(B_k) = 0 \end{cases} \tag{8}$$

Compared with the local complexity, LDC can reflect the characteristic of the transition region is more obvious.

3. The segmentation of transition based on local distinction

When missing colours, the damaged area in the Thangka has the following characteristics: (1) the local difference in the transition regions is larger; (2) There is a difference between the gray value and the local mean; (3) the area which miss colour was smaller; (4) the regions which missed colours, have the features of transition region. Based on the above 4 features, a segmentation algorithm based on local distinction is proposed.

3.1. Local distinction

The bottom of the Thangka is white colour, when the colour missing, the damaged regions have transition features, so the gray value is higher in the serious damaged regions. When original subtract the men filtering image, the result has the feature, the value in the damaged regions is relatively high, which can help us to find the position of the damaged regions, however, there are also some over segmented region, they can be filtered by feature of transition. Local difference can find transition regions, damaged region is in the transition regions, so the local difference value is higher in the damaged regions, we can further filter out some over segmentation regions by formula(9).
The local distinction can be calculated by formula (9).

$$\left\{P(\Omega_k) = LDC \cdot [f(x,y) - \beta\overline{f}], x, y \in [1, N]\right\} \tag{9}$$

Among them, $\overline{f}$ and LDC can be calculated by formula (1) and (7). Because the damaged regions have transition features, the damaged regions colour is white when missing colours, so the term of $f(x,y) - \beta\overline{f}$ denotes getting the regions whose gray value are larger than $\beta\overline{f}$ in the image, β is the weight of $\overline{f}$, $\beta \in (0,1)$, the value of β should be suitable, too smaller will lead to over segmentation, too larger will lead to inadequate segmentation. Experimental verification that segmentation can achieve a very good result when setting $\beta = 0.6$.

The value of the local distinction $P(\Omega_k)$ is larger in the transition and edge. Set the segmentation threshold $segTh = \lambda MAX\{P(\Omega_k)\}$, $\lambda \in (0,1)$, $MAX\{P(\Omega_k)\}$ is the maximum value in the local distinction.

3.2. The step of segmentation

Step 1: calculate the local complexity by formula (4) and normalized, it will be as C in the formula (6).
Step 2: calculate the local difference by formula (7) and normalized, it will be as LDC in the formula (9).
Step 3: calculate the local distinction by formula (9), we will get local distinction image.
Step 4: obtain the segmentation result by setting threshold $segTh$ in the local distinction image.
Step 5: remove the over segmentation in the last segmentation result, we will get final segmentation result.

3.3. *The pseudocode of segmentation methodology*

We give pseudocode of segmentation methodology to explain process of segmentation.

(1) Set size of local window 7×7
(2) Z=0.02, β=0.6, λ=0.25
(3) Calculate local gray mean $\bar{f}(x,y)$
(4) Calculate local complexity C and normalized C
(5) Calculate B_k, $h(B_k)$ and normalized $h(B_k)$ (LDC is the normalized $h(B_k)$)
(6) Calculate local standard deviation R
(7) Calculate P_k
(8) Calculate $SegTh = \lambda MAX(P_k)$
(9) Set $SegTh$ as threshold and get segmentation result
(10) Remove over segmentation and get final segmentation result

4. Experimental results and analysis

4.1. *Parameter setting*

The parameters involved in this paper are verified by experiments. The size of the block is 7×7, and $\lambda \in (0.1, 0.4)$, they can be adjusted according to the experimental results. Segmentation algorithm was accomplished by Matlab.

4.2. *The experiment result*

Figure 1a is a broken Thangka, the size of image is 197×238. Figure 1b shows the result of artificial segmentation, it can evaluate the accuracy of image segmentation and over segmentation rate. Figure 1c shows the segmentation result by Local complexity, it can be seen that there is a fracture in the damaged area, and over segmentation region is large. Figure 1d shows the segmentation result by Local difference, it can be seen that the damaged regions can be segmented, and there is no fracture phenomenon, but it leads to many over segmentation, it is difficult for us to segment by artificial operator. Figure 1e shows the segmentation result by our algorithm, it can be seen that the damaged area can be extracted completely. Compared with local complexity and local difference, our algorithm has less over segmentation, it is easy to separate the damaged region and the over segmentation region when we carry out artificial operator.

In order to verify the effectiveness of our algorithm, we select several traditional segmentation algorithms and compare the result with them. The results are shown in Figure 2. Figure 2a is damaged origin image, its size is 195×156, figure 2b is segmentation result by artificial operator, figure 2c is the segmentation result by our algorithm ($\lambda = 0.25$), we can see that the damaged regions can be segmented well. Figure 2d is segmentation result by Otsu algorithm [15], and it produce much over segmentation region. Figure 2e is the segmentation result by k-mean clustering algorithm [16], the result is bad because of much over segmentation region. Figure 2f is the segmentation result by two-coordinate maximum entropy segment method [17], there has been a shortage of the segmentation, some damaged regions failed to segment. Figure 2g is the segmentation result using local complexity algorithm [13], although the segmentation result has greatly improved than figure 2d, 2e, 2f, but there are still over segmentation. Figure 2h is the segmentation result using fuzzy entropy neighborhood [18], there has been a shortage of the segmentation. Figure 2i is the segmentation result using transition region algorithm based on digital morphology [19], although most of the damaged regions are segmented, but also there are over segmentation badly. Figure 2j is the segmentation result using local difference algorithm [14], but there are over segmentation badly. Through the comparison, we can know that the proposed segmentation algorithm can segment damaged regions which has the feature of the transition region, and create the over segmentation less.

We also choose another broken type Thangka image which are color missing type and scratch type to validate our algorithm, they are shown in Figure 3a and 3d. Figure 3c and Figure 3f is the segmentation result using our algorithm. We can see that the damaged regions which missed color can be segmented from Figure 3c, and scratched region can be segmented from Figure 3f. The segmentation result are very well from the Figure 3c and Figure 3f.

Our algorithm can also applied to segment nature image, Figure 4a is a rice image, and gray distribution in the image is not uniform. Figure 4c and Figure 4d are the segmentation result by Otsu Algorithm and two-dimensional maximum entropy algorithm, segmentation results are not completely. Figure 4b is the segmentation result by our algorithm, we can see that the rice can be segmented very well, so our algorithm is optimal.

4.3. Algorithm performance evaluation

In order to quantitatively evaluate the algorithm, the segmentation accuracy and the over segmentation rate are used as evaluation criteria to measure the processing effect and performance of the algorithm. P_1 is the segmentation regions by algorithm, P_2 is the segmentation regions by artificial operator, $count(P)$ is the total pixels number in the region, the segmentation accuracy is defined as

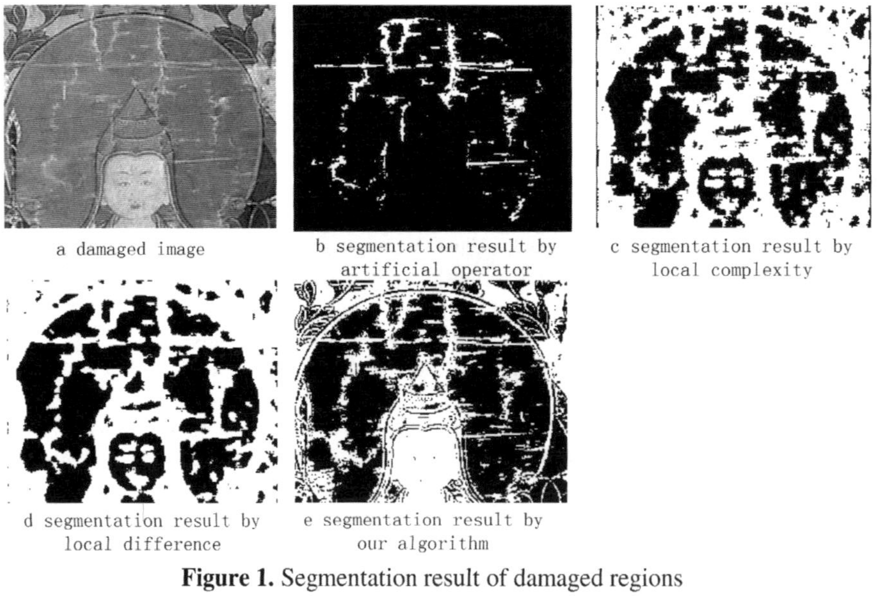

Figure 1. Segmentation result of damaged regions

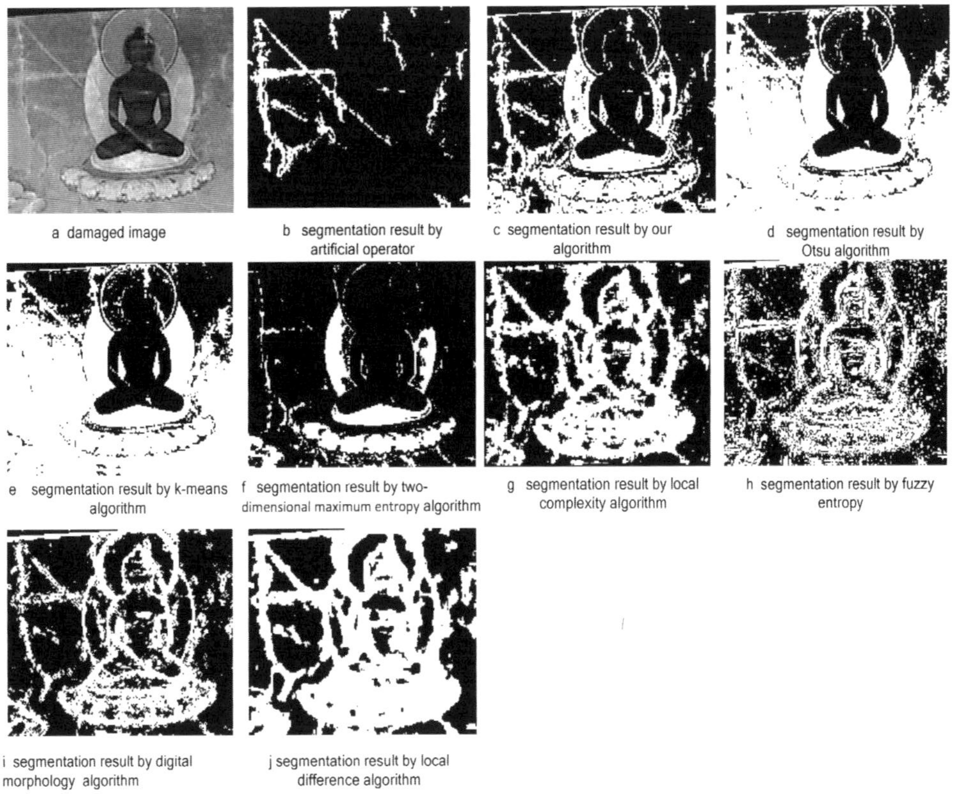

Figure 2. Comparison of segmentation result

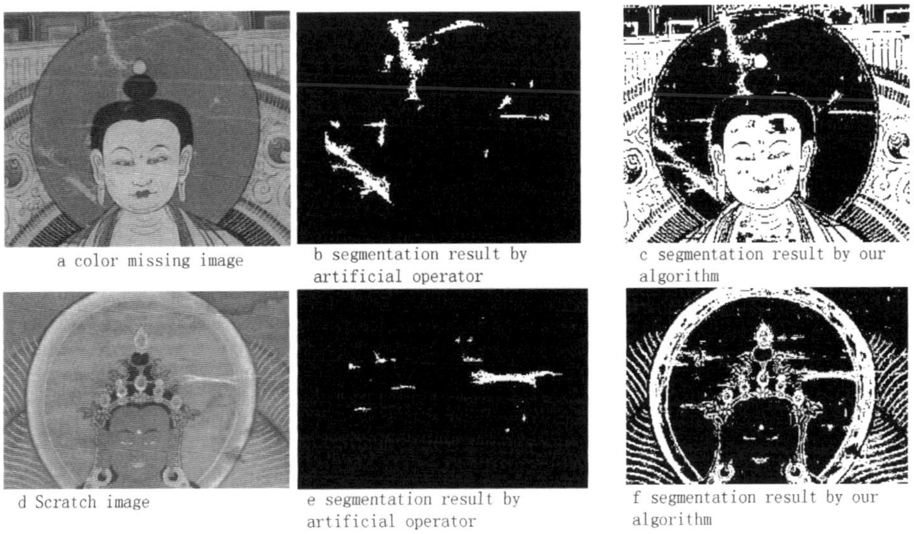

Figure 3. Segmentation result of damaged Thangka

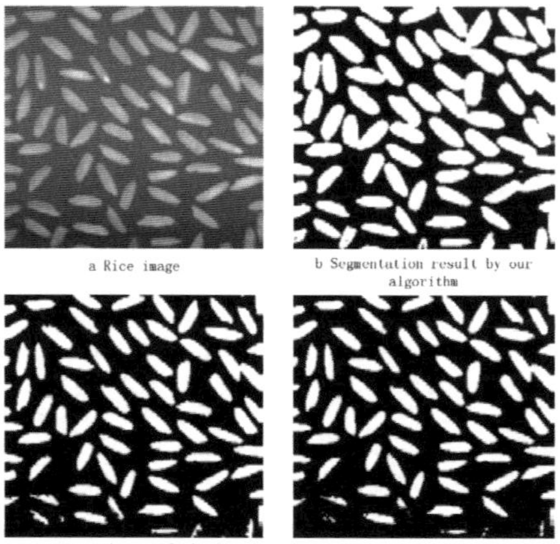

Figure 4. Segmentation result of nature image

$F = ((count(P_1 \cap P_2) / count(P_2))\%$, the over segmentation rate is defined as $F = ((count(P_1 - (P_1 \cap P_2)) / count(P_1))\%$ [20]. Table 1 shows the processing data for the segmentation of Figure 2a using different algorithms, the segmentation accuracy of our algorithm is higher than the other 6 algorithms, the over segmentation rate is significantly lower than the other 6 algorithms, the segmentation effect of our algorithm is better.

Table 1. Comparison of segmentation results for Figure 2a

algorithm	Accuracy rate	over segmentation
Our algorithm	99.05%	69.05%
Otsu algorithm[15]	85.60%	87.10%
Digital morphology	67.88%	81.93%
Two dimensional maximum entropy algorithm[17]	25.60%	85.81%
Local complexity [13]	78.50%	81.70%
k-means[16]	84.47%	86.79%
Fuzzy entropy [18]	69.11%	94.51%

5 Conclusion

Some Thangka damaged regions have the features of transition region, we proposed an algorithm based on distinction after analysing local complexity and local difference. It can segment damaged regions in which colour missed and whose area is small, better than other algorithms. If the gray value of damaged regions is close to the surrounding regions, or the gray value of the damaged regions is lower than the local average, or the area of damaged regions is large, the segmentation result is not ideal. In view of the local distinction algorithm is a transition region segmentation algorithm, segmentation result will lead too much over segmentation for the rich texture image. When Thangka have seriously damaged regions, or have grease regions, or have severe fade regions, the segmentation result is not ideal, it will be the further study focus.

Acknowledgments

Acknowledgements my wok was supported by the National Natural Science Foundation of China(Grant No.60875006, 61162021), Natural Scientific Research of Education department of Anhui province (Grant No.2015KJ012, KJ2015A295), Natural Science Foundation of Fuyang Normal

College (Grant No.2015FSKJ08), and College Students' Innovative & Entrepreneurial (Grant No. 201510371045).

References

[1]Yang X P, Wang S W. Dunhuang Mural Inpainting in Intricate Disrepaired Region Based on Improvement of Priority Algorithm. *Journal of Computer Aided Design & Computer Graphics*, 2011, **23** pp 284-289. (In Chinese)

[2]Liu H M, Wang W L, Xie H. Thangka image inpainting using adjacent information of broken area. *Proceedings of International MultiConference of Engineers and Computer Scientists*. Hong Kong: Newswood Limited, 2008 pp 646-649.

[3]Bertalmio M, Sapiro G, Caselles V, et al. Image Inpainting. *Proceedings of the 27th annual conference on Computer graphics and interactive techniques*. New York: ACM, 2000 pp 417-424.

[4]Criminisi A, Perez P, and Toyama K. Region filling and object removal by exemplar-based image inpainting. *IEEE Trans. on Image Processing*, 2004, **13** pp 1200-1212.

[5]Yang X P, Wang S W, Jia J F, et al. Color recovery of Dunhuang murals with GrabCut segmentation and automatic data acquisition. *Journal of Lanzhou University of Technology*, 2010, **36** pp 114-117. (In Chinese)

[6]Wei B G, Lu D M, Pan Y H, et al. Interactive Image Segmentation Using Multiple Color Spaces. *Chinese Journal Computers*, 2001, **24** pp 770-775. (In Chinese)

[7]Liu H M, Bi X H. Damaged regions segmentation on Thangka for missing colors. *Computer Engineering and Applications*, 2013, **49** pp 219-223. (In Chinese)

[8]Liu H M, BI X H. Segmentation of Faded Regions in Thangka Images. *Journal of Computer-Aided Design & Computer Graphics*, 2013, **25** pp 219-223. (In Chinese)

[9]Zhang Y J. Transition Region and Image Segmentation. *Acta Electronica Sinica*, 1996, **24** pp 12-17. (In Chinese)

[10] Wang F, Zhang T X, Deng J P. A novel method concerning about image transition region extraction and segmentation. *2011 International Conference on Digital Object Identifier*. Hangzhou: IEEE, 2011 pp 95-97.

[11] Li Z, Tang K, Cheng Y, et al. Transition region-based single-object image segmentation. *AEU-International Journal of Electronics and Communications*, 2014, **68** pp 1214-1223.

[12] Li Z, Liu G, Zhang D, et al. Robust single-object image segmentation based on salient transition region. *Pattern Recognition*, 2016, **52** pp 317-331.

[13]Yan C J, Sang N, Zhang T X. Image Transition Region Extraction and Segmentation Based on Local Complexity. *Journal of Infrared and Millimeter Waves*, 2005, **24** pp 312-316. (In Chinese)

[14]Chen C P, Qin W, Fang Z F. Image transition region extraction and segmentation based on LDC. *Laser & Infrared*, 2011, **41** pp 1023-1026. (In Chinese)

[15]Otsu N.A threshold selection method from grey-level histograms. *IEEE Transactions on Systems, Man, and Cybernetics*, 1979, SCM, **9** pp 62–66.

[16] Bezdek J. Pattern Recognition with Fuzzy Objective Function Algorithms. New York: Plenum Press, 1981 pp 25-35.

[17]Abutaleb A S. Automatic thresholding of gray-level pictures using two-dimension entropy. *Computer Vision, Graphics and Image Processing*, 1989, **47** pp 22-32.

[18]Wang Y C, Liang D Q, Wang Y. Transition Region Extraction and Segmentation Based on Image Fuzzy Entropy Neighborhood Unhomogeneity. *Acta Electronica Sinica*, 2008, **36** pp 2445-2449. (In Chinese)

[19]Xu M P, Feng T, Zhou Z A. Analysis of the Image Transition Region Based on Mathematical Morphology. *Journal of Shanghai Second Polytechnic University*, 2008, **25** pp 276-279. (In Chinese)

[20]Zhao S Y, Zhang D, Tan W J, et al. Hippocampus region segmentation method in MR images based on multi-scale level set. *Chinese Journal of Scientific Instrument*, 2012, **33** pp 2286-2291. (In Chinese)

IOP Conf. Series: Journal of Physics: Conf. Series **787** (2017) 012011 doi:10.1088/1742-6596/787/1/012011

A multiple classifier system based on Ant-Colony Optimization for Hyperspectral image classification

Ke Tang, Li Xie and Guangyao Li

No 4800, Cao'an Road, Tongji Universirt, Shanghai, China

E-mail: ke_tang1991@163.com

Abstract. Hyperspectral images which hold a large quantity of land information enables image classification. Traditional classification methods usually works on multispectral images. However, the high dimensionality in feature space influences the accuracy while using these classification algorithms, such as statistical classifiers or decision trees. This paper proposes a multiple classifier system (MCS) based on ant colony optimization (ACO) algorithm to improve the classification ability. ACO method has been implemented on multispectral images in researches, but seldom to hyperspectral images. In order to overcome the limitation of ACO method on dealing with high dimensionality, MCS is introduced to combine the outputs of each single ACO classifier based on the credibility of rules. Mutual information is applied to discretizing features from the data set and provides the criterion of band selection and band grouping algorithms. The performance of the proposed method is validated with ROSIS Pavia data set, and compared to k-nearest neighbour (KNN) algorithm. Experimental results prove that the proposed method is feasible to classify hyperspectral images.

1. Introduction

With the development of remote-sensing image technology, high dimensional data are easily accessed, meaning hyperspectral images are available for characterization, identification, and classification of the land-covers with improved accuracy and efficiency[1]. The information contained in hyperspectral data is performed in more than hundreds of spectral bands, which provide the possibility of distinguishing more classes of the ground type.

A large number of methods have been applied to classify multispectral images over the past decades. Statistical method such as maximum likely hood classifier and Bayesian classifier count on the assumption that the number of each class preforms a normal distribution in the feature space. Decision trees was firstly introduced in remote sensing images in [2,3,4], and provide an acceptable accuracy. Ant colony optimization (ACO) was firstly proposed to mine classification rules by [4]. ACO method is inspired by natural biological systems[5]. By simulating the behavior of ants in their process of searching food, ACO methods can utilize the feedback system of the whole ant colony. The interactional information exchange of ants, which is exactly pheromone system, helps ants learn from past experiences. One advantage of ACO are the easy-understanding rules since the classification rule produced by this induction algorithm is more explicit than mathematical equations. ACO has been demonstrated to be effective and robust when working on multispectral images[6,7].

Unfortunately, these methods above can not be implemented on hyperspectral images directly. Because of the high dimensionality in feature area, the complexity of both traditional classification method and ACO method will dramatically increase which resulting in the curse of dimensionality. [8,9,10] indicated it's beneficial to remove bands with little or no discriminatory information. However, dimensional reduction brings about loss in information of hyperspectral images. Multiple classifier systems(MCS) can overcome this problem. MCS combines same classifiers' results or different classification algorithms[11]. With the help of this system, high dimensionality feature space can be divided into s set of low dimensionality data sets. After a single classifier is produced, MCS combine these classifiers with a specific fusion method. Multiple classifier system can be utilized to improve classification accuracy in hyperspectral images.

In this paper, a new hyperspectral image classification method is developed in this paper. Firstly, a discretization algorithm based on mutual information is implemented to divide each band into several part, which can be regarded as a feature. Then the feature space are split into a set of band groups to overcome high dimensionality problem. Band selection and band grouping method are introduced to reduce the dimensionality. Next ACO method is applied to each group to find rule sets. Finally use fusion method to combine the output of each classifier to produce a classification decision. This paper is composed of four sections. The related algorithms are introduced in Section 2. Section 3 describes the data set of remote sensing image and the designed experiment. The experimental results and discussion are presented in Section 4. Section 5 concludes the paper. Figure 1 presents the flowchart of the proposed system.

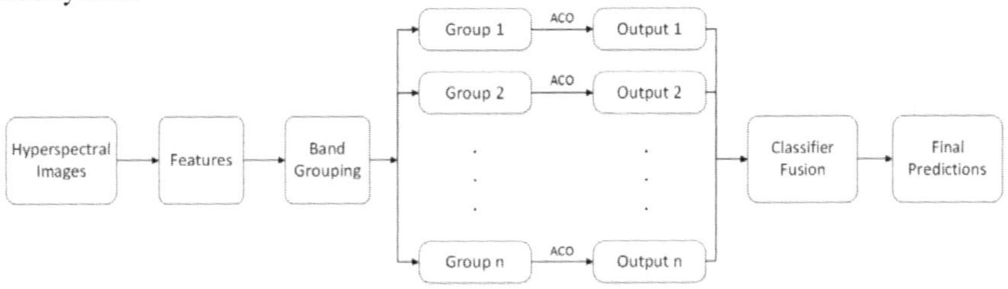

Figure 1. The multiple classifier system based on ACO method

2. Methodology

Ant colony optimization algorithm is innovated by observing food seeking behaviors of ants. When ants are seeking food, they will release pheromone on their route. Ants tend to choose a path that has the largest pheromone. Since pheromone's disappearing with time, the shortest path will collect more pheromone. This process provides a positive feedback, in which the probability of an ant choosing a path is proportional to the number of ants that have passed through that path[7].

First, the hyperspectral remote sensing image pixels are divided into a training data set and a testing data set. Then it comes to data discretization, which is employed to segment the brightness value of each band[7][12]. Since there are more than hundred bands in hyperspectral image and ant colony algorithm can not adapt to high dimensionality, original spectral space can not use this method directly. Besides, not all of the bands are useful to classification, some bands refer to classes those are not interested. This paper proposes a band grouping method based on entropy information to select effective bands and group the remaining bands. After band grouping, ant mining method are used to find the best rules covering the training samples. Then use credibility to describe a rule's confidence degree on classification outputs. Finally, a fusion method based on rules' credibility is applied to establish a multiple classifier system.

2.1. Data discretization

Since the brightness value of each band of a specific pixel in a hyperspectral image ranges from 0 to 65536, conventional classification method will fail when dealing with these continuous values. Continuous attributes must be separated into multiple sections to adapt the decision rules. This will influence the efficiency of the ant miner algorithm and quality of rules. Before a further multi system classifier, a data discretization algorithm is adopted based on information entropy[7].

There are four symbols as (U, R, V, F) to describe the image information. U represents the set of objects, namely the image pixels. R stands for the set of attributes, and $R = C \cup D$, where C and D refer to a condition attribute set and a decision attribute set, respectively. On this condition, C and D refer to the bands of hyperspectral image and the type of lands, respectively. V denotes the domain of the range of value of each band and the value of each class. F is the information function[7].

For a certain condition attribute(a band) $c \in C$, assume its domain is $V_a = [l_a, r_a]$. If there exists a group of points: $l_a < c_1^a < c_2^a < \cdots < c_{m_a}^a < r_a$, then $V_a = [l_a, c_1^a) \cup [c_1^a, c_2^a) \cdots [c_{m_a-1}^a, c_{m_a}^a) \cup [c_{m_a}^a, r_a)$. Attribution a is separated as $m_a + 1$ equivalent classes. c_k^a is called a breakpoint. The goal of discretization is to find a set of breakpoints to represent the data.

In this paper, information entropy is introduced to evaluate the quality of discretization. Supposing $X \subseteq U$ as a subset, $|X|$ is the number of object, k_j is the number of object whose decision attribute is $j(j = 1,2, \cdots, r(d))$, then the information entropy of this subset is defined as:

$$H(x) = -\sum_{j=1}^{r(d)} p_j log_2 p_j, p_j = \frac{k_j}{|X|} \tag{1}$$

X is separated into two sections by a breakpoint b_i^a. The number of sample belonging to X and its value is smaller than value of breakpoint b_i^a is $l_j^X(c_i^a)$, $r_j^X(c_i^a)$ means the number of samples whose value are greater than b_i^a:

$$l^X(b_i^a) = \sum_{j=1}^{r(d)} r_j^X(b_i^a) \tag{2}$$

$$r^X(b_i^a) = \sum_{j=1}^{r(d)} r_j^X(b_i^a) \tag{3}$$

Breakpoint c_i^a divide X into two subsets X_l and X_r, and

$$H(X_l) = -\sum_{j=1}^{r(d)} p_j log_2 p_j, p_j = \frac{l_j^X(b_i^a)}{l^X(b_i^a)} \tag{4}$$

$$H(X_r) = -\sum_{j=1}^{r(d)} p_j log_2 p_j, p_j = \frac{r_j^X(b_i^a)}{r^X(b_i^a)} \tag{5}$$

For a breakpoint b_i^a, information entropy of X is

$$H^X(b_i^a) = \frac{|X_l|}{|U|} H(X_l) + \frac{|X_r|}{|U|} H(X_r) \tag{6}$$

Supposing S is the selected breakpoint set and B is the set of candidate breakpoints. $S = \{X_1, X_2, \cdots, X_m\}$ is considered as the equivalent class of set X. The new information entropy after adding breakpoint $b \notin S$,

$$H(b, L) = H^{X_1}(b) + H^{X_2}(b) + \cdots + H^{X_m}(b) \tag{7}$$

With adding new breakpoints, the decision attribute becomes simplification if $H(b, S)$ decreases.

Let P be the selected breakpoint set, L is the equivalent class of set X, B is the set of candidate breakpoints set. H is the information entropy. The discretization algorithm can be concluded as follows:

1) $P = \phi; S = U; H = H(U)$;

2) For each $b \in B$, calculate $H(b, L)$;

3) If $H \leq minH(b, L)$, terminate the loop;

4) Select breakpoint b_{min}, which contributes the minimum value of $H(b, L)$; $H = H(b, L)$; $B = B - b$;

5) For every $X \in L$, if b_{min} divides equivalent class X into X_1 and X_2, then remove X from S, put equivalent X_1 and X_2 into S;

6) If every sample in equivalent class has the same decision attribute, then end the loop. Turn to step 2 otherwise.

2.2. Band grouping

To split original data space to small groups, information entropy is introduced to evaluate quality of bands. Mutual information can be used to measure the similarity between two random variables[13].

Considering two random variables X and Y, $p(X)$ and $p(Y)$ are marginal probability distributions of X and Y, respectively. $p(X,Y)$ is the joint probability distribution. Then MI is defined as follow:

$$MI(X,Y) = \sum_{X\epsilon \mathbf{X}, Y\in Y} log \frac{p(X,Y)}{p(X)p(Y)} \tag{8}$$

X represents one of candidate bands. Y represents class type in decision attributes. MI can be used to measure the dependency between spectral image and ground type. Higher MI value means more dependency. Such bands with higher MI value have better ability in classification. Fig.2 shows the MI value for all bands.

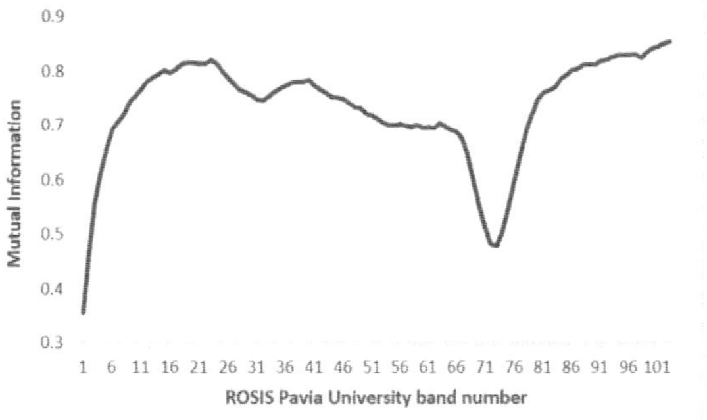

Figure 2. Mutual Information value

Bands with low MI value (lower than a threshold) should be excluded from the condition attributes since these bands hold less information than others. Besides, there are more redundancies between two adjacent bands if the MI is relatively high[9]. Bands with high correlation have weak ability in classification. These two bands need to be separated into different groups. Thus the two steps in this band grouping process can be described as follows:

2.3. Ant Miner

2.3.1. Rule searching. After discretization and band grouping, hyperspectral image are divide into several groups, each group contains several bands. Ant miner algorithm is used to search for classification rules in each group. Rule construction process is similar to the behaviour of ant's seeking food. Ant chooses nodes attributes by attributes until an entire path is constructed. Theoretically, nodes are selected randomly, but this process may take an insufferable time. A heuristic method is designed to help ants find the path in order to shorten the searching time. [7] defines the value of heuristic function of one condition attribute $term_{ij}$ according to statistical data as:

$$\eta_i^j = \frac{\max(\sum freqT_{ij}^1, \sum freqT_{ij}^2, \cdots, \sum freqT_{ij}^k)}{\sum T_{ij}} \tag{9}$$

where T_{ij} is the number of samples which match condition attribute $term_{ij}$, $freqT_{ij}^w$ is the frequency number of class w in T_{ij}. In the process of data mining, data that satisfies rules should be removed from the original data set, which will leading to the change of $\max(\sum freqT_{ij}^1, \sum freqT_{ij}^2, \cdots, \sum freqT_{ij}^k)$ and T_{ij}. Therefore, η needs to be updated after getting a final rule. From the beginning, the information density of each path nodes is initialized as the same value:

$$\tau_{ij}(t=0) = \frac{1}{\sum_{i=1}^{a} b_i} \tag{10}$$

where τ_{ij} is the information density of $term_{ij}$, a is the total condition attributes number, b_i is the potential value that attribute i may hold.

Each node in one attribute are selected out on the probability calculated by following formula:

$$P_{ij} = \frac{\tau_{ij}(t)*\eta_{ij}(t)}{\sum_{i=1}^{a}\sum_{j=1}^{b}\tau_{ij}(t)*\eta_{ij}(t)} \tag{11}$$

Quality of a rule can be defined as follow:

$$Q = \left(\frac{TruePos}{TruePos+FalseNeg}\right) * \left(\frac{TrueNeg}{FalsePos+TrueNeg}\right) \quad (12)$$

where $TruePos$ represents number of samples which entirely match the selecting rule, both condition attributes and decision attributes, $FalsePos$ represents number of samples which satisfy rule's condition attributes, while stand against predict class, $FalseNeg$ represents number of samples which don't match rule's condition attributes while are the same class as prediction, $TrueNeg$ are the number of samples which don't match rule either condition attributes or decision attributes[7].

To provide a basis to the fusion system, credibility of a rule can be defined as follows:

$$C = \left(\frac{TruePos}{TruePos+FalseNeg}\right) \quad (13)$$

2.3.2. Pheromone updating. The next step is updating pheromone. In each ant's searching period, once a complete rule is constructed, the pheromone of nodes which are inside the path should increase, while the others' should decrease. The updating degree can be defined as follows:

$$\tau_{ij}(t+1) = (1-\rho) * \tau_{ij}(t) + \frac{Q}{1+Q} * \tau_{ij}(t) \quad (14)$$

where ρ is pheromone evaporation coefficient, and Q is the quality of a final rule. In every iterations, numbers of ants will find numbers of rules, the rule with highest qualification can be seen as the candidate of final rule, and the other rules are dropped out. If the rule can match specific number of samples, that rule can be added to the final rule sets, otherwise should be dropped out either. Record the credibility of rule. Then it comes to the next iteration, until the number of remaining samples is less than a predetermined value.

2.4. Multi classifier.

The final step is to make a prediction by the rules. [14] utilizes the classifier and confidence measurement to combine prediction results. First, use each classifier to output a unique class so that a vector of classes is produced for each sample. Next, associate a confidence measurement for each class and produces a vector for every classifier and a matrix for ensemble of classifier.

However, this method doesn't consider the differences of abilities of each rule in classification. Besides, there is a possibility that a sample is in accord with multiple rules resulting in a difficulty in making prediction. A multi classifier system based on the credibility of rules is proposed. For each testing sample, testify whether it satisfies the condition attributes of rules in each group of classifier. If so, add up the credibility of the relative rule to the class according to the rule's decision attribute. Finally, the vector of prediction for one sample P is calculated out and forms as follows:

$$P_i = (C_1, C_2, \cdots, C_n), 1 < i < ||test\_size|| \quad (15)$$

where C_i is the sum of credibility of one class, n is the total number of class. Choose the class which has the maximum possibility value from P_i as the final prediction.

3. Experiments

3.1. Data sets

The method proposed in this paper is applied to ROSIS, a well-known hyperspectral data sets. The first data set is acquired by the ROSIS sensor during a flight campaign over Pavia, which is made of 103 spectral bands, covering the wavelength range from 0.43 to 0.86 μm. The geometric resolution is 1.3 meters. This data set holds 610*340 pixels, while some of the samples contain no information and have to be discarded. Pavia University is one of the two scenes in the data set, whose ground-truths differentiate 9 classes. In this paper, totally 42776 pixels were selected out to be made up of the data set. Figure 3 shows the origin view of Pavia University. The types and number of samples are represented in Table 1.

Figure 3. Pavia University

Table 1. Ground-truth of Pavia University data set

#	Class	Samples
1	Asphalt	6631
2	Meadows	18649
3	Gravel	2099
4	Trees	3064
5	Painted metal sheets	1345
6	Bare Soil	5029
7	Bitumen	1330
8	Self-Blocking Bricks	3682
9	Shadows	947

3.2. Results and discussion

In this section, two groups of experiments designed to evaluate the effectiveness of the proposed method. The classification accuracy of the proposed method was evaluated by several experiments in the first part. The purpose of this experiment was to study the effects of discretization iteration and max number of bands when grouping bands. In the second experiment, KNN (k-nearest neighbour) algorithm was a to make a comparison to the proposed method, setting training set size as the control variable.

There are a few parameters in this system: (1) max_bands_per_group: This decides a single classifier's dimensionalit. This amount of condition attributes plus the decision attributes to make up of a sub-training set. (2) discretization_iteration: iteration times of discretization, equalling to the number of intervals in one band. (3) max_iteration: ACO method terminates when the number of iterations exceeds this number. (4) min_cover_per_rule: Abandon one rule if the number of samples covered by this rule is less than the threshold. (5) max_uncovered_sample: ACO method terminates when the remaining training samples is less than this threshold. (6) Min_MI: Bands with MI value lower than this number should be excluded. The default parameters setting are as follows: $max\_iteration = 1000, \min\_cover\_per\_rule = 2, \max\_uncovered\_sample = 20, Min\_MI = 0.55$.

The relationship between classification accuracy and discretization iteration was shown in Figure 4. The classification accuracies are always above 81%. It rises when the iteration increases from 5 to around 12. The highest accuracy is 84.3%. However, there are continuous decreases when iterations keep going up. The reason for this drop is shown in Figure 5. Figure 5 represents the cover rate representing the percentage of covering sample size of training size for each generated rule sets. The cover rate suffers a sharp plunge after the 11th discretization iteration. Lower cover rate indicates that the generated rules can not represent the training set effectively, which finally has an influence on overall accuracy.

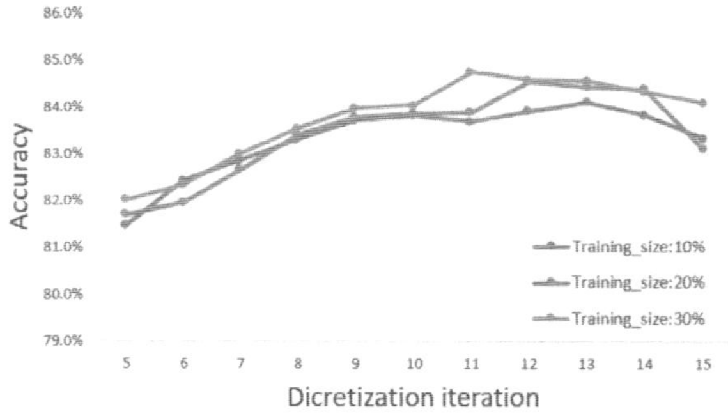

Figure 4. Influence of discretization iteration on accuracy

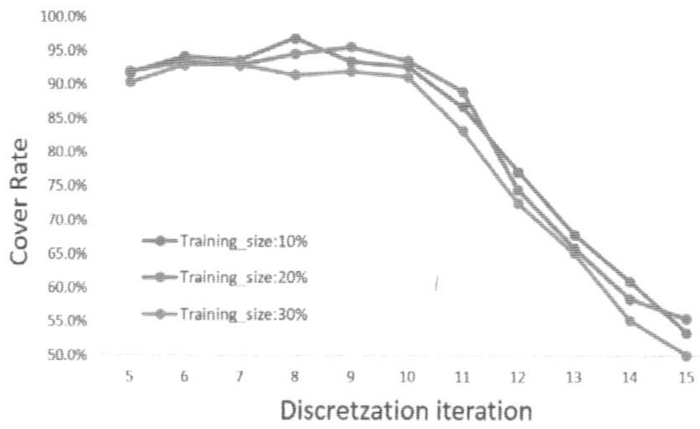

Figure 5. Cover rate per classifier in training set

Figure 6 shows the difference between with or without band selection method before grouping all the bands. This figure illustrates the effectiveness of the proposed band selection method, which will make progress on band grouping and final classification performance. There are improvements on classification accuracy with bands selection.

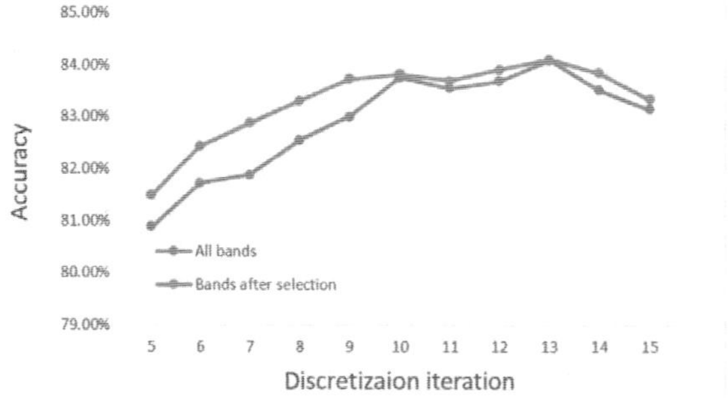

Figure 6. Differences of band grouping on all bands and selected bands

The maximum number of bands in a separate group contributes to classification accuracy and training speed. Figure 7 illustrates the classification ability when setting different size of bands in a group. The classification accuracy will decline with the number of bands in each group increases from 3 to 5. It is reasonable that the number of bands in each group should be more than 2. If the number of bands in each group is too small, the rules are not so believable because it may not be enough to completely represent the sample. On the other hand, too many bands will result in the difficulty on digging rules, and causes either accuracy decrease or time cost rise. The ACO classifiers' performance for this data set are shown in Figure 8 when $max\_bands\_per\_group = 4$ and *discretization_iteration* = 10 on different training size. The classification accuracies of every individual ACO classifier are demonstrated in Figure 8.

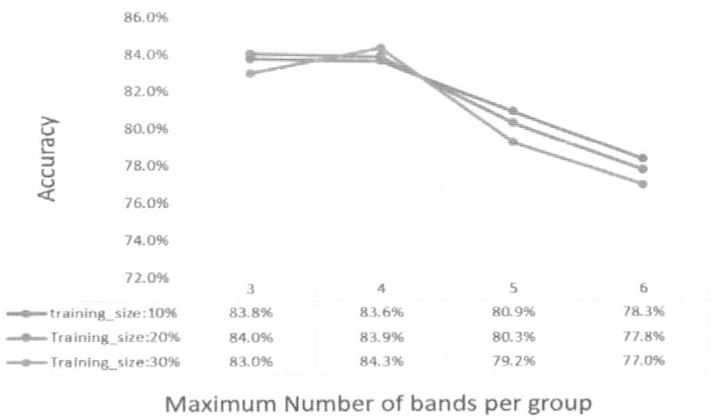

Figure 7. Influence of different size of grouped bands on classification accuracy

The peak of the classification accuracy curve of each classifiers is 78.4% when the lowest is 42.3%. This suggests that the dimensionality reduction method for hyperspectral images in this paper represents an effcacious method of improving the overall classification accuracy.

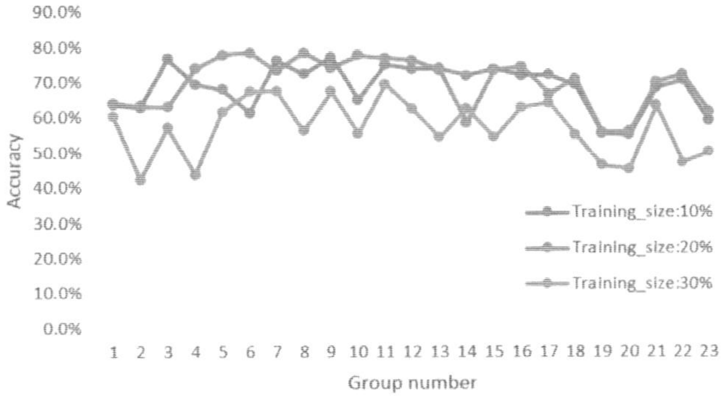

Figure 8. Overall accuracy of ACO classifier on band groups for different training size

To validate the effectiveness of the proposed method, k-Nearest Neighbour (KNN) was introduced to provide a comparison. KNN is a non-parametric approach used for classification and regression[15]. A sample is classified by a majority vote of its neighbours[16,17]. KNN needs a user-defined constant k meaning k nearest neighbours. A drawback of KNN method algorithm is that it is sensitive to the local structure of the training data. The reason to choose KNN as the comparison is that KNN can be directly applied to high dimensionality problem[18] in contrast to the proposed multiple classifier system. In Table 2, the overall accuracy with different training size based on proposed method and KNN method were shown. The results shows that the proposed multiple classifier system outperformed KNN method.

Table 2. Comparison between proposed method and KNN method

Training size(%)	10	20	30
ACO MCS(%)	84.30	83.87	84.32
KNN(%)	83.75	82.90	79.04

4. Conclusion

In this paper, a new multiple classifier system based on ant colony algorithm for hyperspectral images was carried out. A band selection and band grouping system based on mutual information of hyperspectral images was utilized in the proposed method aiming to split the origin images into several band groups. Then ACO method is applied to establish a classifier on each group to generate a multiple classifier system. Finally, a fusion strategy is taken to synthesize all the classifiers. The experiment results demonstrated that the proposed method performed a good classification ability.

Multiple classifier system was introduced to overcome the curse of dimensionality which was the main problem in hyperspectral remote sensing images classification. Feature selection methods were the traditional way of dimensionality reduction for hyperspectral images. It chooses several bands from the original space to represent the data set, which will suffer some information loss. Band grouping method decomposes the whole feature space to subspaces in order to apply ACO method in the later process. ACO method performs better than traditional methods on constructing proper classifier for hyperspectral images.

To combine each single classifier, a fusion method based on the rule confidence is implemented to improve the overall ability. By using the credibility of a specific rule, not just a output class, ACO classifiers' feature are fully employed since the credibility of a rule can be calculated during the rule searching period.

The proposed multiple classifier system has been applied to the classification of Pavia University, ROSIS data set. KNN method is carried out to provide a comparison in terms of classification accuracy. The overall accuracy of the proposed method is 84.11% with Kappa coefficient 80.23% when the KNN method has classification accuracy 83.75% and Kappa coefficient 79.80%.

Further study will focus on a flexible band grouping method that doesn't need a supervision to find the best group size and can adapt to ACO method as well. Moreover, a new fusion strategy considering not only rules' credibility but also single classifiers' accuracy deserves a try. Computation complexity is another possible topic.

References

[1] Camps V G and Lorenzo B 2005 *J. Trans. on Geosci. and Rem. Sens.* **43** 1351-62
[2] Friedl M A and Carla E 1997 *J. Rem. Sens. of Env.* **61** 399-409
[3] Pal M and Paul M 2003 *J. Rem. Sens. of Env.* **86** 554-65
[4] Parpinelli R S, Heitor S and Alex F 2002 *J. Trans. on Evol. Comput.* **6** 321-32
[5] Dorigo M, Mauro B and Thomas S 2006 *J. Comput. Intel. Mag.* **1** 28-39
[6] Dai Q, and Jianbo L 2007 *Int. Symp. on Multispectral Image Processing and Pattern Recognition* vol 7
[7] Xiaoping L 2008 *J. Trans. on Geosci. and Rem. Sens.* **46** 4198-208
[8] Baofeng G 2006 *J. Geosci. and Rem. Sens. Letters* **3** 522-6
[9] Shijin L 2011 *J. Knowledge-Based Sys.* **24** 40-8
[10] Shuang Z, Junping Z and Baoku S 2009 *2nd Int. Cong. On IEEE* vol 10
[11] Bigdeli B, Farhad S and Peter R 2013 *J. Ind. Soc. of Rem. Sens.* **41** 763-76
[12] Li X 2016 *J. Comp.* & Geosci. **89** 252-9
[13] Novovicova J 2007 *Iberoamerican Cong. on Pattern Recognition (Springer Berlin Heidelberg)* vol 13
[14] Kuncheva L I 2004 *Combining pattern classifiers: methods and algorithms* vol 15 (John Wiley & Sons)
[15] Altman N S 1992 *J. the American Sta.* **46** 175-85
[16] Blanzieri E and Farid M 2008 *J. Trans. on Geosci. and Rem. Sens.* **46** 1804-11
[17] Lefei Z 2012 *J. Trans. on Geosci. and Rem. Sens.* **50** 879-93
[18] Li Y and Bo C 2009 *17th Int. Conf. on Geoinformatics* vol 14

A Novel Method for Edge Detection in Images Based on Particle Swarm Optimization

Baby Sherin C and Mredhula L.

Electronics & Communication Engineering, MES College of Engineering, Kuttippuram Malappuram, Kerala, India - 679573.

E-mail: csherin111@gmail.com

Abstract. Edges give important structural information about the images. Edge detection is a process of identifying and locating the edges in an image. Edges are the points where discontinuity of intensity occurs. It also represents the boundaries of objects in images. In this paper a new edge detection method based on Particle Swarm Optimization is discussed. The proposed method uses morphological operations and a thresholding technique to improve the result of edge detector. This algorithm performs better in images comparing to other traditional methods of edge detection. The performance of proposed method is compared with traditional edge detection methods such as Sobel, Prewitt, Laplacian of Gaussian and Canny with parameters Baddeley's Delta Metric. Statistical analysis is performed to evaluate accuracy of edge detection techniques.

1. Introduction

The boundaries between different textures in images are defined as edges. They are always important characteristics of images since edges indicates higher frequency information. Edge detection is the name for a set of mathematical methods which aims at identifying points in a digital image at which the image brightness changes sharply or more formally, has discontinuities in intensity. Edge detection for an image is helpful for many applications in the field of image processing and computer vision, which includes image segmentation, image registration, data compression, pattern recognitions, image classification and so on.

There are large numbers of edge detection algorithms that are widely used for edge detection. Edge detection algorithms are classified into two categories, they are first order derivative based methods and second order derivative based methods. In [1] explains edge detection algorithms which are based on the first derivatives of pixels are used to detect the edges in an image. These methods are simple and fast. In [2] Wang and Xin propose the category of edge detection algorithms uses the second derivatives to detect edges. Its major advantage over the first derivative based edge detectors is it's good localization ability. D. Ziou and S. Tabbone explain comparative analysis of various edge detection techniques in [3]. The evidence for the best detector type is examined by studying the edge maps relative to each other through different parameters. The Canny's edge detection algorithm performs better than all other operators under almost all scenarios.

These methods are simple and typically applied to a particular pixel or its neighbors or a particular area. Images with complex objects or shapes these algorithms cannot effectively detect the edges. These limitations are overcomes by using artificial intelligence (AI). Swarm intelligence (SI) is an artificial intelligence technique based on the study of collective behavior in decentralized, self-organized systems. Examples of these systems found in nature, which includes ant colonies, bird flocking and fish schooling. Particle Swarm Optimization (PSO) is a recent computational intelligence method, introduced by Kennedy and Eberhart in 1995 [4]. PSO is a population-based algorithm for problem solving based on social psychological principles.

From the literature it is well clear that the existing edge detection methods have accuracy up to 60% (It depends upon the complexity of image). To improve accuracy, here we propose a new method of edge detection based on PSO. Obtained simulation results show that proposed method performs better than the traditional edge detection methods. The proposed method of edge detection using PSO uses a morphological operator and a thresholding technique for better result. The edge detection using proposed method has high accuracy and good localization capability comparing to traditional edge detection methods. Our contributions in this work are summarized as follows:

- We propose edge detector that use PSO algorithm for detection of edges in image. The PSO can be used to detect more complex edges in image and hence to increase the localization accuracy.
- From the existing PSO based edge detector, we propose a novel method which uses Morphological and Thresholding algorithm along with PSO edge detector, which improves accuracy of edge detector.
- Performance analysis of each edge detector is evaluated using the parameters Baddeley's Delta Metric (BDM) and accuracy. Both analysis shows that proposed method have better performance than traditional methods of edge detection.

2. Background

In this section, the background theories related to this work is discussed.

2.1. *Edge detection approaches*

Changes in intensity or discontinuities in an image are marked as edges. Otherwise they are the points of significant transitions in an image. Most of the shape information about an image is enclosed in edges. So edge detection is important step in many of image processing applications. Some of the traditional edge detectors are defined here which are used in this work for the comparison of proposed method with some parameters.

Sobel edge detector uses sobel approximation filter, which is used as a gradient with smoothing. A convolution mask of size 3 3 is usually used to detect gradients in X and Y directions [5]. In prewitt edge detector it is similar to sobel only difference is prewitt filter [6] is used. Laplacian of Gaussian (LoG) is another method which detects edges by checking for whether zero crossings occurs or not after filtering with a Gaussian filter. This method is accurate method because it combines the effect of both filtering [7]. Canny edge detector is a multistage algorithm which can find edges in wide range [8].

2.2. *Particle swarm optimization (PSO)*

Particle swarm optimization (PSO) is a universal global optimization, inspired by the social behavior of animals and other biological populations. Recently, PSO has been used in many research areas because of ease of its implementation, fewer operations and high speed of convergence. PSO algorithm is defined in this section, as proposed by Kennedy and Eberhart [4].

In the PSO there is a population of m particles that fly through an n-dimensional search space. The position of the ith particle is represented as the vector $X_i(t)=(X_{i1}(t), X_{i2}(t),\ldots X_{im}(t)\ )$ and is changed according to its own experience and that of its neighbors. Let $X_i(t)$ denote the position of particle P_i at time t. Then X_i is changed at each iteration of PSO by adding a velocity $V_i(t)$. $V_i(t)$ is calculated by the formula:

$$X_i(t+1) = V_i(t) + X_i(t) \quad (1)$$

The velocity is updated based on three components: current motion, particle memory influence, and swarm influence:

$$V_i(t+1) = \omega V_i(t+1) + C_1 r_1 \left(X_{pbesti} - X_i(t)\right) + C_2 r_2 \left(X_{gbesti} - X_i(t)\right) \quad (2)$$

The value of random variables r_1 and r_2 between 0 and 1, ω is inertia weight which controls the impact of the previous velocity; C_1 (self-confidence) and C (population confidence) are learning factors. X_{pbesti} is the best position of ith so far; X_{gbesti} denotes the best position of population so far [9].

3. Proposed method of edge detection

A block diagram of edge detection using PSO is shown in Figure 1. Proposed method consists of morphological operator and thresholding algorithm along with PSO edge detector. Morphological operator is used for image enhancement technique and it also act as noise removal filter. PSO edge detector is to locate the points of discontinuities in image and to obtain edge map of image. Thresholding algorithm is used to classify detected points is edge or not. Single threshold value is used which is obtained by maximum entropy algorithm. The edge detectors performance is analyzed by BDM and accuracy.

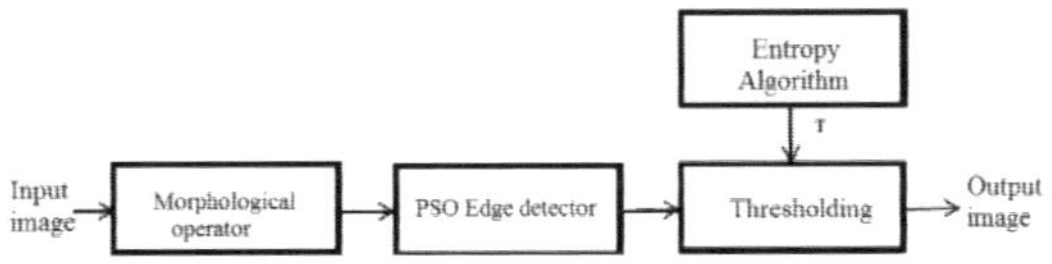

Figure 1. Proposed method of edge detection

3.1. Morphological operator

The morphological operators work with two data's. The original data to be processed and the other is structuring element. Each structuring element has a shape which can be thought as a parameter to the operation. The morphological operator used in our method is used to smoothen the image, to get the final image aggregation of all the extracted edge information. The algorithm is as follows:

$$E1 = \sum_{i=1}^{5}\{(M.B_i) \ominus B_i - (M.B_i)\} \quad (3)$$

$$E2 = \sum_{i=1}^{5}\{(M.B_i) - (M.B_i) \oplus B_i\} \quad (4)$$

$$E3 = \sum_{i=1}^{5}\{(M.B_i) \ominus B_i - (M.B_i) \oplus B_i\} \quad (5)$$

Where B_i is the square structuring element of 3x3. F is the input image then M is defined as:

$$M = (F.B_i) \circ B_i \quad (6)$$

Final output of morphological operator is:

$$E4 = \left(\sum_{i=1}^{3} E_i\right)/3 \quad (7)$$

3.2. PSO edge detector

How PSO can detect edges in image is explained in algorithm 1. After the optimization, obtained fitness curves which indicate the edges and mark these edges in image. Fitness function is dependent upon homogeneity and uniformity factors. The fitness function for PSO is given by maximizing f_c to find the best fitting curve which passes through a pixel.

$$f_c = H_c - U_c \quad (8)$$

H_c and U_c are homogeneity and uniformity factors respectively.

Algorithm 1 : PSO Edge Detection Algorithm

- Initialize PSO population randomly(position and velocity of each particle)
- Repeat
- For all particle do
- Evaluate the fitness value of the particle
- Find in the particle neighborhood, the leader particle
- Calculate particle velocity according to equation (2)
- Update particle position according to equation (1)
- End
- Update leader
- Select best particle in the population

3.3. Thresholding

Once we have computed a measure of edge strength using PSO edge detector, the next stage is to apply a threshold, to decide whether edges are present or not at an image point. So setting threshold value is important procedure. Here maximum entropy algorithm is used to set threshold. It is a histogram based method to find optimum threshold. After thresholding the image is transformed into edge map of the input image. Transformation to edge map according to threshold T is explained in equation (9).

$$f(x,y) = \begin{cases} 255, g(x,y) \geq T \\ 0, g(x,y) < T \end{cases} \quad (9)$$

Where g(x,y) is input image and f(x,y) is the edge map obtained.

4. Experimental results & analysis

The edge detection in images is done using traditional edge detection methods and proposed method. In proposed method morphological operator initially smooth's the image, then PSO edge detector find the edge strength magnitude by locating the points of discontinuities in image. Thresholding is applied to locate all the edge points in image. The performance comparison is made between proposed method and traditional methods using parameters BDM and Accuracy.

We use one image for the experiment in which they are natural image which are commonly used for the evaluation of edge detection algorithm. The artificial ground truth image is provided by University of Cordoba (UCO) [10]. Localization accuracy of edge detectors are evaluated by the comparison with the ground truth. Ground truth images contain thin and single pixel width edges and they are generated by a specified algorithm. Figure 2 shows the original images and their outputs using various edge detectors, from these figure it is clear that edges are more clear and continuous in proposed method of edge detection. The number of broken edges is less in proposed method comparing with other methods of edge detection.

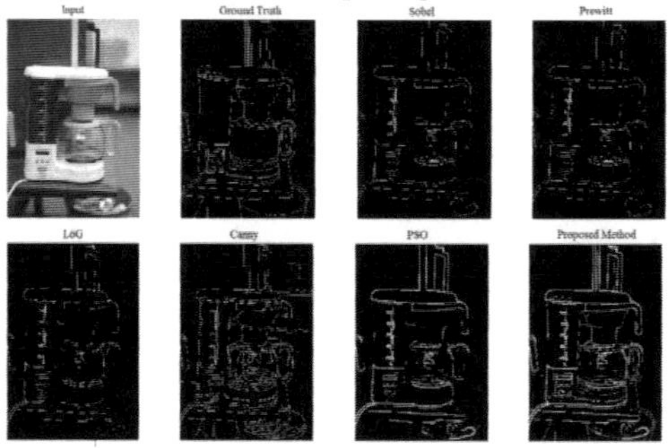

Figure 2. Outputs of various edge detectors for 'Coffee Maker' image.

Performance of proposed method of edge detection is analyzed by comparing with traditional edge detectors with parameters BDM and Accuracy. BDM is a parameter used to calculate the dissimilarity measure between two binary images. i.e., more the value, more the detected image varies from ideal image [11]. Accuracy is a statistical measure which can be calculated by the difference between edge pixels obtained by edge detection methods and a reference map (Ground truth) can be inferred.

Table 1 shows the performance analysis of different edge detector for Coffee Maker image with parameters BDM and accuracy. From the tables, it is clear that accuracy of proposed method is high when comparing to other methods. It is higher than that of PSO edge detector. Accuracy of proposed method is greater than 70%. Among the traditional edge detection techniques Canny is better, even though it does not have accuracy above 60%.

BDM is dissimilarity measure between two images. Ideally for an edge detector it is zero, in practical for a good edge detector it is smaller as possible. Here, the proposed method shows small value of BDM for image. It shows that output of proposed method have more similarity to ground truth image. The above figure and table shows that proposed method of edge detection based on PSO have better performance in quantitative analysis; it also shows better result in visual analysis also.

Table 1. Performance analysis for 'Coffee Maker' image.

Technique	BDM	Accuracy
Sobel	18.121	45.4
Prewitt	17.543	46.82
LoG	16.89	51.21
Canny	13.079	58.5
PSO	9.612	63.2
Proposed Method	7.11	72.254

5. Conclusion

In this paper, we proposed a novel method of edge detection based on PSO. Here the work done is to extract all edge patterns of an image with minimum computation time. Experimental results are observed for traditional edge detection methods, PSO edge detector and proposed method. Among traditional edge detectors canny edge detector performs better than other techniques. On visual analysis we have concluded that the proposed method gives a good quality of edge map relative to other methods discussed here. The output of PSO edge detector has good quality than the canny edge detector; it shows that optimization algorithms are also suitable for edge detection. Morphological operator helps to remove unwanted information's and makes the edge pixels more clearly. Thresholding technique has increased the localization accuracy of the proposed method of classifying the each edge pixel. Proposed method gives better performance in quantitative analysis also, proposed method have smaller value of BDM relative to other methods discussed here. Accuracy of the proposed method is above 70%, it is the best result for an edge detector.

References

[1] R. Maini and H. Aggarwal, "Study and comparison of various image edge detection techniques," *Int.J. of Image Process.(IJIP),* vol. 3, no. 1, pp. 1-11, 2009.
[2] X. Wang, "Laplacian operator-based edge detectors," *IEEE Trans. Pattern Anal. Mach. Intell.*, vol. 29, no. 5, pp. 886-890, 2007.
[3] D. Ziou, S. Tabbone et al., "Edge detection techniques-an overview," *Pattern Recognition and Image Anal.*, vol. 8, pp. 537-559, 1998.
[4] J. Kennedy, "Particle swarm optimization," in *Encyclopedia of Mach. Learning.,* Springer, 2011, pp. 760-766.
[5] I. E. Abdou and W. K. Pratt, "Quantitative design and evaluation of enhancement/thresholding edge detectors," *Proc. of the IEEE.*, vol. 67, no. 5, pp. 753-763, 1979.
[6] R. C. Gonzalez and R. E. Woods, "Digital image processing," vol. 2, 2007.
[7] R. M. Haralick, "Digital step edges from zero crossing of second directional derivatives," *IEEE Trans. Pattern Anal. Mach. Intell.*, no. 1, pp. 58-68, 1984.
[8] J. Canny, "A computational approach to edge detection," *IEEE Trans. Pattern Anal. Mach. Intell*, no. 6, pp. 679-698, 1986.
[9] M. Setayesh, M. Zhang, and M. Johnston, "Detection of continuous, smooth and thin edges in noisy images using constrained particle swarm optimisation," in *Proc. of the 13th Ann. Conf. on Genetic and Evol. Computation.*, ACM, 2011, pp. 45-52.
[10] N. L. Fernandez-Garc, A. Carmona-Poyato, R. Medina-Carnicer, and F. J. Madrid-Cuevas, "Automatic generation of consensus ground truth for the comparison of edge detection techniques," *Image and Vision Computing.*, vol. 26, no. 4, pp. 496-511, 2008.
[11] Rue, Håvard. "Delta Metric, Baddeley's." *Encyclopedia of Statistical Sciences(1999).*

Micro-Expression Recognition based on 2D Gabor Filter and Sparse Representation

Hao Zheng[1,2,3]

[1] Key Laboratory of Trusted Cloud Computing and Big Data Analysis, School of Information Engineering, Nanjing XiaoZhuang University, Nanjing, China
[2] MOE Key Laboratory of Computer Network and Information Integration, School of Computer Science and Engineering, Southeast University, Nanjing, China
[3] State Key Laboratory for Novel Software Technology, Nanjing University, Nanjing, China

E-mail: zhh710@163.com

Abstract. Micro-expression recognition is always a challenging problem for its quick facial expression. This paper proposed a novel method named 2D Gabor filter and Sparse Representation (2DGSR) to deal with the recognition of micro-expression. In our method, 2D Gabor filter is used for enhancing the robustness of the variations due to increasing the discrimination power. While the sparse representation is applied to deal with the subtlety, and cast recognition as a sparse approximation problem. We compare our method to other popular methods in three spontaneous micro-expression recognition databases. The results show that our method has more excellent performance than other methods.

1. Introduction

Expression recognition has been widely studied in the pattern recognition. However, micro-expressions recognition has three characteristics with poor performance. Firstly, micro-expressions are rapid facial movements with less than 0.55 second. Secondly, the intensity of the micro-expressions is very low according to facial muscles' movement. Thirdly, micro-expressions involve a fragment of the facial region. Therefore micro-expressions are difficult to recognize in the aspect of its short duration and occurrence. The flourishing of expression recognition methods depend on the well-established facial expression databases, such as JAFFE, CK+, BU-3DFE, Oulu-CASIA and Multi-PIE. However, there are few well-established micro-expression databases due to the difficulty of eliciting micro-expressions. In current literature, there are only three spontaneous micro-expression databases, i.e. SMIC [1], CASME [2], CASME II [3], while there are some works to date on automatic recognition of spontaneous micro-expressions.

Among all the available work, Li published the SMIC dataset and released a baseline performance of up to 48.78% accuracy for 3 classes, adopting LBP-TOP for feature extraction and SVM polynomial kernel for classification with cross validation method. Yan also reported a baseline performance of up to 63.41% accuracy for a 5-class classification task on CASME II database, adopting LBP-TOP and SVM for feature extraction and classification respectively. Then Wang et al. [4] proposed LBP-SIP method, which incorporated a multi-resolution Gaussian pyramid by concatenating the feature histograms of all four pyramid levels, and obtained the best recognition accuracy of 67.21%. In addition, Polikovsky et al. [5] used 3D-gradient descriptor for micro-expression recognition. Wang et al.[6] treated a micro-expression gray-scale video clip as a 3rd-order tensor and used Discriminant Tensor Subspace Analysis and Extreme Learning Machine to recognize micro-expressions.

But above methods have not achieved ideal results, we need to further improve the performance. 2D Gabor functions and wavelets have achieved impressive results when used for pattern recognitions. Many researchers have used Gabor wavelets in order to improve recognition performance [7]. Gabor wavelets show desirable characteristics of spatial locality and orientation selectivity, and are localized in the space and frequency domains [8]. In addition, sparse representation has been proved to improve the robustness of recognition, and wildly used in face recognition, expression recognition, etc. Motivated by above idea, in this paper, a novel method named 2D Gabor filter and sparse representation (2DGSR) is proposed to improve robustness for spontaneous micro-expression recognition. In the proposed method, 2D Gabor filter is used for enhancing the robustness of the variations due to increasing the discrimination power. While the sparse representation is applied to deal with the subtlety, and cast recognition as a sparse approximation problem. Our extensive experiments on spontaneous micro-expression databases show that the proposed 2DGSR method has very competitive performance with state-of-the-arts.

The rest of the paper is organized as follows. Section 2 presents the proposed 2DGSR method. Section 3 conducts experiments. Section 4 concludes the paper.

2. 2D Gabor filter and sparse representation (2DGSR)

The Gabor transform adds the Gaussian window to Fourier transform in order to realize the local analysis in spatial domain and spatial frequency domain. 2D Gabor filter can achieve optimal joint localization properties in the spatial domain and in the spatial frequency domain. So 2D Gabor filter is introduced in the proposed method.

2.1. 2D Gabor filter

When 2D Gabor filters are used for micro-expression recognition, the face image should be filtered by 2D Gabor filters firstly, then feature extraction is employed, finally downsample can be performed. The 2D Gabor filtering for face image is that an image $I(x, y)$ is convolved with a 2-D Gabor function $g(x, y)$ to obtain a Gabor feature image $u(x, y)$ as follows:

$$u(x, y) = \iint_{\Phi} I(\xi,\eta)g(x-\xi, y-\eta)d\xi d\eta \tag{1}$$

where $(x, y) \in \Phi$, Φ denotes the set of image points.

So the 2D Gabor function $\psi(x, y)$ can be obtained by a product of a Gaussian and a cosine function as following:

$$\psi_{\lambda,\theta,\phi}(x, y) = e^{-((x'^2+\gamma^2 y'^2)/2\sigma^2)} \cos(2\pi \frac{x'}{\lambda} + \phi) \tag{2}$$

where $x' = x\cos\theta + y\sin\theta, y' = -x\sin\theta + y\cos\theta, \sigma = 0.56\lambda, \gamma = 0.5$. σ represents the scale parameter, the small σ means high spatial resolution, the image filtering coefficients reflect local properties in fine scale, while the large σ means low spatial resolution , the coefficients reflect local properties in coarse scale. θ specifies the orientations of 2D Gabor filter, λ denotes the wavelength of the cosine factor, γ is the spatial aspect ratio and specifies the ellipticity of Gaussian factor, ϕ specifies the phase offset of the cosine factor.

Suppose $I(x, y)$ is the gray level distribution of an image. The Gabor representation is generated from convoluting the image $I(x, y)$ and the Gabor kernel $\psi_{\lambda,\theta,\phi}(x, y)$. The Gabor representation is defined as:

$$\Theta_{\lambda,\theta,\phi}(x, y) = I(z) * \psi_{\lambda,\theta,\phi}(x, y) \tag{3}$$

To reduce the space dimension, we downsample each $\Theta_{\lambda,\theta,\phi}(x, y)$ by a factor ρ and normalize it to zero mean and unit variance.

$$\Theta^{\rho}{}_{\lambda,\theta,\phi}(x, y) = \frac{\Theta_{\lambda,\theta,\phi}(x, y) - m_{\lambda,\theta,\phi}}{\sigma_{\lambda,\theta,\phi}} \tag{4}$$

where $m_{\lambda,\theta,\phi}$, $\sigma_{\lambda,\theta,\phi}$ denote the mean and the standard deviation of the sums, respectively. So the new Gabor representation $\Theta^{\rho}{}_{\lambda,\theta,\phi}(x, y)$ is obtained. Then we concatenate the 40 Gabor representations together and get an augmented feature vector

$$\tau^{\rho}{}_{\Xi,\Xi} = \Xi_{\mu=0}^{7} \Xi_{v=0}^{4} \Theta^{\rho}{}_{\lambda,\theta,\phi} \tag{5}$$

Where Ξ denotes the junction symbol. So the argument Gabor feature vector which encompasses all the elements of the Gabor wavelets representation set is obtained.

2.2. 2D Gabor filter and sparse representation

The augmented Gabor face feature vector τ, which is a local feature descriptor, can not only enhance the face feature but also tolerate face image local deformation to some extent. Suppose we denote

$$T(A) = [T(A_1)\ T(A_2)\ \cdots\ T(A_K)], T(A_i) = [\tau(s_{i,1})\ \tau_i(s_{i,2})\ \cdots\ \tau(s_{i,n_i})],$$

$$x = [x_1; x_2; \cdots; x_m] \tag{6}$$

The 2D Gabor feature sparse representation is as following:

$$\tau(y) = T(A_1)x_1 + T(A_2)x_2 + \cdots + T(A_m)x_2 = T(A)x \tag{7}$$

So the object function of 2DGSR is

$$< \hat{x} >= \arg\min_{\hat{x}}(\| \tau(y) - T(A)x \|_2^2 + \lambda \|x\|_1) \tag{8}$$

By using 2D Gabor features for micro-expression recognition, the feature dictionary A will be transformed into 2D Gabor feature dictionary.

Thus the proposed 2DGSR method is summarized in Algorithm 1.

Algorithm 1 (2D Gabor filter and Sparse Representation (2DGSR))

Step 1 obtain 2D Gabor feature dictionary $T(A)$

Step 2:Normalize the columns of $T(A)$

Step 3: Solve the l_1 minimization problem:

$$< \hat{x} >= \arg\min_{\hat{x}}(\| \tau(y) - T(A)x \|_2^2 + \lambda \|x\|_1)$$

Step4:Compute label of the test image:

$$r_i(y) = \| y - T(A)\delta_i(\hat{x}) - \hat{\varepsilon} \|_2, \ for\ i = 1,...,k.$$

3. Experiments and discussions

In this section, we compare our proposed 2D Gabor feature with two baseline features in micro-expression recognition, i.e., LBP-TOP[9],[10] and HOOF[11]. We use two versions of the HOOF feature such as HOOF-whole, HOOF-ROIs. HOOF-whole is the original HOOF feature applied to the entire facial region, four to ten bins are used in the HOOF histogram. HOOF-ROIs is a combinatorial HOOF feature which created by applying the HOOF feature in each of the 36 ROIs, four to ten bins are used in each ROI's HOOF histogram. We test our proposed method on three micro-expression database, including SMIC, CASE, and CASME II to verify its effectiveness for micro-expression recognition. These databases consist of spontaneous micro-expressions which appear in real life. In all experiments we applied the SRC to classify and employed Leave-one-subject-out (LOSO) cross validation.

3.1 CASME database:

The Chinese Academy of Sciences Micro-Expression (CASME) database includes 195 spontaneous facial micro-expression recorded by two different 60 fps cameras. Two expert coders were recruited in the work to code the duration and AU combination in these micro-expressions. The selected micro-expressions either have a total duration less than 500ms or onset duration less than 250ms. These samples are coded with the onset, apex and offset frames, furthermore tagged with AUs. In this database, micro-expressions are classified into 7 categories (happiness, surprise, disgust, fear, sadness, repression and tense).

CASME database is divided into two classes: Class A and Class B. The samples in Class A were recorded by the BenQ M31 consumer, camera with 60fps, with the resolution set at 1280*720 pixels. The participants were recorded in natural light. The samples in Class B were recorded by the Point Grey GRAS-03K2C industrial camera with 60 fps, with the resolution set to 640*480 pixels. The participants were recorded in a room with two LED lights. In the experiments, we merged the 7 categories into 4 classes: Positive, Negative, Surprise, Others. We selected 97 samples from Class B.

From Table 1, it can be seen that 2DGSR outperform than other methods, especially 2DGSR is almost 20% higher than HOOF-whole feature extraction. This is because 2DGSR makes full use of gabor feature information about spatial and orientation, and enhance the robustness of recognition. Meanwhile from Table 2, we can see that the results of 2DGSR had a good recognition rates in all four classes.

Table 1: The best recognition rates of four features for micro-expression recognition in CASME database.

Feature	LBP-TOP	HOOF-whole	HOOF-ROIs	2DGSR
Accuracy	66.34%	51.89%	57.92%	**71.19%**

Table 2: Confusion matrix of micro-expression recognition obtained by 2DGSR on CASME database

	Positive	Negative	Surprise	Others
Positive	48.6%	4.1%	3.6%	6.6%
Negative	5.4%	57.6%	5.2%	13.2%
Surprise	3.9%	4.8%	69.1%	2.5%
Others	42.1%	33.5%	22.1%	77.7%

3.2 CASME II database

The CASME II database includes 246 spontaneous facial micro-expressions recorded by a 200 fps camera. These samples were selected from more than 2,500 facial expressions. The selected micro-expressions in this database either had a total duration less than 500 ms or an onset duration (time from onset frame to apex frame) less than 250 ms. In CASME II, the samples in which the facial feature points in the first frame cannot be correctly detected by using the DRMF method. We used 236 samples from 26 subjects, categorized into four classes: positive (31 samples), Negative(65 samples), Surprise(21sample), and Others(119 samples).

Table 3 show that 2DGSR outperformed other three methods, and also achieved the best recognition rates in CASME II database. We can see that the recognition rates in CASME II database is worse than those in CASME database, because compared with CASME, the database is improved in increased sample size, fixed illumination, and higher resolution(both temporal and spatial).

Table 3. The best recognition rates of four features for micro-expression recognition in CASME II database.

Feature	LBP-TOP	HOOF-whole	HOOF-ROIs	2DGSR
Accuracy	59.12%	44.36%	54.65%	**64.88%**

4. Conclusion

In the paper, we apply the 2D Gabor features and sparse representation to micro-expression recognition. 2D Gabor feature and sparse representation make our robust to its subtlety. Experimental results show that our method has better performances than other three popular methods. Although our method performs well, recognition speed is not fit for real time. In the future, we will improve the method to speed up Gabor wavelet transformations.

References

[1] Li, X., Pster, T., Huang, X., Zhao, G., Pietikainen, M.: A spontaneous micro-expression database: inducement, collection and baseline. In: 10th IEEE International Conference and Workshops on Automatic Face and Gesture Recognition (FG), pp.1-6 (2013)

[2] Yan, W.J., Wu, Q., Liu, Y.J., Wang, S.J., Fu, X.: CASME database: a dataset of spontaneous micro-expressions collected from neutralized faces. In: 10th IEEE In-ternational Conference and Workshops on Automatic Face and Gesture Recognition (FG), pp. 1-7 (2013)

[3] Yan, W.J., Li, X., Wang, S.J., Zhao, G., Liu, Y.J., Chen, Y.H., Fu, X.: CASME II: an improved spontaneous micro-expression database and the baseline evaluation. PLoS One 9(1), e86041 (2014)

[4] Wang Y., See J, Phan RC-W, Oh Y-H.: Efficient Spatio-Temporal Local Binary Patterns for Spontaneous Facial Micro-Expression Recognition. PLoS ONE 10(5): e0124674 (2015)

[5] Polikovsky S., Kameda Y., and Ohta Y.: Facial micro-expressions recognition using high speed camera and 3D-gradient descriptor. In 3rd International Conference on Crime Detection and Prevention. IET, pp. 1-6 (2009)

[6] S.J. Wang, H.L. Chen, W.J. Yan, Y.H. Chen, and X. Fu: Face recognition and micro-expression based on discriminant tensor subspace analysis plus extreme learning machine. Neural Processing Letters (2013)

[7] Ilkka Autio, Tapio Elomaa, Flexible View Recognition for Indoor Navigation Based On Gabor Flters and Support Vector Machines, Pattern Recognition, vol. 36, 2003, pp.2769-2779.

[8] Chengjun Liu, and Harry Wechsler, Gabor Feature Based Classification Using the Enhanced Fisher Linear Discriminant Model for Face Recognition, IEEE Transactions on Image Processing, vol. 11, 2002, pp.467-476.

[9] T. Pfister, X. Li, G. Zhao, and M. Pietik¨ainen, "Recognising spontaneous facial micro-expressions," in 2011 IEEE International Conference on Computer Vision (ICCV). IEEE, 2011, pp. 1449–1456.

[10] G. Zhao and M. Pietik¨ainen, "Dynamic texture recognition using local binary patterns with an application to facial expressions," IEEE Transactions on Pattern Analysis and Machine Intelligence, vol. 29, no. 6, pp. 915–928, 2007.

[11] R. Chaudhry, A. Ravichandran, G. Hager, and R. Vidal, "Histograms of oriented optical flow and binet-cauchy kernels on nonlinear dynamical systems for the recognition of human actions,"in 2009 IEEE Conference on Computer Vision and Pattern Recognition (CVPR). IEEE, 2009, pp. 1932–1939.

New Denoising Method Based on Empirical Mode Decomposition and Improved Thresholding Function

Wahiba Mohguen and RaïsEl'hadiBekka

Université de Sétif 1,Faculty of Technology,Department of Electronics, LIS Laboratory ,19000 Sétif, Algeria

E-mail: wasotel@yahoo.fr

Abstract. This paper presents a new denoising method called EMD-ITF that was based on Empirical Mode Decomposition (EMD) and the Improved Thresholding Function (ITF) algorithms. EMD was applied to decompose adaptively a noisy signal into intrinsic mode functions (IMFs). Then, all the noisy IMFs were thresholded by applying the improved thresholding function to suppress noise and improve the signal to noise ratio (*SNR*). The method was tested on simulated and real data and the results were compared to the EMD-Based signal denoising methods using the soft thresholding. The results showed the superior performance of the new EMD-ITF denoising over the traditional approach. The performance were evaluated in terms of *SNR* in *dB*, and Mean Square Error (*MSE*).

1. Introduction

The Empirical Mode Decomposition (EMD) method has recently been pioneered by Huang and al. [1] for analyzing the nonlinear and non-stationary signals. The aim of the EMD method is to adaptively decompose any signal into oscillatory components called intrinsic mode functions (IMFs) using a sifting process. EMD denoising can be based on partial reconstruction or inspired by standard wavelet thresholding [2-8]. The basic principle of wavelet thresholding is to cancel all the coefficients that are lower than a threshold related to the noise level. However, the performance of wavelet approach relies the basis wavelet which is predetermined in advance and on the threshold. A more flexible alternative was performed by EMD approach. Several EMD-Based denoising methods using thresholding were proposed in [6]. Indeed, it was shown that the direct application of wavelet thresholding to IMFs can lead to very bad results for the continuity of the reconstructed signal. Therefore, an EMD interval thresholding was proposed (EMD-IT) in which the thresholding was performed to the zero-crossing interval as a whole. EMD-IT procedure resembles wavelet thresholding more than directEMD thresholding because wavelet thresholding is applied to the wavelet coefficients [6]. The main factors affecting the quality of Wavelet Threshold Denoising are denoising threshold and selection of the suitable wavelet threshold function. The hard threshold function does not change the local properties of the signal, but it can lead to some fluctuation in the reconstruction of the original signal. The hard threshold function leads to a loss of some high frequency coefficients above the threshold. In order to overcome the drawbacks of the classical threshold functions, Lu Jing-yi et al. proposed an improved threshold function by increasing the adjustment factor [9]. A direct application of wavelet improved threshold function in the EMD case (EMD-ITF) was considered in this paper. Numerical simulation and real data test were performed to evaluate this method, and the results were compared to EMD soft threshold function in terms of *SNR* and *MSE*.

The paper is organized as follows. Section II introduces the EMD algorithm. Section III describes the EMD-Soft thresholding and the new proposed method EMD-ITF. The simulation results are illustrated in section IV. Finally, section V presents the conclusion.

2. EMD algorithm

EMD is an adaptive method to decompose a signal $x(t)$ into a series of IMFs. The IMFs must satisfy the following two conditions: (i) The number of maximum must equal the number of zeros or differ at most by one. (ii) In each period, it is necessary that the signal average is zero.

The EMD algorithm consists of the following steps [1]:

1. Find local maxima and minima in $x(t)$ to construct the upper and lower envelopes respectively using cubic spline interpolation.
2. Calculate the mean envelope $m(t)$ by averaging the upper and lower envelopes.
3. Calculate the temporary local oscillation $h(t) = x(t) - m(t)$.
4. Calculate the average of $h(t)$, if average $h(t)$ is close to zero, then $h(t)$ is considered as the first IMF, named $c_i(t)$ otherwise, repeat steps (1)–(3) while using $h(t)$ for $x(t)$.
5. Calculate the residue $r(t) = x(t) - c_i(t)$.
6. Repeat steps from (1) to (5) using $r(t)$ for $x(t)$ to obtain the next IMF and residue.

The decomposition process stops when the residue $r(t)$ becomes a monotonic function or a constant that no longer satisfies the conditions of an IMF.

$$x(t) = \sum_{i=1}^{N} c_i(t) + r_N(t) \tag{1}$$

3. EMD Based denoising

3.1. EMD Soft thresholding

Having a noisy signal $y(t)$ given by:

$$y(t) = x(t) + \eta(t) \tag{2}$$

Where $x(t)$ is the noiseless signal and $\eta(t)$ is independent noise of finite amplitude. In EMD-Soft thresholding method, the noisy signal $y(t)$ was first decomposed into noisy IMFs: $c_{ni}(t)$. These noisy IMFs was thresholded by soft function in order to obtain an estimation of the noiseless IMFs $\hat{c}_i(t)$ of the noiseless signal. In this work the universal threshold is used proposed in [10] and it identified as follows:

$$\tau_i = C\sqrt{E_i 2\ln(n)} \tag{3}$$

Where C is a constant depending of the type of signal that was set to 0.5 in this work, n is the length of the signal and E_i is given by [4]:

$$\hat{F}_k = \frac{E_1^2}{0.719} 2.01^{-k}, \qquad k = 2,3,4,N \tag{4}$$

Where E_1^2 is the energy of the first IMF defined by:

$$E_1^2 = \left(\frac{median(|c_{n1}(t)|)}{0.6745} \right)^2 \tag{5}$$

A direct application of wavelet soft thresholding [11] in the EMD case:

$$\hat{c}_i(t) = \begin{cases} c_{ni} - \tau_i & if \quad c_{ni} \geq \tau_i \\ 0 & if \quad |c_{ni}| < \tau_i \\ c_{ni} + \tau_i & if \quad c_{ni} \leq -\tau_i \end{cases} \tag{6}$$

A reconstruction of the denoised signal is given by:

$$\hat{x}(t) = \sum_{i=1}^{N} \hat{c}_i(t) + r_N(t) \tag{7}$$

3.2. *EMD-ITF Denoising*

A direct application of wavelet improved threshold function [9] in the EMD case:

$$\hat{c}_i(t) = \begin{cases} \operatorname{sgn}(c_{ni}(t)) \left[|(c_{ni}(t)| - \dfrac{\tau_i}{\exp^{3\alpha\left(\frac{c_{ni}(t)-\tau_i}{\tau_i}\right)}} \right] & if \quad |c_{ni}| \geq \tau_i \\ & if \quad |c_{ni}| < \tau_i \end{cases} \tag{8}$$

A reconstruction of the denoised signal is given by equation (7).

4. Simulation results

In this section, we assess our proposed denoising algorithm compared to EMD-soft thresholding denoising. The new EMD-ITF approach was applied to five test signals (Doppler, Blocks, Bumps, Heavysine and Piece-Regular). The size of the signals was equal to 2048. The method was also tested on real *ECG* signal using the MIT-BIH database [12]. For simulated signals, the variance of the additive white Gaussian noise was set so that the *SNR* before denoising was maintained at 15 *dB*. The original signals and the corresponding noisy versions are depicted in Fig. 1. The *SNR* before denoising of the real *ECG* signal was *20 dB*. Each noisy signals were decomposed into IMFs using EMD process and all IMFs are thresholded by the Improved Thresholding Function and soft thresholding.The performance of the proposed method was affected by the choice of the α value as shown in Fig.2 that depicts the *SNR* after denoising as function of α. The α values for which the the *SNR* after denoising are maximum are 0.4, 0.2, 0.2, 0.1, 0.1 for ECG, bumps, heavysine, Piece Regular, blocks, and Doppler signals respectively. Indeed, the maximum *SNR* after denoising depends on the signal and α. A comparative study of two methods considered in this work is presented in Table 1.This table shows that the proposed method provides the best results in terms of *SNR* and *MSE* for all test signals. Therefore, the proposed EMD-ITF outperforms totally the conventional EMD-soft thresholding denoising. Fig.3 shows the denoising results of applying EMD Soft and EMD- ITF to simulated signals. Fig.4 displays the denoising results of real *ECG* signal using EMD-Soft and EMD- ITF.

Table 1. *SNR* before and *SNR* after denoising of the different signals.

	SNR (dB)(after denoising)					
Methods	***Blocks* 15*dB***	***Bumps* 15*dB***	***Heavysine* 15*dB***	***Doppler* 15*dB***	***Piece-Regular* 15*dB***	***ECG* 20*dB***
EMD-Soft	20.1370	20.4058	26.4670	22.2368	21.6292	26.2561
EMD-ITF	20.6421	23.3905	26.6830	24.4025	22.3292	28.1154
Methods	*MSE*					

IOP Conf. Series: Journal of Physics: Conf. Series **787** (2017) 012014 doi:10.1088/1742-6596/787/1/012014

EMD-Soft	0.0854	0.0295	0.0214	0.0005	0.0220	0.0023
EMD-ITF	0.0760	0.0148	0.0204	0.0003	0.0187	0.0015

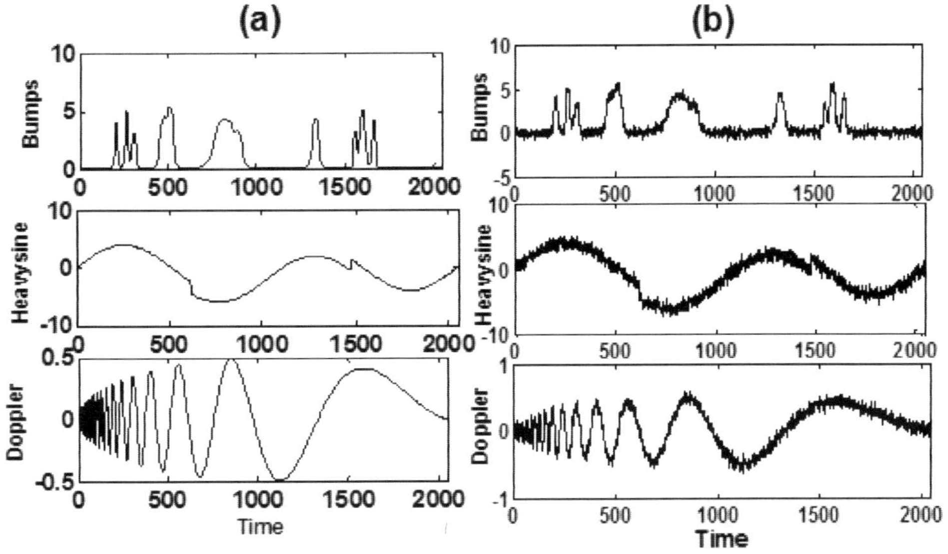

Figure 1. (a). Test signals with n=2048. (b).Noisy test signals *SNR=15dB*

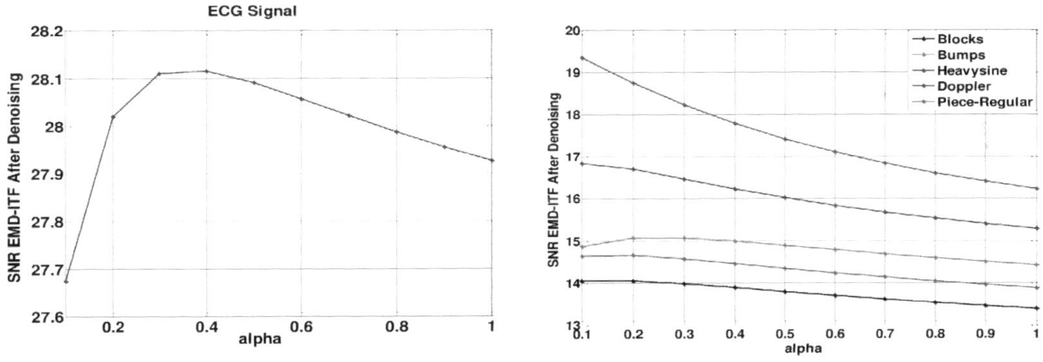

Figure 2. Performance evaluation of EMD-ITF denoising method.

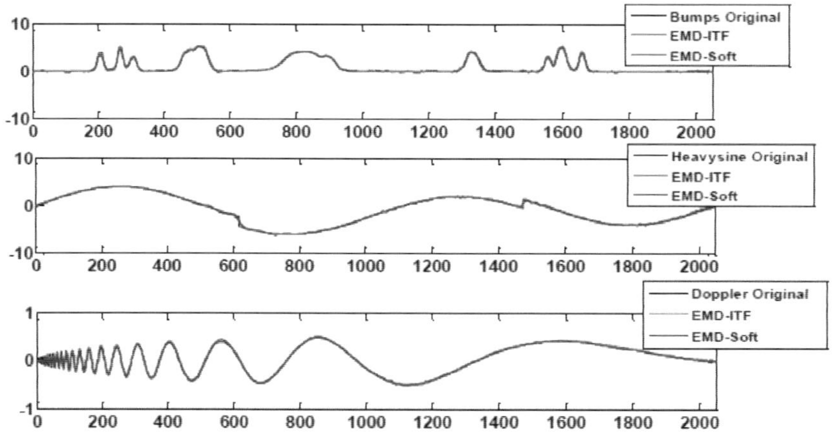

Figure 3. Denoising results in $SNR = 15dB$ of test signals corrupted by Gaussian noise.

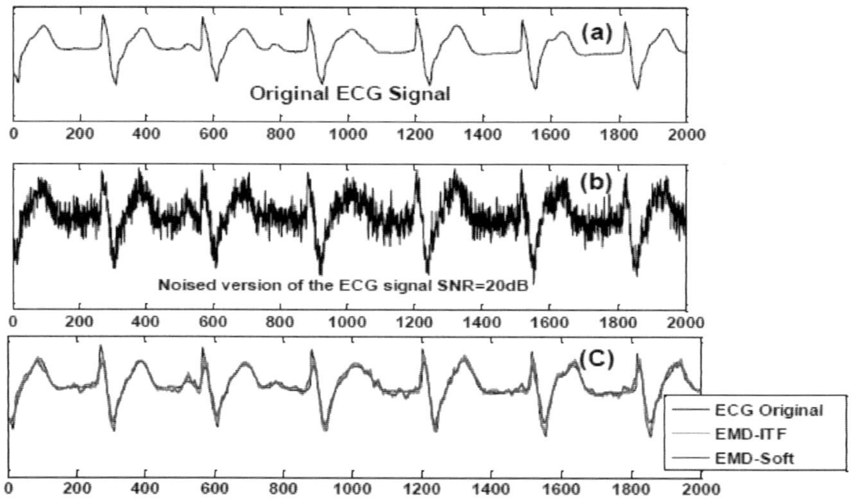

Figure 4. Denoising results in *SNR* (20dB) of real ECG signal

5. Conclusion

A new signal denoising method is proposed based on empirical mode decomposition and the Improved Thresholding Function to suppress noise in the signal and improve the output *SNR*. The proposed method was tested on real *ECG* signal and simulated signals (Doppler, Blocks, Bumps Heavysine, and Piece- regular) corrupted by white Gaussian noise. Based on *SNR* and *MSE*, the simulation results are in favour of the new EMD-ITF denoising method. We showed that the new approach is useful for removing noise

References

[1] N.E. Huang, Z. Shen, S.R. Long, M.C. Wu, H.H. Shin, Q. Zheng, N.C. Yen, C.C. Tung and H.H. Liu, "The Empirical Mode Decomposition and the Hilbert spectrum for nonlinear and non-stationary time series analysis,"*Proceedings of the Royal Society ofLondon*,454:903–995, 1998.

[2] A .O. Boudraa, J. C. Cexus, Z. Saidi, "EMD-based signal noise reduction, Int. J. Signal Process. 1(1), 33–37, 2004.

[3] A. O. Boudraa, J. C. Cexus, "Denoising via empirical mode decomposition", in Proceedings of the IEEE International Symposium on Control, Communications and Signal Processing (ISCCSP '06), p. 4, Marrakech, Morocco, March 2006.

[4] P. Flandrin, G. Rilling, P. Goncalces, "EMD equivalent filter banks, from interpretations to application," in *Hilbert-Huang Transform and its Application*, N. E. Huang and S. Shen, Eds., 1st ed. Singapore: World Scientific, 2005.

[5] A. O. Boudraa, J. C. Cexus, "EMD-Based Signal Filtering", IEEE Transactions on Instrumentation and Measurement, Vol. 56, NO. 6,pp.2196-2202 ,2007.

[6] Y. Kopsinis, S. Mclanglin, "Development of EMD-based denoising methods inspired by wavelet thresholding ", IEEE Trans.Signal Process.57(4) 1351–1362, 2009.

[7] Y. Kopsinis, S. McLaughlin, "Empirical Mode Decomposition Based Soft-Thresholding," ," in *Proc. 16th Eur. Signal Process. Conf.(EUSIPCO)*, Lausanne, Switzerland, Aug. 25–29, 2008.

[8] G.S. Tsolis and T. D. Xenos, ,"Signal Denoising Using Empirical Mode Decomposition and Higher Order Statistics ," International Journal of Signal Processing, Image Processing and Pattern Recognition Vol. 4, No. 2, pp.91-106, 2011.

[9] L. Jing-yi, L. Hong, Y. Dong,and Z. Yan-sheng"A New Wavelet Threshold Function and Denoising Application, , "
Mathematical Problems in Engineering; Vol. 2016, 8 pages.

[10] D. L. Donoho and I. M. Johnstone, "Ideal spatial adaptation by wavelet shrinkage,"*Biometrika*, vol. 81, no. 3, pp. 425 – 455, Aug. 1994.

[11] Donoho DL, "De-noising by soft-thresholding, " IEEE Trans Inform Theory, Vol.14, No.3, pp. 612-627,1995.

[12] A.L. Goldberger, L.A.N. Amaral, L. Glass, J.M. Hausdorff, P.C. Ivanov, R.G. Mark, J.E. Mietus, G.B.Moody, C.-K. Peng, H.E. Stanley, PhysioBank, PhysioToolkit, and PhysioNet: components of a new research resource for complex physiologic signals, Circulation, Vol. 101, N° 23, pp. 215–220, 2000.

Identity Recognition Algorithm Using Improved Gabor Feature Selection of Gait Energy Image

LIANG Chao[1,a], JIA Ling-yao[1,b] and SHI Dong-cheng[1,c]

[1]School of Computer Science & Engineering, Changchun University of Technology, Changchun 130012, China

[a]liangchao@mail.ccut.edu.cn, [b]ayaoabc@yeah.net, [c]dcshi@mail.ccut.edu.cn

Abstract. This paper describes an effective gait recognition approach based on Gabor features of gait energy image. In this paper, the kernel Fisher analysis combined with kernel matrix is proposed to select dominant features. The nearest neighbor classifier based on whitened cosine distance is used to discriminate different gait patterns. The approach proposed is tested on the CASIA and USF gait databases. The results show that our approach outperforms other state of gait recognition approaches in terms of recognition accuracy and robustness.

1. Introduction

Gait recognition [1] is the process of identifying and individual by the way they walk. This is less unobtrusive biometric, which offers the possibility to identify a person at a distance. Moreover, gait recognition offers great potential for recognition of low-resolution videos, where other biometrics technologies may be invalid because of insufficient pixels.

2. Previous Work

Since Han [2] proposed the gait energy image as a new spatio-temporal gait representation, a large number of improved approaches have emerged. Liang [3] present a novel approach to extract gait features based on gait energy image, and P Theekhanont [4] uses gait energy image transformed to a trace transform image for human identification. S Sivapalan [5] demonstrate the applicability of compressed sensing in the field of gait recognition as a very effective dimensionality reduction technique. And Xue H [6] apply PCA and LDA to extract the main vector of gait feature of GEI. In this paper, the Gabor magnitude of GEI and an innovative reduced dimensions method is proposed to select distinctive features.

3. The Gait Recognition System

The structure diagram of the gait recognition system based on GEI is shown in Figure 1.

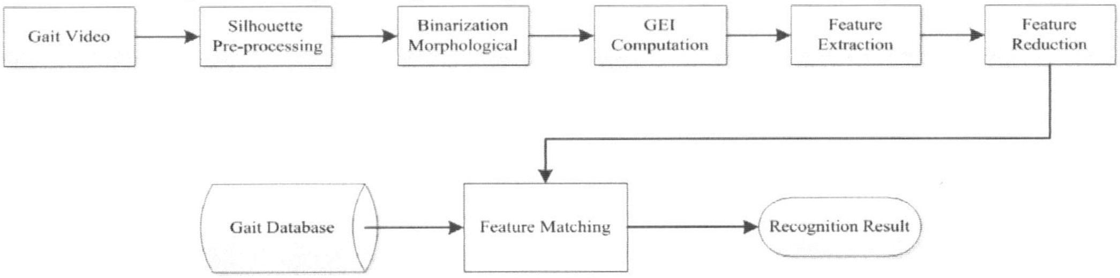

Figure 1. The flow chart of gait recognition system based on gait energy image

Given a size-normalized and horizontal-aligned human walking binary silhouette sequence B(x, y, t), the grey-level GEI A(x, y) is defined as follows

$$A(x,y)=\frac{1}{N}\sum_{i=1}^{N}B(x,y,t) \quad (1)$$

where N is the number of frames in complete cycles of the sequence, t is the frame number of the sequence, x and y are values in the 2D image coordinate. The examples of normalized and aligned silhouette frames in different human walking sequences and GEIs are shown in Figure 2.

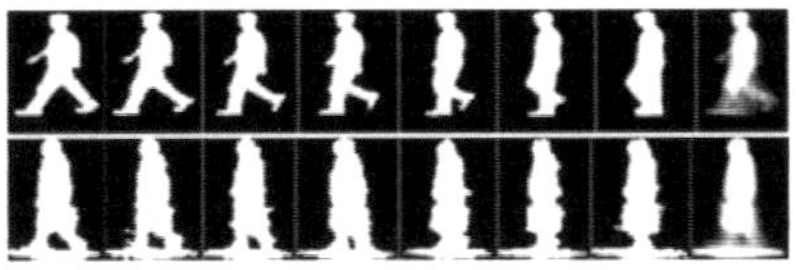

Figure 2. The average silhouette image over the whole sequence-Gait Energy Image (GEI)

3.1. Feature extraction algorithm based on the combination of gait energy map and Gabor wavelet

Gabor filters have proven themselves to be a powerful tool for gait feature extraction and robust gait recognition. They represent complex band-limited filters with an optimal localization in both the spatial as well as the frequency domain. In general, the family of 2D Gabor filters can be defined in the spatial domain as follows:

$$\psi(x,y)=\frac{f_u^2}{\pi k\eta}e^{-((f_u^2/k^2)x'^2+(f_u^2/\eta^2)y'^2)}e^{j2\pi f_u x'}, \text{ u=0,1,...,scales-1, v=0,1,...,orientions-1} \quad (2)$$

where $x'=x\cos\theta_v+y\sin\theta_v$, $y'=-x\sin\theta_v+y\cos\theta_v$, $f_u=f_{max}/2^{(u/2)}$ and $\theta_v=v\pi/8$.

3.2. Feature extraction of GEI using gabor filters

Let $I(x,y)$ stand for a grey-scale image of GEI and, moreover, let $\psi_{u,v}(x,y)$ denote a Gabor filter given by its center frequency f_u and orientation θ_v. The feature extraction procedure can then be defined as a filtering operation of the given GEI $I(x,y)$ with the Gabor filter $\psi_{u,v}(x,y)$ of size *u* and orientation v. In the Equation(3), $G_{u,v}(x, y)$ denotes the complex filtering output, that is

$$G_{u,v} = I(x, y) * \psi_{u,v}(x, y) \tag{3}$$

3.3. Gait feature dimension reduction based on Kernel Fisher Analysis

In this part, the improved kernel Fisher discriminant and Gabor feature based kernel Fisher analysis, which are main contributions of this paper. The idea of kernel Fisher discriminant analysis is to yield a nonlinear discriminant analysis in the higher space as F. In this paper, the F space for the Fisher method just as the Equation (4).

$$J(w) = \frac{w^T S_B^{\Phi} w}{w^T S_W^{\Phi} w} \tag{4}$$

where the S_B^{Φ} and S_W^{Φ} are the between and within class scatter matrices respectively. In this paper, the polynomial kernel is adopted. And then, the gait features of the training data in the kernel space is got by the Equation (5).

$$\arg\max(J(w)) \bullet \Phi(x) = \sum_{i=1}^{l} \alpha_i k(x_i, x) \tag{5}$$

Lastly, the features of test data are got by the following steps:

$$K = (X^T Y + 1)^2 \tag{6}$$

$$V = P^T * K \tag{7}$$

where the $\boldsymbol{K}$ denotes the kernel matrix, $\boldsymbol{X}$, $\boldsymbol{Y}$ are the test data and training data respectively. **P** is just the transformation matrix calculated by training data. $\boldsymbol{V}$ is just the features of test data.

3.4. Nearest neighbor classifier based on whitened cosine similarity measure

In this paper, the GEIs are normalized and aligned, so they can be clustered by the distance measure. According to related experimental results, the nearest neighbor classifier based on the white cosine similarity [7] is used to discriminate different gait patterns.

4. Experimental Results and Analysis

Our experiments are carried out on the CASIA DatasetB gait database and the USF HumanID gait database.

4.1. Experiment on CASIA gait database

On the CASIA gait database, a dataset consisting of 124 subjects in the 90° view is constructed. Each person has 10 GEIs with the resolution 240 * 240 and they are divided into training and test set with the ratio of 1:1. The parameters of Gabor filter banks are set as follows: nscales=6, norientions=8, $k=\eta=\sqrt{2}$, f_{max}=0.25. Give one gait energy image, the filter response is shown in Figure 3.

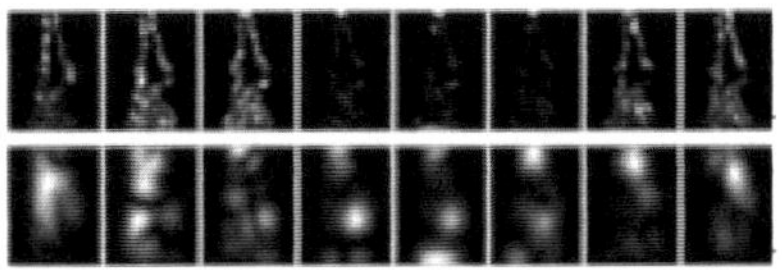

Figure 3. Filter response of GEI

To investigate the performance of gait recognition on the gait databases, the cumulative matching characteristic (CMC) and receiver operating characteristic (ROC) are used. On the CASIA, the experimental results as well as comparison with other approaches of individual recognition by gait are shown in Figure 4 ~5. The experimental data are shown in Table 1 and Table 2.

4.2. Experiment on USF gait database

A database consisting of 122 subjects based the USF HumanID database is constructed. For each individual, there are 12 gait energy images with a resolution of 88 x 128, in which the CALNB and CALBF are 6 respectively. They are divided into training and test set with the ratio of 1:1. The experimental results are shown in Figure 6 ~7, and the related data are shown in Table 3~4.

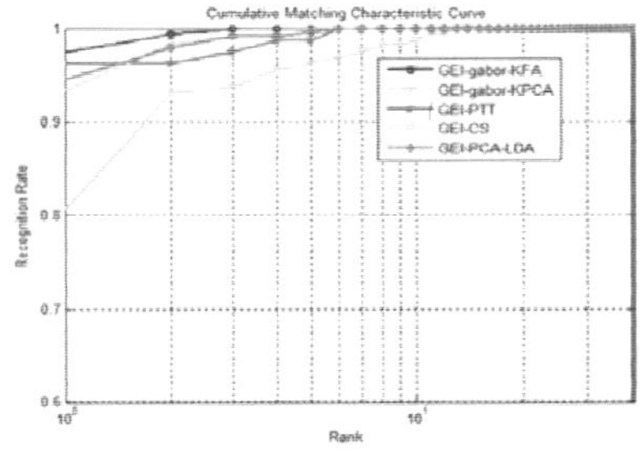

Figure 4. Recognition performance of CMC curves on CASIA

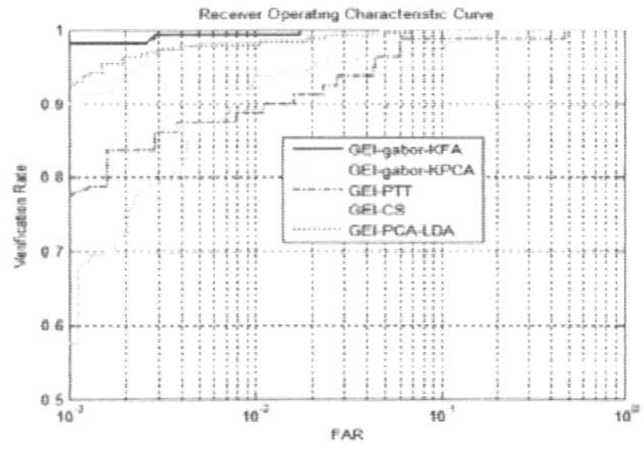

Figure 5. Verification performance of the ROC curve on CASIA

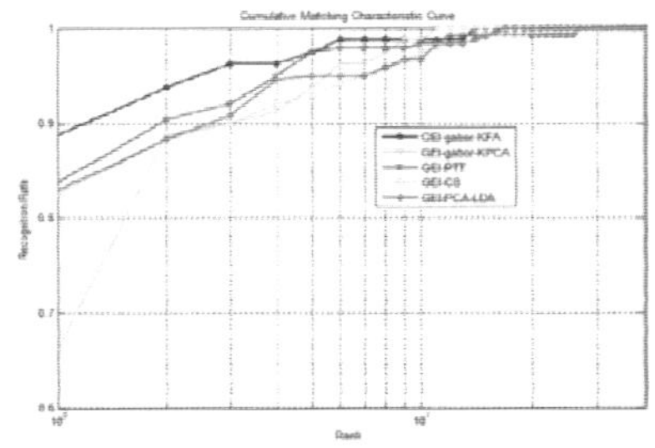

Figure 6. Recognition performance of CMC curves on USF

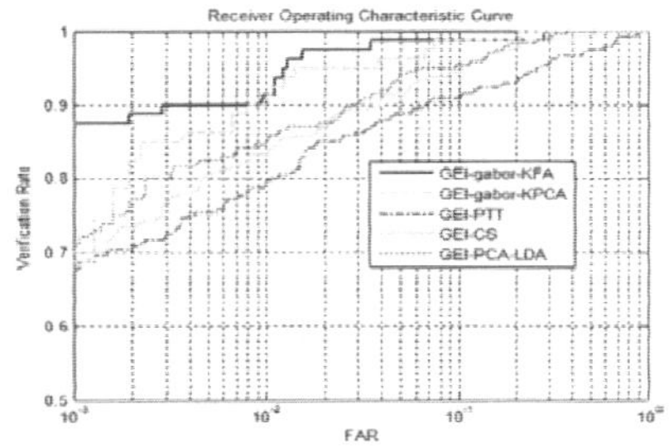

Figure 7. Verification performance of the ROC curve on USF

Table 1. CMC curve identification performance data on CASIA

Approach	GEI-gabor-KPCA[3]	GEI-PTT[4]	GEI-CS[5]	GEI-PCA-LDA[6]	Proposed
Rank1	80.63%	96.25%	93.33%	94.58%	97.50%
Rank5	96.25%	98.75%	99.58%	99.58%	100.00%

IOP Conf. Series: Journal of Physics: Conf. Series **787** (2017) 012015 doi:10.1088/1742-6596/787/1/012015

Table 2. ROC curve verification performance data on CASIA

Approach	GEI-gabor-KPCA[3]	GEI-PTT[4]	GEI-CS[5]	GEI-PCA-LDA[6]	Proposed
FAR=1%	93.13%	88.75%	98.33%	97.92%	99.38%
FAR=0.1%	56.88%	76.25%	89.17%	90.42%	98.13%

Table 3. CMC curve identification performance data on USF

Approach	GEI-gabor-KPCA[3]	GEI-PTT[4]	GEI-CS[5]	GEI-PCA-LDA[6]	Proposed
Rank1	66.25%	83.75%	82.50%	82.92%	88.75%
Rank5	93.75%	97.50%	92.92%	95.00%	97.50%

Table 4. ROC curve verification performance data on USF

Approach	GEI-gabor-KPCA[3]	GEI-PTT[4]	GEI-CS[5]	GEI-PCA-LDA[6]	Proposed
FAR=1%	90.00%	79.58%	83.33%	84.58%	91.25%
FAR=0.1%	66.25%	67.50%	68.75%	69.58%	87.50%

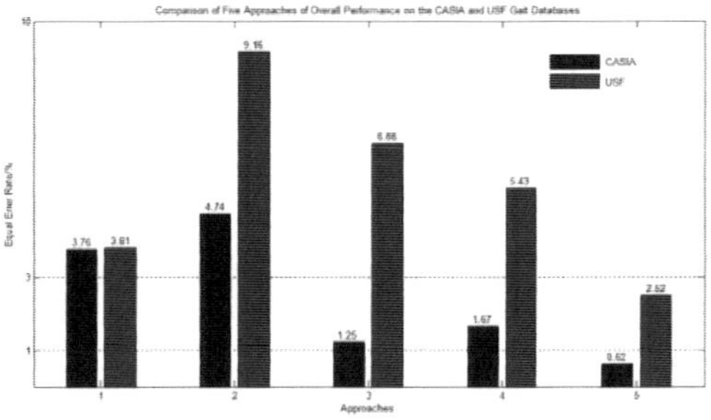

Figure 8. Equal error rate of five approaches on the CASIA and USF gait databases

In the CMC and ROC curves, the best performance of one approach should be close to the upper left corner as near as possible. As shown in Fig. 4~5 that our approach is closest to the upper left corner. The values of Rank1 and Rank5 of the proposed are the largest of the five approaches, and our approach is the only one converging to 1 in rank 5. The experimental data are shown in Table 1 ~2 as well as comparison with other approaches of individual recognition by gait.

It can be seen from the data in Table 2 that the recognition rate of the proposed approach is the highest when the false acceptance rate (FAR) is equal to 1% and 0.1%, and at the low FAR, the advantages of our approach are more prominent. With the decrease of false acceptance rate, the correct recognition rate just dropped by 6.25%, which is the smallest of the five approaches. This exhibits that the recognition performance of the proposed approach in this paper is relatively robust.

The same conclusions, as experiment on CASIA gait database, can be drawn from the experimental results on USF shown in Figure 6~ 7 and the related data in Table 3~ 4. The equal error rate in Figure 8 reflects the overall performance of one approach. The smaller the equal error rate, the better the performance of the algorithm.

5. Conclusions and Future Work

In this paper, the amplitude of Gabor of gait energy image is extracted and the kernel Fisher discriminant analysis is used to reduce the dimensionality of gait features of training set. The gait features of test set are selected in a novel way. According to the kernel matrix combined with transformation matrix, the features of test set are selected. Experimental results show that the approach proposed has the characteristics of high recognition rate and strong robustness. Further research will focus on the following aspects: a) explore more effective method to extract gait features; b) seek more effective method of similarity measurement; c) seek more effective classifier.

Acknowledgement
This work was supported by the Education Department of Jilin province science and technology research project in 13th Five-Year in 2016 under contract No.349, wish to express the authors' gratitude here.

References
[1] Chai Y, Ren J, Han W, et al. Human gait recognition: Approaches, datasets and challenges[C]// International Conference on Imaging for Crime Detection and Prevention. IET, 2011:3-3.
[2] Han J, Bhanu B. B.: Individual recognition using gait energy image[J]. IEEE Transactions on Pattern Analysis & Machine Intelligence, 2006, 28(2):316-322.
[3] Liang S C, Zhou M, An-An L I. GEI based gait recognition by using KPCA and SVM[J]. Application Research of Computers, 2010, 27(7):2798-2800.
[4] Theekhanont P, Miguet S, Kurutach W. Gait recognition using GEI and pattern trace transform[C]// International Symposium on Information Technology in Medicine and Education. IEEE, 2012:936-940.
[5] Sivapalan S, Rana R K, Chen D, et al. Compressive Sensing for Gait Recognition[C]// International Conference on Digital Image Computing: Techniques and Applications. IEEE Computer Society, 2011:567-571.
[6] Xue H, Hao Z. Gait recognition based on gait energy image and linear discriminant analysis[C]// IEEE International Conference on Signal Processing, Communications and Computing. IEEE, 2015.
[7] Loog M. On distributional assumptions and whitened cosine similarities.[J]. IEEE Transactions on Pattern Analysis & Machine Intelligence, 2008, 30(6):1114-5.

Normal Vector Projection Method used for Convex Optimization of Chan-Vese Model for Image Segmentation

W B Wei, L Tan, M Q Jia and Z K Pan
College of Computer Science &Technology, Qingdao University, Qingdao 266071, China

E-mail: njustwwb@163.com

Abstract. The variational level set method is one of the main methods of image segmentation. Due to signed distance functions as level sets have to keep the nature of the functions through numerical remedy or additional technology in an evolutionary process, it is not very efficient. In this paper, a normal vector projection method for image segmentation using Chan-Vese model is proposed. An equivalent formulation of Chan-Vese model is used by taking advantage of property of binary level set functions and combining with the concept of convex relaxation. Threshold method and projection formula are applied in the implementation. It can avoid the above problems and obtain a global optimal solution. Experimental results on both synthetic and real images validate the effects of the proposed normal vector projection method, and show advantages over traditional algorithms in terms of computational efficiency.

1. Introduction

The variational method and partial differential equation method have already become main methods in fields of image processing and computer vision[1-5]. With advantages such as universality of multi-model integration, expression flexibility of complex topology and stability of numerical computation method, variational level set method has been one of the basic methods of image segmentation. Chan-Vese model[6] is a combination of Munford-Shah model[7] and variational level set method[8]. It is a variational method of image segmentation which could implement multi-phase image segmentation[9], motion segmentation, texture image segmentation[10] and implicit surface segmentation[11].

Traditional method which uses signed distance function as level sets has certain disadvantages:(1) Image segmentation method, which based on the variational level set method, generally uses gradient descent method to solve the energy functional, so that the minimum obtained from energy functional is probably a local minimum[12], which leads to a high requirement of function initialization. Nevertheless, results acquired sometimes are not really ideal, and even wrong; (2) Complicated level set evolution equation needs to be computed in order to get solution of energy functional. This will slow down computation speed. To this end, Lie [13] and Bresson [14] replaced signed distance functions with binary tag function, transformed Chan-Vese model of two phase image segmentation into global optimal model by constraint relaxing, and finally a threshold method can be applied to the result to acquire global optimal solution of the original problem. This method avoids impact on segmentation results due to initialization of different level sets. Four methods were used to solve this model respectively: Gradient Descent Method(GDM) [15], Dual Method (DM) [16], Split-Bregman Method (SBM) [17] and Alternating Direction Method of Multipliers(ADMM) [18]. On this basis, a new fast segmentation method called Normal Vector Projection Method(NVPM) is proposed. The effectiveness and efficiency of the proposed method are validated by comparison of experimental results. Also, the proposed method can be generalized to other variational model of image processing.

The rest of this paper is organized as follows. In Section 2, the binary level set formulation of the functional of Chan-Vese model used in our paper is reviewed along with its traditional solution method. Our proposed method is discussed in Section 3 and its iterative discrete formulas for implementation will be presented in detail. In Section 4, some numerical experiments are given to illustrate the effectiveness of our method by comparing with other methods. Finally a conclusion is given in section 5.

2. Convex optimization of Chan-Vese model

Chan-Vese model, which is based on variational level set method, is a two-phase image segmentation Munford-Shah model approximated by piece-wise constant. It divides the image into two types of heterogeneous area. Chan-Vese model can be transformed into the following energy functional minimization problem through combining the method that use binary tag functions as level sets which is proposed by Lie and Bresson.

$$\underset{c,\phi\in\{0,1\}}{Min}\left\{E(c,\phi)=\alpha_1\int_\Omega(c_1-f)^2\phi dx+\alpha_2\int_\Omega(c_2-f)^2(1-\phi)dx+\gamma\int_\Omega|\nabla\phi|dx\right\} \tag{1}$$

$f(x)$: $\Omega\to R$ is the image that is yet to be segmented which is defined on Ω. α_1, α_2, γ are penalty parameters, $c_1, c_2 \in c$ are piece-wise constants which are used for segmenting the image into different areas, $\phi(x)$ is level set function. We assume that there is a closed region $\Omega_1 \subset \Omega$, then the definition of binary tag function is as follows.

$$\phi(x)=\begin{cases}1, & x\in\Omega_1\\ 0, & x\notin\Omega_1\end{cases} \tag{2}$$

Eq.(1) is the energy function contains more than one variable, we solve it through alternating optimization method generally, i.e. solve the function under fixed ϕ with respect to c.

$$c_1=\int_\Omega f\phi dx\Big/\int_\Omega\phi dx,\ c_2=\int_\Omega f(1-\phi)dx\Big/\int_\Omega(1-\phi)dx \tag{3}$$

Then, fixing c and solving the function with respect to ϕ. By Eq.(1), the formula of energy functional minimization about ϕ can be written as

$$\underset{c,\phi\in\{0,1\}}{Min}\left\{E(\phi)=\int_{\Omega}Q_{12}(c_1,c_2)\phi dx+\gamma\int_{\Omega}|\nabla\phi|dx\right\} \tag{4}$$

where $Q_{12}(c_1,c_2)=\alpha_1(c_1-f)^2-\alpha_2(c_2-f)^2$. Eq.(4) can be transformed into following problem via convex relaxation and threshold on binary tag function ϕ

$$\underset{\phi\in[0,1]}{Min}\left\{E(\phi)=\int_{\Omega}Q_{12}(c_1,c_2)\phi dx+\gamma\int_{\Omega}|\nabla\phi|dx\right\} \tag{5}$$

The according gradient descent function about Eq.(5) is

$$\begin{cases}\partial\phi/\partial t=\left(\gamma\nabla\cdot\left(\nabla\phi/|\nabla\phi|\right)-Q_{12}(c_1,c_2)\right) & in\Omega\\ \partial\phi/\partial\vec{n}=0 & on\partial\Omega\end{cases} \tag{6}$$

After getting ϕ, then project it into the interval between 0 and 1.

$$\tilde{\phi}^{k+1}=Max\left(Min\left(\phi^{k+1},1\right),0\right) \tag{7}$$

Finally, ϕ is obtained through threshold technology.

$$\phi^{k+1}=\begin{cases}1 & \tilde{\phi}^{k+1}>\alpha=0.5\\ 0 & otherwise\end{cases} \tag{8}$$

3. Normal vector projection method of Chan-Vese convex model and its optimization method

Through introduction of four methods above, the method[19][20] which uses low-order term instead of high-order term and optimize alternately, not only improves computation efficiency, but also reduces complexity of energy functional. On the basis of this thought, the normal vector projection method is proposed, which uses approximated variable to replace normal vector of curvature term in gradient descent function. This will simplify solving process, and then conduct the projection of normal vector to constrain the introduced approximated variable directly. This method is more concise on the structure of algorithm, moreover, it can guarantee the fast convergence of the energy functional. This chapter not only put forward NVPM of Chan-Vese model, but also the optimization algorithm of NVPM.

3.1 Vector projection method

It is easy to get Euler-Lagrange equation according to Eq.(5) which is depicted as follow.

$$Q_{12}(c_1,c_2)-\gamma\nabla\cdot\left(\nabla\phi/|\nabla\phi|\right)=0 \tag{9}$$

Here the four-order curvature term which is difficult to calculate and the primary source of low iteration speed. According to method that has been introduced which utilizes low-order term to replace high-order term and alternately optimize each variable so that calculation efficiency is improved. Introducing instrumental variable $\vec{p}=\left[p_1,p_2\right]^T$ to approximately represent normal vector of ϕ, so

$\vec{p} \approx \nabla\phi/|\nabla\phi|$ can be obtained, and the above equation is converted as

$$Q_{12}(c_1,c_2) - \gamma\nabla\cdot\vec{p} = 0 \tag{10}$$

Now fix $\vec{p}$ to find ϕ^{k+1}, the formula above should be calculated by using gradient descent method.

$$\partial\phi/\partial t = \gamma\nabla\cdot\vec{p} - Q_{12}(c_1,c_2) \tag{11}$$

The time step is τ_p, now we obtain the iteration function on ϕ.

$$\phi^{k+1} = \phi^k + \tau_p\left(\gamma\nabla\cdot\vec{p}^k - Q_{12}(c_1,c_2)\right) \tag{12}$$

After obtained ϕ^{k+1}, we need to make a projection of ϕ^{k+1} according to Eq.(7). Then fix ϕ^{k+1}. We can obtain $\vec{p}$ with normal vector projection method directly based on $\vec{p} \approx \nabla\phi/|\nabla\phi|$.

$$\vec{p}^{k+1} = \nabla\phi^{k+1}/max\left(\left|\nabla\phi^{k+1}\right|, 1\right) \tag{13}$$

Finally, threshold ϕ^{k+1} according to Eq.(8).

In summary, we can list the NVPM in a pseudo code as follows.

Algorithm 1: NVPM for CV model

Step 1: Initialize unknown values as $\phi^0, \vec{p}^0$

Step 2: For $k \geq 1$, solve the following problems alternatively

2.1 Sub-problem 1 for ϕ^{k+1}: $\phi^{k+1} = Arg\min\left\{\varepsilon_1(\phi) = E\left(\phi, \vec{p}^k\right)\right\}$,

$\phi^{k+1} = Max\left(Min\left(\phi^{k+1},1\right),0\right)$.

2.2 Sub-problem 2 for $\vec{q}^{k+1}$: $\vec{p}^{k+1} = Arg\min\left\{\varepsilon_2(\vec{p}) = E\left(\phi^{k+l}, \vec{p}\right)\right\}$.

2.3 Threshold ϕ^{k+1}: $\phi^{k+1} = \begin{cases} 1 & \tilde{\phi}^{k+1} > \alpha = 0.5 \\ 0 & otherwise \end{cases}$.

Step 3: The overall loop will be terminated if the stopping criteria (described in section 3) are satisfied.

3.2 Optimization algorithm of normal vector projection method

The method proposed by Nesterov[21] could accelerate and optimize the process of gradient descent through over relaxation. A kind of optimization algorithm of the normal vector projection method is given to further improve computation efficiency. An over relaxation parameter α and an intermediate variable $\tilde{\phi}$ are introduced before we solve the gradient descent process of Eq.(24).

$$\tilde{\phi}^{k+1} = \phi^k + \tau_p\left(\gamma\nabla\cdot\vec{p}^k - Q_{12}(c_1,c_2)\right) \tag{14}$$

After getting $\tilde{\phi}^{k+1}$, we need to update the over relaxation parameter and then obtain α^{k+1}.

$$\alpha^{k+1} = \left(1+\sqrt{4\alpha^k+1}\right)/2 \tag{15}$$

Then, update ϕ through intermediate parameter $\tilde{\phi}^{k+1}$ and over relaxation α^k and α^{k+1}.

$$\phi^{k+1} = \tilde{\phi}^{k+1} + \left(\alpha^k - 1\right)\left(\tilde{\phi}^{k+1} - \tilde{\phi}^k\right)/\alpha^{k+1} \tag{16}$$

We remain need to make a projection of ϕ^{k+1} according to Eq.(7). The rest of the steps are similar as the non-optimized NVPM. Via normal vector projection presented by Eq.(26) $\vec{p}^{k+1}$ can be got, and according to Eq.(8) we obtain ϕ^{k+1} by threshold.

The pseudo code is shown as follows.

Algorithm 2: NVPM+ for CV model

Step 1: Initialize unknown values as $\phi^0, \vec{p}^0$

Step 2: For $k \ge 1$, solve the following problems alternatively

2.1 Sub-problem 1 for ϕ^{k+1}: $\tilde{\phi}^{k+1} = Arg\min\left\{\varepsilon_1(\phi) = E\left(\phi, \vec{p}^k\right)\right\}$.

Update over relaxation parameter α : $\alpha^{k+1} = \left(1+\sqrt{4\alpha^k+1}\right)/2$.

Update ϕ: $\phi^{k+1} = \tilde{\phi}^{k+1} + \left(\alpha^k - 1\right)\left(\tilde{\phi}^{k+1} - \tilde{\phi}^k\right)/\alpha^{k+1}$,

$\phi^{k+1} = Max\left(Min\left(\phi^{k+1}, 1\right), 0\right)$.

2.2 Sub-problem 2 for $\vec{q}^{k+1}$: $\vec{p}^{k+1} = Arg\min\left\{\varepsilon_2(\vec{p}) = E\left(\phi^{k+l}, \vec{p}\right)\right\}$.

2.3 Threshold ϕ^{k+1}: $\phi^{k+1} = \begin{cases} 1 & \tilde{\phi}^{k+1} > \alpha = 0.5 \\ 0 & otherwise \end{cases}$.

Step 3: The overall loop will be terminated if the stopping criteria (described in section 3) are satisfied.

4. Numerical experiments and analysis

In this section, the numerical results of our proposed methods are applied on some real cases and they will be compared with different methods (Gradient Descent Method (GDM), Dual Method (DM), Split-Bregman Method (SBM), Alternating Direction Method of Multipliers (ADMM)) to demonstrate the effectiveness and efficiency of our methods. All the experiments are operated on the same platform (Matlab7.8) on a PC(Intel(R) , Core(TM) , CPU i7 2. 60GHz) . The same initial contours and initiations of variables for all the methods in each experiment are used in order to have a relatively neutral criterion for comparison.

As described in [22] the iterations need to be terminated when the following criteria are satisfied:

i We need to monitor the constraints errors in iterations:

$$\left(R_1^k\right) = \left(\left\|\vec{R}_1^k\right\|_{L^1} / \left\|\vec{R}_1^0\right\|_{L^1}\right) \tag{17}$$

with

$$\vec{R}_1^k = \vec{p}^k - \nabla\phi^k / |\nabla\phi^k|$$

ii During iterating, the relative errors of the solution ϕ^k should be noticed. They should reduce to a sufficiently small level:

$$\|\phi^k - \phi^{k-1}\|_{L^1} / \|\phi^{k-1}\|_{L^1} \quad (18)$$

iii The relative energy error can be chosen as stopping criterion

$$|E^{k+1} - E^k| / |E^k| \le \varepsilon \quad (19)$$

where, E^k is the energy value. The computation stops automatically when $|E^{k+1} - E^k| / |E^k|$ is less than a predefined tolerance, which means the energy approaches its steady state. Until now, the proposed NVPM and NVPM+ are completed and the effectiveness will be given with extensive experiments in the following.

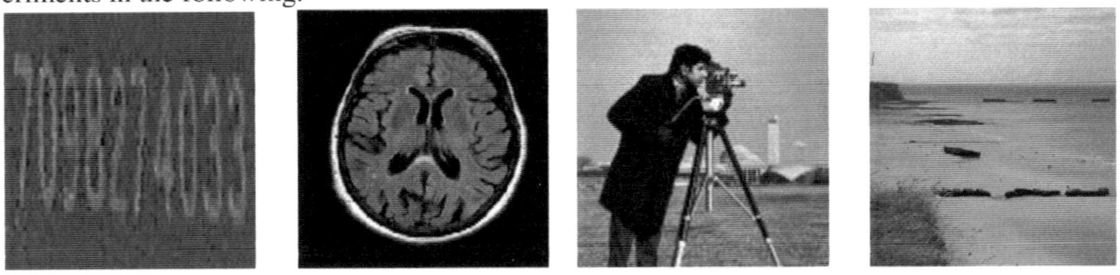

Figure 1. Images for segmentation

Four images to be segment are presented in Fig.1. There are blurred image, MRI image, character image and nature scene image.

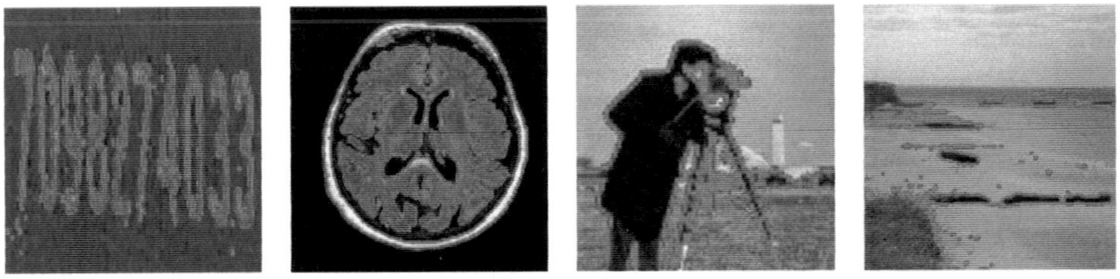

(a) GDM

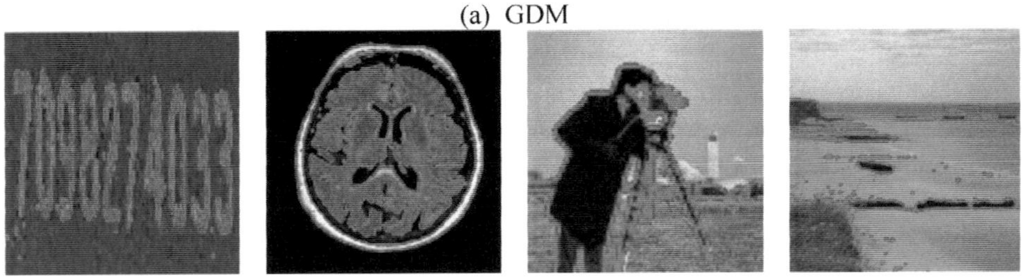

(b) DM

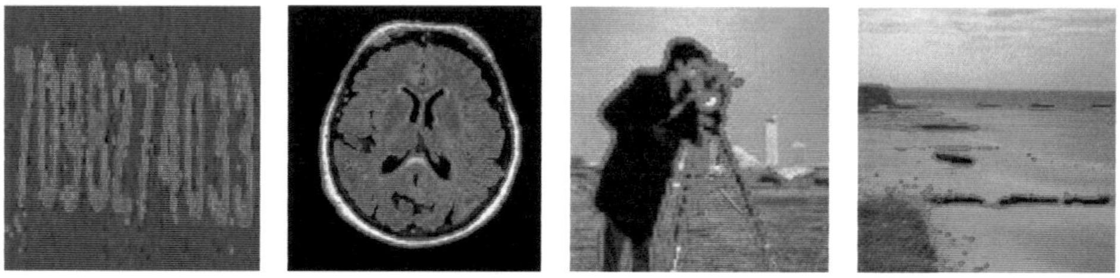

(c) SBM

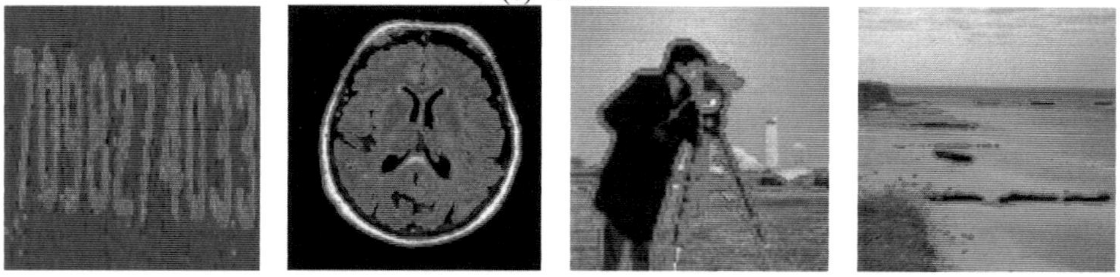

(d) ADMM

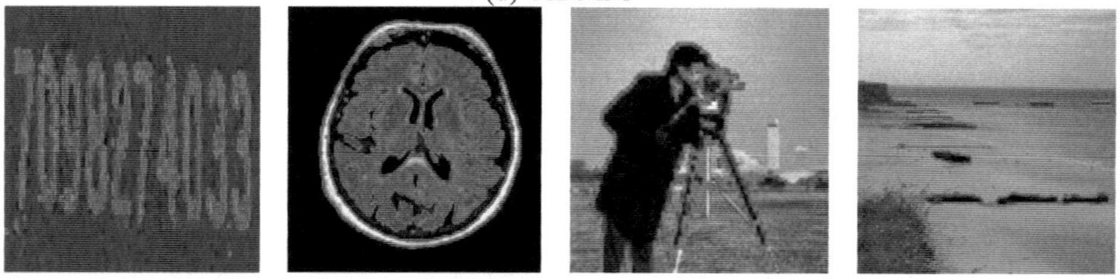

(e) NVPM

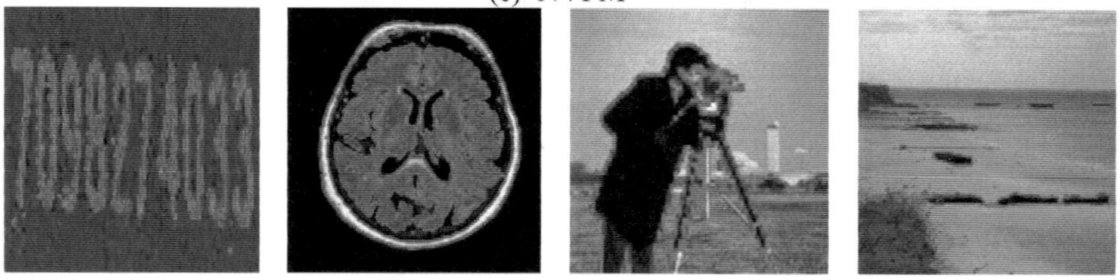

(f) Optimization Algorithm of NVPM

Figure 2. Segmentation results with different methods

Fig.2 (a)-(f) shows the segmentation results of GDM, DM, SBM, ADMM, NVPM model and its optimization. From the experimental results, we can see that each method can achieve better segmentation results.

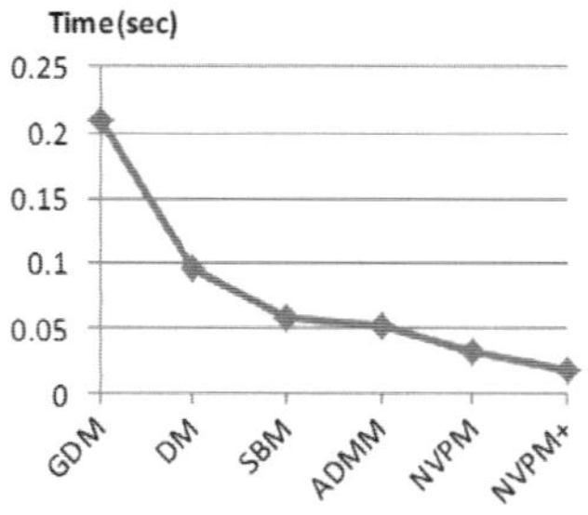

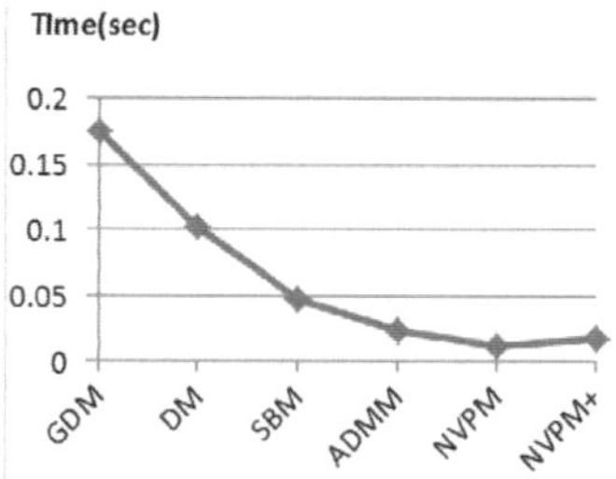

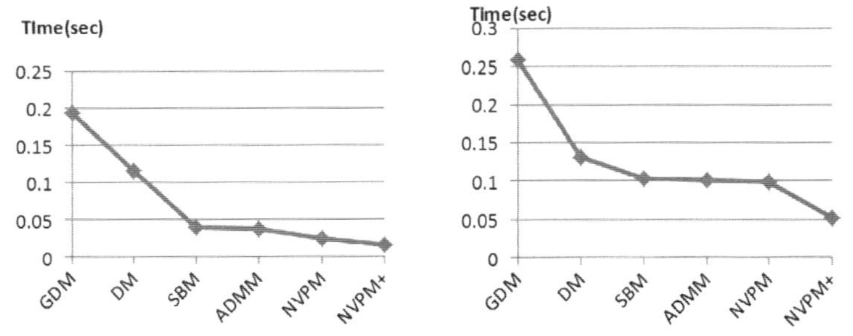

Figure 3. Comparisons of computational time with different methods

Comparisons of computational time using different methods for four kinds of images are given in Fig.3. According to the results of numerical experiments and comparing to other algorithm, no matter optimization or not, the computation rate of our method is always better than DM, GDM, SBM and ADMM. Among them, the NVPM optimization algorithm is the fastest, followed by the non-optimized NVPM, SBM and ADMM are equivalent and better than DM and GDM, the lowest calculation rate of GDM. The advantages of our method in terms of computational speed are verified through multiple sets of experiments.

5. Conclusions

In this paper, normal vector projection method used for global convex active contour model is proposed. This method with a simple structure uses the properties of introduction of auxiliary variables to solve energy functional directly. The calculation procedure is simplified to a large extent, and the computational efficiency is accelerated at the same time. The comparison of results indicate that our approach owns good enough effects and it is a good way to efficiently minimize the difficult functional. In addition, the method can also be optimized by acceleration algorithm and good results have been obtained. It is supposed to yield shorter running time while the quality of results is identical. Our method can also be applied into surface restoration and segmentation, motion segmentation and medical image 3D reconstruction in the future work.

References

[1] Aubert G 2001 *Mathematical Problems in Image Processing: Partial Differential Equations and the Calculus of Variations.* New York: Springer

[2] Osher S 2003 *Geometric level set methods in imaging, vision, and graphics.* New York: Springer-Verlag

[3] Paragios N 2006 *Handbook of mathematical models in computer vision.* New York: Springer

[4] Chan T 2005 *Image Processing And Analysis: Variational, PDE, Wavelet, And Stochastic Methods.* Philadelphia: SIAM Society for Industrial and Applied Mathematics.

[5] Tai X C 2007 *Image processing based on partial differential equations. Proceedings of the international conference on PDE-based image processing and related inverse problems* CMA Oslo Norway August 8–12 2005 Mathematics & Visualization

[6] Chan T F 2001 *Active contours without edges.* IEEE Transactions on Image Processing A Publication of the IEEE Signal Processing Society, 10(2), 266-277

[7] Mumford D 1989 *Optimal approximations by piecewise smooth functions and associated variational problems*. Communications on Pure and Applied Mathematics, 42(5):577-685
[8] Zhao H 1996 *A variational level set approach to multiphase motion*. Journal of Computational Physics, 127(1):179-195
[9] Vese L 2002 *A multiphase level set framework for image segmentation using the mumford and shah model*. International Journal of Computer Vision, 50(3):271-293
[10] Liu J 2011 *A fast segmentation method based on constraint optimization and its applications: Intensity inhomogeneity and texture segmentation*. Pattern Recognition, 44(9):2093–2108.
[11] Bertalmio M 2001 *Variational problems and partial differential equations on implicit surfaces*. Journal of Computational Physics 174(2):759-780
[12] Chan T F 2001 *Active contours without edges*. IEEE Transactions on Image Processing 10(2): 266-277
[13] Lie J 2006 *A binary level set model and some applications to Mumford-Shah image segmentation*. IEEE Transactions on Image Processing 15(5): 1171-1181
[14] Bresson X 2007 *Fast global minimization of the active contour/snake model*. Journal of Mathematical Imaging and Vision 28(2):151-167
[15] Rudin L 1992 *Nonlinear total variation based noise removal algorithms*. Physica D: Nonlinear Phenomena 60(1): 259-268.
[16] Chambolle A 2004 *An algorithm for total variation minimization and applications*. Journal of Mathematical Imaging and Vision, 20(1-2): 89-97
[17] Goldstein T. 2008 *The split Bregman method for L1 regularized problems*. SIAM Journal on Imaging Sciences, 2(2), 323-343
[18] Goldstein T 2014 *Fast Alternating Direction Optimization Methods*. SIAM Journal on Imaging Sciences, 7(3):1588-1623.
[19] Pan Z K 2009 *A Variational Level Set Method of Multiphase Segmentation for 3D Images*. Chinese Journal of Computers 32(12):2464-2474.
[20] Wang Q 2010 *Split-Bregman Method and Dual Method for Multiphase Image Segmentation*. Journal of Computer-aided Design & Computer Graphics 22(9): 1561-1569
[21] Nesterov Y 1983 *A method of solving a convex programming problem with convergence rate o(1/k^2)*. Soviet Math(27):372-376
[22] Tai X C 2014 *Fast numerical schemes related to curvature minimization: a brief and elementary review*. UCLA CAM Report: 14-40

Video Stabilization Using Feature Point Matching

Shamsundar Kulkarni[1], D.S.Bormane[2] and S.L.Nalbalwar[3]

[1] Research Scholar, JNTU Kakinada, India.
[2] Principal,JSPM's Rajarshi Shahu College of Engg. ,Savitribai Phule Pune University , Pune,
[3] Professor & Head (E&Tc), Dr. Babasaheb Ambedkar Technological University, Lonere, Raigad, India.

Abstract. Video capturing by non-professionals will lead to unanticipated effects. Such as image distortion, image blurring etc. Hence, many researchers study such drawbacks to enhance the quality of videos. In this paper an algorithm is proposed to stabilize jittery videos. A stable output video will be attained without the effect of jitter which is caused due to shaking of handheld camera during video recording. Firstly, salient points from each frame from the input video is identified and processed followed by optimizing and stabilize the video. Optimization includes the quality of the video stabilization. This method has shown good result in terms of stabilization and it discarded distortion from the output videos recorded in different circumstances.

1. Introduction

Recently, the market of handheld camera has grown rapidly. However, video capturing by non-professional user normally will lead to unanticipated effects, such as image distortion, image blurring etc. Hence, many researchers focused on such drawbacks to enhance the quality of casual videos. [1]

Digital video stabilization is the process of removing unwanted movement from a video stream. Generally the processes of stabilization goes through three phases namely 1) Motion estimation 2) Motion smoothing 3) Motion compensation. The purpose of first phase is to estimate the motion between frames. After that the parameters of estimated motion which are obtained from the first phase will be sent to motion smoothening, where it removes the high-frequency distortion and calculates the global transformation, which is very important to stabilize the current frame [2]. Next, warping will be done by image composition for the motion compensation. These three-step frameworks are the essential steps in most of the video stabilization algorithms.

2. Literature Survey

Initial step in the video stabilization is selection of feature points. These feature points are corners.Corner detectors were developed in the late 1970's. Labeeb Mohsin Abdullah et.al. [1] had proposed an algorithm to stabilize the image sequence by using Harris corner detection technique. Harris Corner Detection is one of the fastest algorithms to find corner values [3]. Aleksandr Shnayderman et.al. had proposed SVD (Signature Value Decomposition)based gray scale image quality measurement. Edward Rosten et.al proposed faster and better machine learning approach to corner detection. This algorithm is used for

IOP Conf. Series: Journal of Physics: Conf. Series **787** (2017) 012017 doi:10.1088/1742-6596/787/1/012017

identifying Interest points in an image [4]. Yue Wang et.al. had proposed Real-Time Video Stabilization for Unmanned Aerial Vehicles. Elmar Mair et.al.proposed technique of Adaptive and Generic Corner Detection Based on the Accelerated Segment Test. C. Harris and M.J. Ephens proposed combined corner and edge detection to cater for image regions containing texture and isolated features, which is based on the local auto-correlation function. Mohammed A. Alharbi, had proposed Fast Video Stabilization Algorithm. [5] In this affine motion model is utilized to determine the parameters of translation and rotation between images. The determined affine transformation is then exploited to compensate for the abrupt temporal discontinuities of input image sequences. [6]

3. Methodology

Features from accelerated segment test (FAST) is a corner detection method which could be used to extract feature points and later used to track and map objects in many computer vision tasks. FAST is an algorithm proposed originally by Rosten and Drummond for identifying Interest points in an image. It is fast than many other well-known feature extraction methods, such as Difference of Gaussians (DoG) used by SIFT, SUSAN and Harris. An interest point in an image is a pixel which has a well defined position and can be robustly detected [7]. Interest points have high local information content and they should be ideally repeatable between different images.

3.1 Algorithm &Implementation

The Flow chart of video stabilization algorithm is shown in Fig.1. The total process of video capturing and stabilization has been performed in several steps. These steps are elaborated with simulation results.

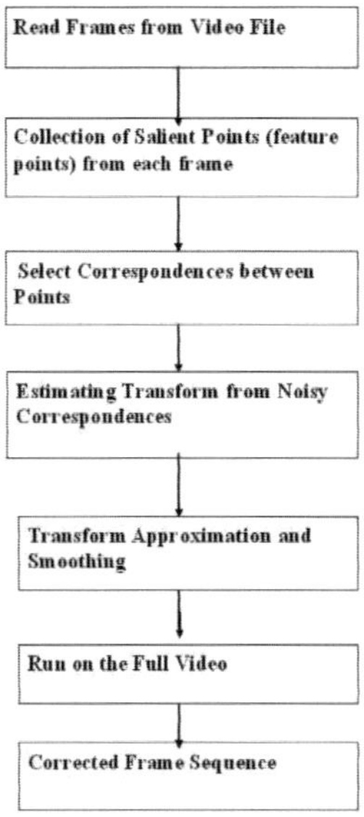

Figure.1 Flow chart of the of video stabilization algorithm

Step 1. Read Frames from Video File

In this step, first two frames of a video sequence are read. We read them as intensity images since color is not necessary for the stabilization algorithm and also using grayscale images improve the speed. Fig.2 shows both frames side by side. The data related intensity of images has been separated from color.

Figure 2. Reading the two frames

Step2 .Collection of Salient Points (feature points) from each frame

After reading two frames, Red-Cyan color composite is produced to illustrate the pixel wise difference between them as shown in figure 3. Our goal is to determine a transformation that will correct for the distortion between the two frames, for this Geometric Transform function can be used, which return an affine transform [8]. The input for this function shall comprise a set of point correspondences between the two frames. To generate these correspondences, collection of points of interest from both frames is performed [9]. For this purpose, FAST is corner detection algorithm is used. Fig.4 shows Image showing the interest point under test and the 16 pixels on the circle. It gives high measured quality video and less computational time. The detected points from both frames are shown in the fig.5 (a) &(b). Observe how many of them cover the same image features, such as points along the tree line, the corners of the large road sign, and the corners of the cars.

Figure 3. Red-Cyan color composite

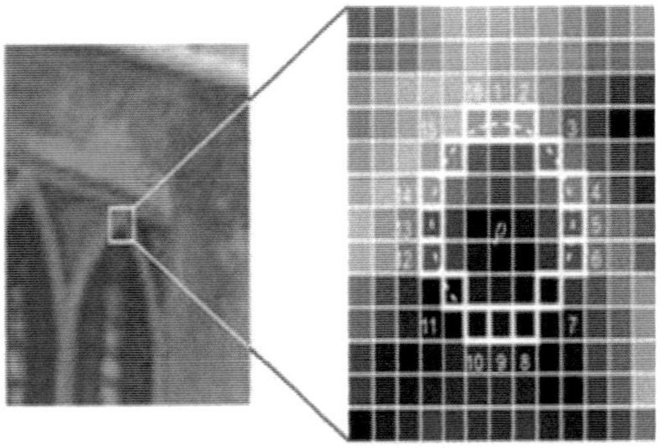

Figure 4. Image showing the interest point under test and the 16 pixels on the circle

Figure 5. (a) corners in frame A

Figure 5. (b) Corners in frame B

Step 3. Select Correspondences between Points

In this step, correspondences between the points derived in the step 2 should be established. For each point, we extract a Fast Retina Key point (FREAK) descriptor centered around it [10]. The matching cost we use between points is the Hamming distance since FREAK descriptors are binary. Points in frame A and frame B are matched putatively. Note that there is no uniqueness constraint, so points from frame B can correspond to multiple points in frame A. Fig.6 shows the same color composite given above, in this points from frame A are shown with red colour, and the points from frame B in green colour. Yellow lines are drawn between points to show the correspondences. Many of these correspondences are correct, but still there is significant number of outliers.

Step 4. Estimating Transform from Noisy Correspondences

Many of the point correspondences obtained in the previous step are identified with limited accuracy. To rectify this problem, Random Sample Consensus (RANSAC) algorithm is used which is implemented in the Geometric Transform function. It is observed that the inliers correspondences in the image background are not aligned with foreground. The reason behind this is the background features are far enough those act as if they were on an infinitely distant plane. We can assume that background plane is static and will not change dramatically between the first and second frame, instead, this transform is capturing the motion of the camera. Thus correcting process will stabilize the video. Furthermore, as long as the motion of the camera between frame A and frame B is minimize or the time of sampling the video is high enough, this condition is maintained. The RANSAC algorithm is repeated multiple times and at each run the cost of the result is calculated by projecting frame B onto frame A via Sum of Absolute Differences between the two image frames. Fig.7 shows a color composite showing frame A overlaid with the re-projected frame B, along with the re-projected point correspondences. It is clear from this figure that the results are favorable, with the inliers correspondences nearly exactly coincident. The cores of the images are both well aligned, such that the red-cyan color composite becomes almost purely black-and-white in that region.

Figure 6. Correspondences between Points

Figure 7. color composite showing frame A overlaid with the re-projected frame B

Step 5. Transform Approximation and Smoothing

We could use all the six parameters of the affine transform, but, for numerical simplicity and stability, we choose to re-fit the matrix as a simple scale-rotation-translation transform. This has only four free parameters as compared to the full affine transform which are one scale factor, one angle, and two translations. Finally reconstructed S-R-t transform output is shown in Figure 8.

Step 6. Run on the Full Video

The above procedure of estimating the transform between two images has been performed in the MATLAB® function cvexEstStabilizationTform. The function cvexTformToSRT also converts a general affine transform into a scale-rotation-translation transform. The smoothed video frame is shown in fig.9 by using Video Player in Matlab.

Figure 8. Color Composite of Affine and s-R-t Transform output

Figure 9. Camera movement to stabilize video

Step 7. Corrected Frame Sequence:

During computation, we computed the mean of the raw video frames and mean of the corrected frames. These mean values are shown side-by-side in fig.10. The left image shows the mean of the raw input frames, proving that there was a great deal of distortion in the original video. The mean of the corrected frames on the right, however, shows the image with almost no distortion. Foreground details have been blurred because of the car's forward motion. This shows the efficacy of the stabilization algorithm.

Figure.10 Mean of the raw video frames

4. Result and conclusion

In this experimentation two shaky videos are considered. FAST detection method is applied for feature point detection. Quality factor and computational time are analyzed. Video 1(shaky _car) has the following specifications: the length of 4 seconds, width x height of 320x240, frame rate of 30 frames/second and size of 1.2 MB. . Video 2(shaky _Highways) has the following specifications: the length of 5 seconds, width x height of 320x240, frame rate of 30 frames/second and size of 1.4 MB. This technique provides logical and computationally efficient approach in terms of stabilizing high jitter videos suffered from distortion. The FAST detection algorithm presented in the work provides a fast and robust stabilization system and improves real-time performance. The quality and computational time are measured and compared for sample videos as shown in Table 1.

Table 1. Performance parameters

Sr.No	**Sample Video**	**Computational Time (s)**	**Quality Value**
1	Shaky _video1	8.81	22.20 %
2	Shaky _video2	9.18	39.21%

5. Conclusion and Future scope

The use of FAST detection method shows improved result in terms of stabilization. FAST detection technique is useful in enhancing the quality of low-grade video surveillance cameras. FAST detection technique is particularly helpful in identifying people, license plates, etc. from low-quality video surveillance cameras. In future, we can find better feature detector to overcome the consequences of extreme shaking of handheld camera in real time implementation for video stabilization.

References

[1] Aleksandra Shnayderman, Alexander Gusev, and Ahmet M. Eskicioglu"An SVD-Based Grayscale Image Quality Measure for Local and GlobalAssessment ",IEEE 15(2), 2006.

[2]B.-Yu. Chen, K.-Yi. Lee, W.-T. Huang, J.-S. Lin, "Capturing Intention-based Full-Frame Video Stabilization," Computer Graphics Forum, Vol. 27, No. 7, p.1805 - p.1814, 2008.

[3] C. Harris and M.J. Stephens, "A combined corner and edge detector", Proc of Alvey Vision Conference, pp 147–152, 1988.

[4] Edward Rosten, Reid Porter and Tom Drummond, "FASTER and better: A machine learning approach to corner detection" in IEEE Trans. Pattern Analysis and Machine Intelligence, 2010, vol 32, pp. 105-119

[5] C. Morimato and R. Chelleppa, "Evaluation of Image Stabilization algorithms", Proc. of IEEE int. Conf. on Acoustics, speech and signal processing, Vol-5,pages 2789-2792, 1995.

[6]Jie Xu, Hua-wen Chang, Shuo Yang, and Minghui Wang "Fast Feature-Based Video Stabilization without Accumulative Global Motion Estimation" IEEE 2012.

[7]Ken-Yi Lee Yung-Yu Chuang Bing-Yu Chen Ming Ouhyoung "Video Stabilization using Robust Feature Trajectories" 2009 IEEE 12th International Conference on Computer Vision (ICCV).

[8]Keng-Yen Huang, Yi-Min Tsai, Chih-Chung Tsai, and Liang-Gee Chen "Feature-based Video Stabilization for Vehicular Applications" 2010 IEEE 14th International Symposium on Consumer Electronics.

[9] Labeeb Mohsin Abdullah, Nooritawati Md Tahir & Mustaffa Samad , "Video Stabilization based on Point Feature Matching Technique", 2012 IEEE Control and System Graduate Research Colloquium (ICSGRC 2012).

[10]Xie Zheng, Cui Shaohui, Wang Gang, Li Jinlun "Video Stabilization System Based on Speeded-up Robust Features" International Industrial Informatics

GPU accelerated Foreign Object Debris Detection on Airfield Pavement with visual saliency algorithm

Jun Qi, Guoping Gong and Xiaoguang Cao

Image Processing Center, Beihang University, Beijing, China

E-mail: xgcao@buaa.edu.cn

Abstract. We present a GPU-based implementation of visual saliency algorithm to detect foreign object debris(FOD) on airfield pavement with effectiveness and efficiency. Visual saliency algorithm is introduced in FOD detection for the first time. We improve the image signature algorithm to target at FOD detection in complex background of pavement. First, we make pooling operations in obtaining saliency map to improve recall rate. Then, connected component analysis is applied to filter candidate regions in saliency map to get the final targets in original image. Besides, we map the algorithm to GPU-based kernels and data structures. The parallel version of the algorithm is able to get the results with 23.5 times speedup. Experimental results elucidate that the proposed method is effective to detect FOD real-time.

1. Introduction

A clean and flat airport pavement provides a safe environment for airplane's take-off and landing. It is vital to detect foreign object debris(FOD) timely, such as little stones and screws in pavement, which could make damages to airplane tires and even lead to heavy accident.

Li [1] designed a multi-sensor system and applied combined methods to make FOD detection. Liu [2] proposed a traversal method cutting images into small pieces and extracting features, such as Harris corner, GLCM of each piece for classifying. However, these two methods are complicated or time consuming. The detection of FOD with effectiveness and efficiency becomes a challenging task.

Recently, visual saliency has drawn much attention in many areas, including neurobiology and computer vision [3]. It is suggested that human's attention is directed to visually distinctive regions in an image rapidly without much guidance. This capability inspires us to make saliency detection, aiming at highlighting visually salient regions or objects in an image. Considering the background characteristics of airport pavement, where FOD would be visually conspicuous, we apply visual saliency algorithm to make FOD detection in airfield pavement.

At the same time, in order to meet the need of real-time detection, parallelization is required to reduce the computation time. Nowadays, GPU has been widely used in various fields, especially in computer vision and image processing [4]. GPU contains thousands of cores to process parallel workloads efficiently. Using GPU can make full use of hardware performance and improve the speed of detection algorithm obviously.

Our main contributions are summarized as follows:

First, an effective visual saliency algorithm for FOD detection is proposed.

Second, we accelerate the algorithm using GPU with highly parallel structure, and achieve real-time detection, which will be of great significance in real scene.

The paper is organized as follows. Section 2 introduces our visual saliency algorithm. The GPU implementation details are presented in Section 3. In Section 4, experiments and analysis are made to show the performance and speed of proposed algorithm. Conclusions are drawn in Section 5.

2. Visual saliency algorithm

In existing works, visual saliency algorithms can be generally divided into two parts, spatial domain and frequency domain. Considering the speed of FOD detection in airfield pavement, we take frequency domain algorithms into account to compute saliency, which can be efficiently mined through simple spectrum modulation.

Our visual saliency algorithm is shown in Fig 1. Firstly, preprocessing methods including frequency and spatial filter are employed to eliminate interference of complex background. Secondly, we use an improved algorithm based on image signature to obtain saliency map. Finally, saliency blocks are segmented with connected component analysis and the locations are mapped to original image.

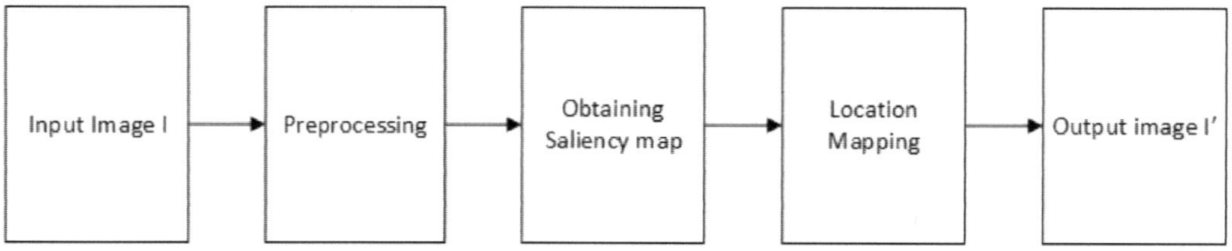

Figure 1.visual saliency algorithm.

2.1. preprocessing

Actually, airfield pavement image has complex background with road groove, maker line, underground lamp, which can make obstruction to FOD detection. Firstly, we make Discrete Fourier Transform(DFT) to the original image, and do band pass in frequency domain. Road groove was filter without losing FOD detection object. Then, Inverse Discrete Fourier Transform(IDFT) and normalization are made to get image in spatial domain back. At last, average filter is used to eliminate noisy points.

2.2. obtaining saliency map

Image signature [5] is defined as follows:

A grey image combines with foreground and background.

$$\mathbf{x} = \mathbf{f} + \mathbf{b} \quad \mathbf{x}, \mathbf{f}, \mathbf{b} \in \mathbb{R}^N \tag{2.1}$$

$\mathbf{f}$ represents the foreground and is assumed to be sparsely supported in the standard spatial basis. $\mathbf{b}$ represents the background, and is assumed to be sparsely supported in the basis of the Discrete Cosine Transform (DCT). In airfield pavement image, our detection target FOD can be treated as foreground. The image signature is defined as:

$$\mathbf{x}\text{=ImageSignature}(\mathbf{x}) = sign(DCT(\mathbf{x})) \tag{2.2}$$

We can approximately isolate the support of $\mathbf{f}$ by taking the sign of the mixture signal $\mathbf{x}$ in the transformed domain and then inversely-transform it back into the spatial domain.

$$\bar{\mathbf{x}} = IDCT(sign(\mathbf{x})) \tag{2.3}$$

Pooling is widely used in deep neural network, which can maintain spatial invariance. The pooling window size can be arbitrary, and windows can be overlapping. We improve Hou's saliency model by applying pooling methods.

In order to improve speed, we first make subsampling and overlapping pooling to the image to reduce the dimension to from (2048, 2048) to (202, 202). Then, image signature is calculated to the image.

At last, we apply non-overlapping mean pooling and non-overlapping max pooling to improve recall. Max pooling is a way of taking the most responsive node of the given interest region. The saliency map size we obtained finally is (64, 64), where 1 pixel corresponds to 32 pixels in original image. Max pooling helps find the most likely node to avoid missing true targets in location mapping stage.

2.3. location mapping

To get final saliency regions, we make connected component analysis of the saliency map. And rules are made to map location in original image according to the area of connected component. For example, large area of connected component is excluded, which is not our detection target, such as small stones and screws. Small area is judged as effective saliency regions and its location will be mapped to a box in original image.

The results of our algorithm are shown in Fig 2. Fig 2(a) is the original image. Fig 2(b) is image after preprocessing. Fig 2(c) is the saliency map and Fig 2(d) is the location mapping results.

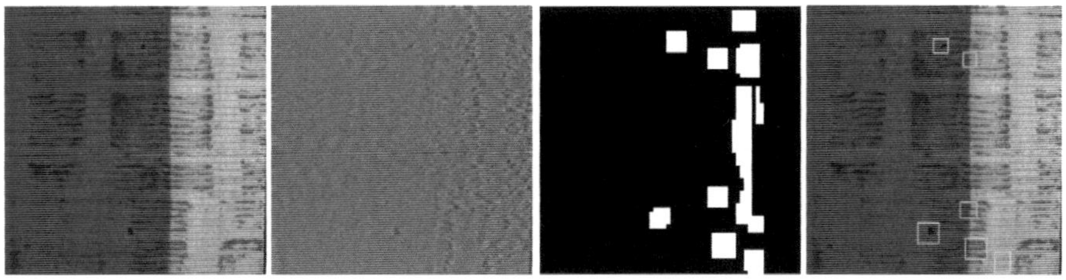

(a) (b) (c) (d)

Figure 2. Results of visual saliency algorithm.

3. Implementing with GPU

Compute Unified Device Architecture(CUDA) is a parallel computing platform and a software environment for parallel computing created by Nvidia [6]. It allows software developers to use a CUDA-enabled graphics processing unit(GPU) which can launch thousands of threads in parallel for general purpose processing. In order to achieve higher performance, we make parallel optimizations to some parts of the algorithm in consideration of memory model and threads scheduling of CUDA.

3.1. convolution

Convolutions are used widely by many applications in image processing task, including blur filters, mean pooling and so on. For instance, each pixel has to sum up its own value and others values in neighbourhood in an average filter. Then, an average application will be made to get the output pixel value.

It is noticed that the calculation of adjacent output pixel value shares most input pixel value. Considering shared memory's high bandwidth and low latency, we can load pixels to be filtered and the corresponding apron pixels [7] into shared memory first. Then, in next step, data to be calculated would be read from shared memory other than global memory, which could reduce repeated and time-consuming communications between global memory and processing units.

3.2. reduction

To do normalization of an image, we have to get minimum and maximum pixel value of an image first. Sequential computation complexity of the operation reduction is over the number of input data $O(N)$, whereas parallel computation complexity can be $O(\log N)$. Following is the parallel reduction algorithm implemented in CUDA.

For a (2048, 2048) size image, we set gridDim (64, 64), and blockDim (32, 32). We first make reduction in every block, which has 1024 threads. Threads allocation details are shown in Fig 3.

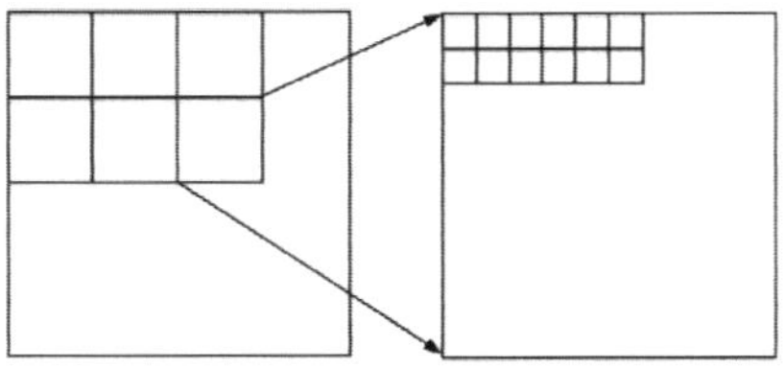

Figure 3. Threads allocation.

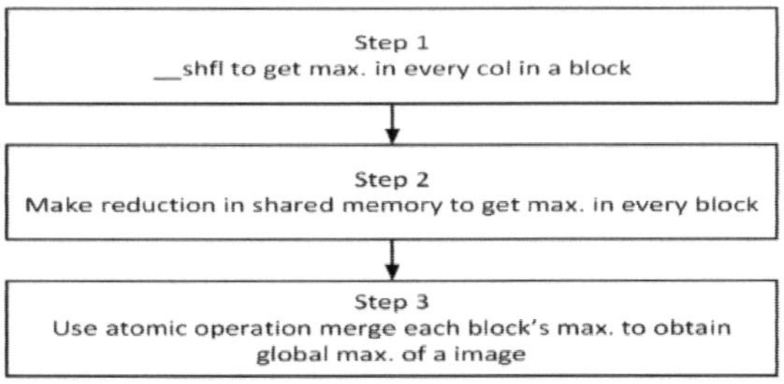

Figure 4. Reduction steps.

SM3.0 introduced the "shuffle" instruction, which can be used to perform a reduction across the 32 threads in a wrap [8]. In a block which has 32*32 threads, we get maximum pixel value in every row using __shfl instruction and copy the value to an array with size 32, which is allocated in shared memory. Then, we only have to make reduction in the shared memory to get the maximum of each block. In order to eliminate the need to invoke multiple kernels, we use atomic operations atomicMax to merge maximum value in every block. The reduction details are show in Fig 4.

3.3. image signature

The main operations in image signature are making DCT and IDCT transformation. Take DCT for example, a two dimension DCT defines as follows:

$$F(u,v) = c(u)c(v)\sum_{i=0}^{N-1}\sum_{j=0}^{N-1} f(i,j)\cos\left[\frac{(i+0.5\pi)}{N}u\right]\cos\left[\frac{(j+0.5\pi)}{N}v\right] \tag{3.1}$$

$$c(u) = \begin{cases}\sqrt{1/N}, & u=0\\ \sqrt{2/N}, & u\neq 0\end{cases} \quad u = 0,1,\ldots,N-1 \tag{3.2}$$

The two dimension approach performs DCT on input sample X by subsequently applying DCT to rows and columns of the input signal. In matrix notation this can be expressed using the following formula:

$$F = AXA^T \quad A(i,j) = c(i)\cos(\frac{j+0.5\pi}{N}i) \tag{3.3}$$

As DCT 8×8 algorithm [9] will loss accuracy of the result, we make DCT transform using matrix multiply instead. CUDA provides cuBLAS library to make matrix multiplying operations with high efficiency. Coefficient matrix are calculated only once and load to device memory in initialization stage. The image signature of each image can be computed very fast with CUDA in later stage.

3.4. multi-streams and multi-GPUs

Stream is a sequence of operations that execute in issue-order on the GPU [10]. SM 2.x-class and later GPUs are capable of concurrently running multiple kernels. To improve the utilization of hardware, we can process a batch of images with starting multiple streams, which would be concurrently running. Comparing to processing one by one, batch operation will reduce average processing time of a single image.

At the same time, multiple GPUs can operate in parallel for enough computational density task. CUDA has supported multiple GPUs since the beginning, and each GPU can be controlled by a separate CPU thread. An amount of images is assigned to multiple GPUs, which highly improve the processing speed.

4. Experiments and analysis

Experiments were performed using two Nvidia TITAN X GPU and one Intel i7-4790K CPU. In order to evaluate the detection performance of our algorithm, we set up a pavement image dataset containing 1000 images. The FOD on the airfield pavement is mainly stones and screws.

The recall and precision are defined as follows in Fig 5.

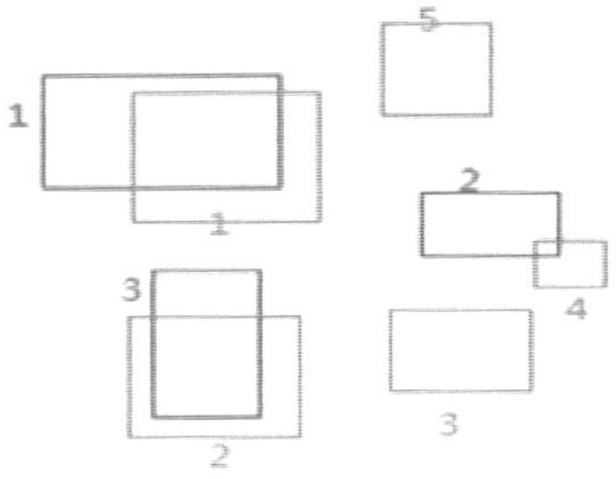

Figure 5. The definition of recall and precision.

Red boxes are real target and green boxes are generated by saliency algorithm. It is regarded as correct detection that the generated box covers the real target mostly. The recall of Fig 5 is 2:3, and the precision is 2:5.

Table 1 gives the detection precision and recall of the saliency detection algorithm.

Table 1. Recall and precision of Hou's model and ours.

	recall	*precision*
Image Signature	87.83%	64.76%
Ours	96.46%	62.93%

The result shows that our algorithm gets higher recall in FOD detection, which guarantees that almost all FOD will be detected in the original image.

As saliency map segmentation and connected domain analysis need complicated logical operation, which are not suitable for transplanting to GPU, we make parallel accelerations on the remaining parts of the detection framework.

We experimented with algorithms by varying number of concurrent streams. Fig. 6 shows the decrement in average executional time as stream size gets varied from 1 to 50. It is often better to set more streams to make full utilization of hardware.

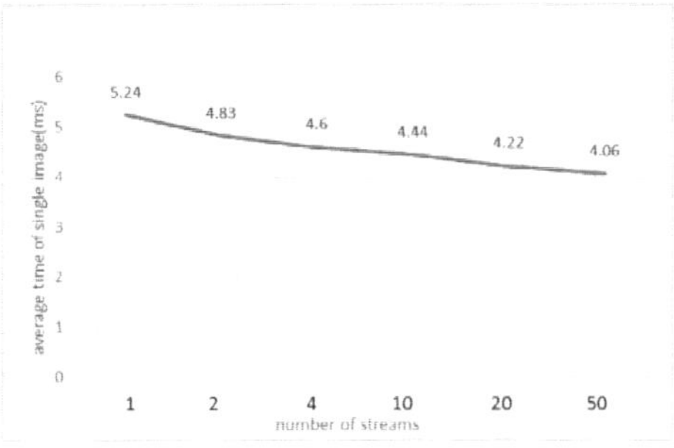

Figure 6. Average processing time of single image with multi-streams

Table 2 shows the average execution time of single image in CPU, one GPU and multiple GPUs.

Table 2. Speed up with GPU.

	Preprocess and image signature time(ms)	Speed up	Saliency map segmentation and connected component analysis time(ms)	Total time	Total Speedup
CPU	287.94	--	8.54	296.48	--
GPU	4.06	70.9	--	12.60	23.5
Two GPUs	2.15	133.9	--	10.69	27.7

Saliency map segmentation and connected component are executed in CPU only. The experiment results show that the GPU implementation parts obtained 70.9 times speedup and get approximately linear speedup in multiple GPUs with 133.9 times speedup. In conclusion, 23.5 times speedup is achieved using single GPU in total algorithm.

5. Conclusion

The paper presents a GPU accelerated visual saliency algorithm, which performs effectiveness and efficiency in FOD detection on airfield pavement. We improve image signature algorithm with pooling operations in generating saliency map and connected component analysis in locating saliency blocks. We make parallel optimizations to some parts of the algorithm with GPU. Experiments result shows that recall rate gets increased and the GPU implementation achieves 23.5 times speedup. The processing time of single image reduce to 12.60ms, which is significant in real-time detection.

References

[1] Y Li and G Xiao 2011 A new FOD recognition algorithm based on multi-source information fusion and experiment analysis *International Symposium on Photoelectronic Detection and Imaging Advances* vol 34 (Infrared Imaging and Applications) pp 112-112

[2] D Liu, X Cao, B Xue and H Li 2013 Feature analysis and detection of FOD under complex background of airport pavement vol 21 (Electronic Design Engineering) pp 12-15

[3] Itti Laurent and Christof Koch 2000 A saliency-based search mechanism for overt and covert shifts of visual attention vol 40(Vision research) pp 1489-1506.

[4] Vinith B, Akhila M K, Naik N, and Rathna G N 2015 GPU Accelerated Face Recognition System with Enhanced Local Ternary Patterns Using OpenCL *Digital Image Computing: Techniques and Applications (DICTA), 2015 International Conference on IEEE* pp 1-7

[5] Hou X, Harel J and Koch C 2012 Image signature: Highlighting sparse salient regions *IEEE transactions on pattern analysis and machine intelligence* pp 194-201

[6] *Cuda C. Programming guide* 2012 pp 1-175

[7] Podlozhnyuk V 2007 *Image convolution with CUDA* (NVIDIA Corporation white paper) p 6

[8] Harris M 2014 *CUDA Pro Tip: Do The Kepler Shuffle* Parallel Forall. Np

[9] Obukhov A and Kharlamov A 2008 *Discrete cosine transform for 8x8 blocks with CUDA* NVIDIA white paper

[10] Wilt Nicholas 2013 *The cuda handbook: A comprehensive guide to gpu programming* Pearson Education

Adaptive classifier for steel strip surface defects

Mingming Jiang[1, a], Guangyao Li[1, b], Li Xie[1, c], Mang Xiao[1, d] and Li Yi[1, e]

[1] College of Electronics and Information Engineering, Tongji University, Shanghai, 201804, PR China

E-mail: [a]stupidhod@hotmail.com, [b]lgy@tongji.edu.cn, [c]robert3443@vip.qq.com, [d]mangxiao@163.com, [e]5yili@tongji.edu.cn

Abstract. Surface defects detection system has been receiving increased attention as its precision, speed and less cost. One of the most challenges is reacting to accuracy deterioration with time as aged equipment and changed processes. These variables will make a tiny change to the real world model but a big impact on the classification result. In this paper, we propose a new adaptive classifier with a Bayes kernel (BYEC) which update the model with small sample to it adaptive for accuracy deterioration. Firstly, abundant features were introduced to cover lots of information about the defects. Secondly, we constructed a series of SVMs with the random subspace of the features. Then, a Bayes classifier was trained as an evolutionary kernel to fuse the results from base SVMs. Finally, we proposed the method to update the Bayes evolutionary kernel. The proposed algorithm is experimentally compared with different algorithms, experimental results demonstrate that the proposed method can be updated with small sample and fit the changed model well. Robustness, low requirement for samples and adaptive is presented in the experiment.

1. Introduction

With the growth competition of the steel strip, the quality of the steel strip has become seriously important. In course of time, steel strip quality and surface quality of structural products have assumed significant importance [1]. Importance of surface quality which demands the effective and efficient methods to replace the conventional artificial visual inspection that the expert only inspect about 0.05% of the total steel surface and easily impacted by fatigue and other unfavorable conditions. The artificial inspection cannot satisfy the quality requirements. Therefore, an automatic accurate real-time steel surface inspection system has become an essential section of the production system. Boukouvalas et al. [2] have researched on ceramic tile automatic inspection using computer vision in early 1995. They used Wigner distribution and spatial frequency methods to detect defects. Boukouvalas et al. [3] described an integrated system developed for the detection of defects on color ceramic tiles in early 1997.

The inspection system for surface defect mainly consists of two parts, defect segmentation and defect processing. Defect segmentation mainly used spatial domain. Spatial domain techniques are fast and easy to implement, and they have been used in most of the papers where suitability for real-time operation. Yun et al. [4] has used first order gradient filter followed by an edge pair search and used morphological dilation and erosion to join edge pair. Finally, they used a defect size thresholding to discard pseudo defects. Liu et al. [5] used gradient operators and region growing to identify defects. JM Blackledge et al. [6] first used a Wiener filter for noise reduction to detect six types of defects in cold strip. Then, they extracted rough locations of defects with Sobel edge detector. Finally, defect locations were identified by a specially developed region growing method. Choi et al. [7] used Laplacian gradient operator for defect detection.

The defect processing contains feature extraction and defect classification. In recent years, abundant researches have been taken by scholars. Different kinds of feature extraction and classification methods have been introduced to classify the surface defect of steel strip. In [8,9], the defects are classified by K-nearest neighbor (KNN) methods with Co-occurrence matrix. Santanu Ghorai et al. [10] described an automated visual inspection system with discrete wavelet transform (DWT) features and support vector machine. Wu et al. [11] described an algorithm with undecimated wavelet transform (UWT) and mathematical morphology to detect geometric defects that achieved a 90.23% accuracy that is hard for real industrial application in 2008. However, in 2013 a noise robust method based on completed local binary proposed by Song [12] has arrived average 98% accuracy and was effective enough applied to real production. From a review of vision-based steel surface inspection system [1], most of systems have achieved higher than 90% accuracy.

To apply to real production, there are still some difficulties for defect detection of steel surface. In real production environment, one product line may produce multiple products, one product may be manufactured in multiple production lines at the same time. As the time gone, the aged equipment and the detection device will change, and the classifier trained by the original database will work with accuracy deterioration. It is hard for the steel company to train every new product with a single database. To train a new model for a new built production line, the production line have to produce without a trained inspection system for a long time until it get enough samples.

Unfortunately, most of the researches focus on the algorithm with enough samples, there are few researches research on train new classifier based on old one. Solly et al. proposed a rapidly evolution system with the expert's feedback. The research describes an adaptive interactive evolution methodology for determining parameters to control segmentation of surface defects on images, but the algorithm do not change the classifier.

Therefore, to solve these issues we proposed an evolutionary classifier with a Bayes kernel (BYEC) evolve with latest batch of samples. To update the classifier, we can concern either modifying the structure of ensemble (weaker components are replaced by base classifiers trained on the most recent data), or updating the aggregation technique (e.g., updating weights).

In this paper, we generate n component SVM classifier with different feature sub-space, then we focus on the topic to generate Bayes kernel sequentially from fixed size blocks of training examples (data chunks). In such ensembles, when a new block arrives, the base SVMs are evaluated and their combination weights are updated. This window techniques provide a simple forgetting mechanism by selecting examples introduced to the learning algorithm, thus eliminating those examples that come from an old distribution.

The rest of the paper is organized as follows: Section 2 depicts the procedures of the surface defects inspection system. Section 3 presents the features we used in this paper. The method to build SVM sub-classifiers on random subspace of features is provided in Section 4. The method to build Bayes kernel and evolve the kernel are discussed in Section 5. Section 6 gives comparative experimental results of this algorithm and research some factors that affect the results. Finally, we conclude this paper in section 7 and give some suggestions for future work.

2. Structure of the evolutionary classifier for steel surface defects

For this evolutionary inspection system for surface defects, it would be trained as the classic classifier. We can update the classifier based on old one with latest samples. The structure of the inspection system is presented in Fig. 1. Our system mechanism mainly contains three processes: defect acquisition, feature extraction and defect classification.

The image acquisition system has been discussed in many literature [2,3,4,5,6,7,13,14,15], it has been a mature field and we will not discuss it in this paper again.

As it illustrated in the Fig. 1, five kinds of features were introduced in this system. We integrated the uniform local binary patterns (ULBP) [16], Gray-level co-occurrence matrix (GLCM) features [17], and Histogram of Oriented Gradient (HOG) feature, Gray level histogram and Gabor filter features. These abundant features guarantee enough information for the classifier. Multiple classifier system was imported to utilize these features.

The classification part has two components described in Fig. 1: base SVM classifiers and Bayes kernel classifier. The fuser Bayes kernel is the key contribution of this system compared with the classic steel surface inspection system. It fuses the result from the base classifiers and update the aggregation parameters.

Support vector machine (SVM) [18] is a popular small sample set learning method, which has a very good performance for pattern classification problems by minimizing the Vapnik-Chervonenkis (VC) dimension and achieving a minimal structural risk. In light of the small sample set requirement, fast learning and good performance, SVM is a good choice for this method. Every SVM classifiers are trained by a subspace of the feature space, and we can change the weights of the base SVM classifiers.

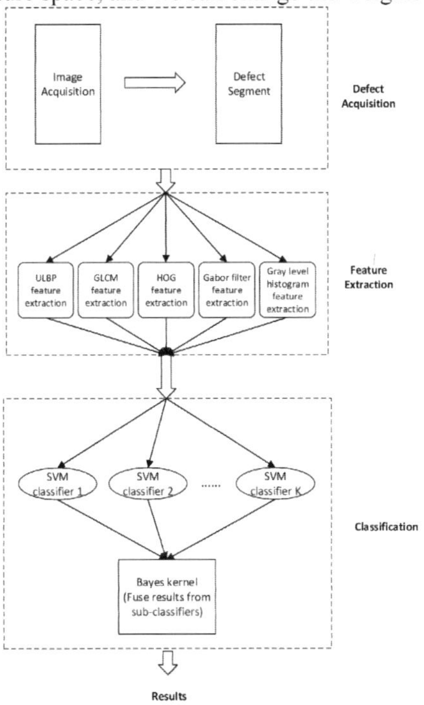

Figure 1. Structure of the evolutionary classifier for steel surface defects with Bayes kernel

Multiple classifier system (MCS) offers many alternatives for unorthodox handing for realistic increasing complex problems [19], it allows us to exploit the potential of the individual classifiers and get an enhanced performance by the combination. It can perform the best individual classifier and easily implemented on parallel, multithreaded and distributed architecture, which is very important for the real time production environment. It is well suited to deal with concept drift that means the statistical dependencies between object features and its classification may change in time that leads accuracy deterioration. A dynamic weighted majority algorithm were described in [20]. We also use an evolvable weighted method to change our classifier with time.

The method we aggregate the base classifiers must be adjusted with small sample. The Bayes classifier is a good candidate as it build a reference from prior probability to posterior probability. We can score the performance of the classifiers on the latest sample. With a small sample, we can get the approximate posterior of the base classifiers and use it to re-weight the integrated classifier to get new one. A Bayes kernel was built as these reasons.

3. Features extraction

The feature extraction is one of the important tasks of surface inspection system, which is mainly to describe the defect characteristics. To avoid the lack of information after integration, we introduce superfluous features to overcome this weakness. Five different kinds of features are extracted in the inspection system, they dedicate to describe the property of texture, color and shape respectively. The feature space consists of Gray-level co-occurrence matrix, uniform local binary pattern, Histogram of Oriented Gradient, Gray level histogram and Gabor filter.

Gray level co-occurrence matrix $N_g \times N_g$ is a matrix where $\mathrm{N_g}$ is the number of gray levels in the image, this matrix reflects the direction, adjacent distance and change range of the image. We used the GLCM feature to cover the texture feature of the defect.

The histogram is a key tool in image processing, it is one of the most useful techniques in gathering information about a matrix. Dalal Navneet and Triggs Bill proposed the Histogram of oriented gradient (HOG) in 2005 [21], this feature is used in computer vision and image processing for object detection, which counts the occurrences of the oriented gradient in an image which have been used in .

Local binary pattern is one of the most successful statistical approaches for texture classification due to its gray-scale and rotation invariance, this feature reflects the local texture of the image.

The Gabor filter is used for edge detection in image processing, the frequency and orientation representations of Gabor filters are similar to those of the human visual system and efficient to describe the texture and shape.

4. The SVM sub-classifiers on random subspace of features

After the feature extraction, we get a feature vector of 212 dimensions. In general, the feature dimensions are too large compared with the scale of the sample that may lead to over-fitting problems. So instead of training one classifier to cover all the feature space, we separate the features with random sampling scheme without replacement and keep almost equal dimensions for every subspace.

In order to overcome the small sample, we introduce support vector machine, which has been widely used in many areas, such as computer vision, natural language processing and Neuroimaging, for the good performance, fast training and the small requirements for labeled samples.

There are six kinds of defects in the database which we used for experiment, as SVM is a binary classifier, we implemented the multiclass classification with the "one-against-one" scheme which is usually applied for binary classifier [22,23]. For every base SVM classifier in this paper, the number of binary SVM classifiers we need is defined as Eq. 1, k is the number of classes exists in data set.

$$n_{SVM} = k(k-1)/2\,,\ where\ k \geq 1. \tag{1}$$

We combine the results from the binary classifier with voting scheme: every binary classifier has a vote and a data point is classified to the class with the maximum number of votes. To solve the clash of same votes, we simply choose the class with greater sequence number. Besides this, we circle increase the sequence number for the classes with every multiclass SVMs to avoid the accumulation deviation, a simple example indicates the risks without this strategy, assume we have three classes, A, B and C, they all have the same accuracy and scale, finally samples classified to A are least.

5. Combination of sub-classifiers by the Bayes kernel

5.1. Bayes aggregation kernel

The combination of the sub-classifiers is very important to this evolutionary classifier because it not only response for improving the performance of the final integrated classifier but also for the ability to update itself to fit the new changed model. Many ensemble methods have been presented [24,25,26], nevertheless, these methods do not suite for this adaptive inspection system. This is because: (1) these methods are sensitive to the size of the training sample set, however, even in a mature production line there are not too many labeled samples, (2) the fusion of classifier may be biased as the combination of samples from changed model.

As these reasons, a new fusion strategy is proposed, native Bayes classifier, which is a highly practical Bayesian learning method which deduce from the Bayesian theorem (Eq. 2), this theorem has been proposed for about three hundred years by Thomas Bayes and developed into a great branch of machine learning, in some domains, it has presented comparable to neural network and the other machine learning methods. The naive Bayes classifier $f(x)$ is described by a conjunction of attribute values when $f(x)$ is limited to a finite set v.

In Bayes learning, the training examples are described by a feature vector $(a_1, a_2, a_3 \dots a_m)^{\mathrm{T}}$, the Bayes classifier make decision based on the probability for every possible value and select the most portable target.

In this model, we assign every base classifier as a feature and described them as a probability matrix. The feature vector of the Bayes vector is defined as $= (D_1, D_2, D_3 \dots D_m)^{\mathrm{T}}$, the D_{i} is the decision made by the i th classifier, and the decision of the Bayes classifier is taken from a finite set $V(v_1, v_2, \dots v_n)$. The prior probability that the ith classifier make decision v_k and the real class is v_j is defined as Eq. 3. $R(r_1, r_2, \dots r_n)$ is the real label set, and c is the number of the samples.

$$P(B_i \mid A) = \frac{P(B_i)P(A \mid B_i)}{\sum_{i=1}^{n} P(B_i)P(A \mid B_i)} \tag{2}$$

$$P(R = v_j \mid D_i = v_k) = \frac{1}{c}\sum_{l=1}^{c}\begin{cases}1 & \textit{if } D_i = v_k \textit{ and } r_l = v_j \\ 0 & \textit{otherwise}\end{cases} \tag{3}$$

We can deduce the post probability of the decision v_j, that the i th classifier make decision v_j, defined by Eq. 4.

$$PM(D_i = v_k \mid R = v_j) = \frac{P(R = v_j \mid D_i = v_k)}{\sum_{l=1}^{n} P(R = v_j \mid D_i = v_l)} \tag{4}$$

The advisable decision number the classifier can make is defined by variable n . With post probability for every individual classifier, the Bayes classifier can take decision based on multinomial model [27]. The naive Bayes classifier simplify assumption that the attribute values are conditionally independent, individual classifiers are independent in this model. The probability of observing the conjunction of classifiers $D_1, D_2, D_3 \dots D_m$ is just the product of the post probability of the individual classifier.

For every class, the probability is calculated with Eq. (5):

$$PV(R = v_i) = \frac{1}{c}\sum_{l=1}^{c}\begin{cases}1 & r_l = v_j \\ 0 & \textit{otherwise}\end{cases} \tag{5}$$

The r_i is the instance of the sample set and the size of sample set is defined as c. For every decision v_i belongs to V, the post probability that v_i is the current answer when the decision vector of sub-classifiers is $d_1 d_2 \dots d_m$ defined as Eq. 6.

$$P(v_i \mid d_1 d_2 \dots d_m) = PV(R = v_i)\prod_{i=0}^{m} PM(d_i \mid v_i) \tag{6}$$

The v_j is one possible label of the finite result set V. We calculate the probability of every $v_i \in \mathrm{V}$ when the decision vector of sub-classifiers is $d_1 d_2 \dots d_m$. The result v of the classifier is the most portable v_i. The Bayes classifier not only takes advantage of the true positive result but also the true negative result from base classifiers.

5.2. Kernel update method

The key structure in this model is the post probability matrix for base classifiers that we trained on labeled samples. To adjust our model, we changed the post probability matrix.

For the application enviroment, the accuracy loss are categorized into sudden and gradual loss. The sudden lose happens as the technological updating, equipments updating or new production line appplied, and the gradual loss happens along the time. For these two scenes, we have different updating schemes.

Let S be the sample set labeled by expert, this BYEC algorithm maintains a slide window and a pool of base SVM classifiers. The S_0 is the origin sample set, and the pool of base SVMs is trained with the origin sample set S_0. After each data chunk of samples labeled, the size of the sample set S increases and the slide window slide to cover the latest samples contain c examples described as Fig. 2. We let B_i to describe the sample set contained by the ith moved slide window, $B_1, B_2, \ldots, B_n$ each containing c examples. For each sample set B_i, the post matrix PM_i is calculated with Eq. 4 and class probability matrix PV_i is calculated with Eq. 5.

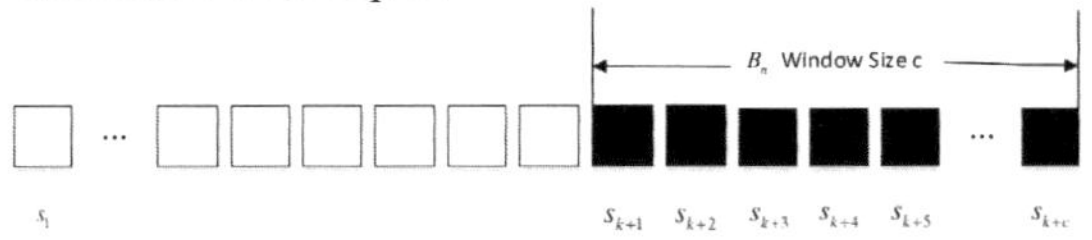

Figure 2. sample set slide window

$PM_i(D_t = v_k | R = v_j)$ denotes the probability that defect instance x is an instance of v_j and classifier D_t predicts its class label v_k. The result of BYEC classifier is generated by Eq. 6 with PM_i and PV_i.

As the B_i is the latest chunk of sample at the slide time, the PM_i most refect the possibility between the base classifiers and the real instance of the defect. For the gradual accuracy lose scene, every time, a chunk of labeled defects inputed to adjust the classifier, the BYEC slide the slide window and generate the new PM and PV and substitutes the old one. For the sudden accuracy lose scene, we clear the slide window and only take the inputed chunk of samples to generate the new PM and PV.

The base classifiers predicts the class of instance x and output the prediction vector $(D_{x1}, D_{x2}, D_{x3} \ldots D_{xm})^T$, the Bayes aggregation kernel predict with Eq. 6 to get the most possible result.

6. Experimental results

To evaluate the effectiveness of the inspection system for surface defects, a surface defect data set was used and described in the subsection. Meanwhile we compared this approach with some other classification methods. In addition, some factors were examined to present how they affecte the classification accuracy.

6.1. Experiment implementation details

The accuracy of this evolutionary classifier has been compared with other classifiers such as SVM [28], NN-BP [29]and KNN [9]. To reveal the fairness of the classifier, a surface defect database, NEU surface defect database [30] was used. There are six kinds of typical defects of the hot-rolled steel strip surface in the database and 1800 grayscale images, 300 samples for every defect: rolled in scale(RS), crazing(CR), inclusion(IN), patches(PA), scratches(SC) and pitted surface(PS).

As Fig. 3 present, the defect image is collected and sampled as 200×200 resolution. The bias between different production lines are mainly caused by electronic circuit noise and sensor noise due to poor illumination and/or high temperature and these factors often lead to Gaussian noise in image acquisition [31], so we add Gaussian noise to the NEU database with different variance to simulate the defect images from different changed production lines. Fourteen contrasting data sets are used with standard deviations $(0 - 13)$, the data with 0 deviation is the original data set, the paired photographs are indicated in Fig. 4.

6.2. Adaptive of the classifier

To evaluate the adaptive of this classifier, one original NEU defect data set and thirteen defect data sets were used. The standard deviations of the noise added to the NEU defect set were $(0 - 13)$, with an equal mean of 0. 212 features were extracted from every defect image.

The BYEC classifier that composed by 25 SVM sub-classifiers were trained by 70% of the original NEU defect data set and the remaining 30% data were used to evaluate the accuracy of the BYEC on the original data set. Then we randomly sampled 10% of the processed data sets to adjust our BYEC classifier. Finally, the accuracy of the adjusted BYEC classifier and the original classifier on the processed data set were tested by the reminder 90% of the processed data. Same as the BYEC, 70% of the original data set were used to train the KNN, BPNN and SVM classifier and 30% were used to evaluate the accuracy on the original data set, then the accuracy of these classifiers on processed data set were test by 90% of the processed data set. The parameters of BPNN and KNN are determined by cross-validation testing. The average accuracy of the classifiers on every data set were run 100 times and the data set were sampled individually.

Fig. 5 depicts the accuracy obtained by KNN, BPNN, SVM, Original BYEC and BYEC classifier. The standard deviation of the defect samples ranges from 0 to 13 and increases by 1 at each iteration, so the first data set is the original data set, the second is added Gaussian noise with standard deviation 1, and so on. The accuracy associated with the SVM is higher than the corresponding values for the KNN, original BYEC and BYEC classifier for the original data set, but the accuracy of the SVM, KNN, BPNN and Original BYEC classifier decline as the standard deviation increases. The accuracy of KNN, SVM and BPNN decline with a nearly 30 degree and the Original BYEC decline a little slower than them.

However, the increase of stand deviation has a little impact on our BYEC classifier. It can be observed that this proposed method performs more adaptive to different standard deviations in comparison with other classifiers. The accuracy of BYEC classifier on the original data is lower than the SVM, the possible reason of this may be that information loss are leaded by the combination of SVM classifiers. The Original BYEC without adjustment process also suggests a relatively high adaptive compared with other classifiers.

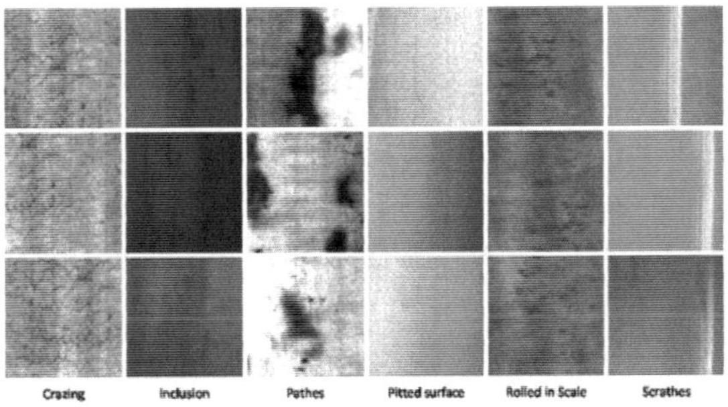

Figure 3. Each row is one of the six typical surface defect on the NEU database that sampling from 300 samples for one class

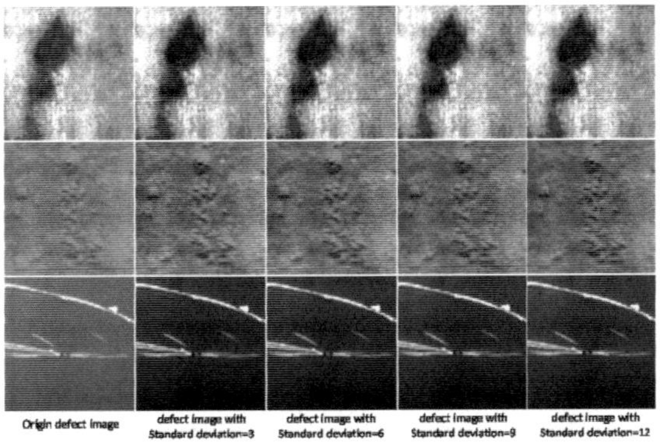

Figure 4. Each row if one defect image

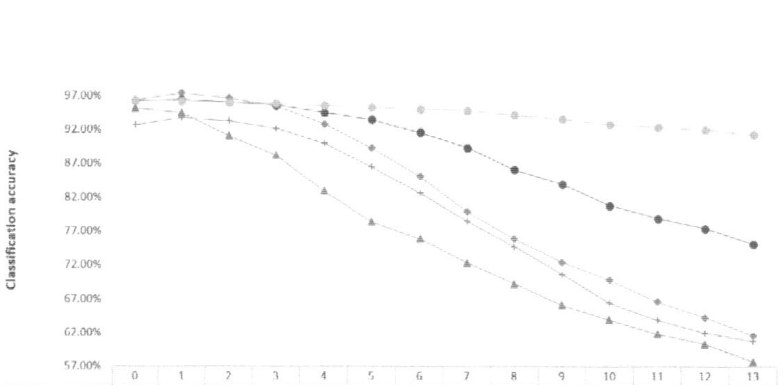

	0	1	2	3	4	5	6	7	8	9	10	11	12	13
Knn	92.68%	93.74%	93.25%	92.14%	89.99%	86.50%	82.63%	78.43%	74.72%	70.66%	66.44%	63.94%	62.02%	60.77%
BPNN	95.16%	94.43%	91.09%	88.22%	82.91%	78.34%	75.86%	72.35%	69.27%	66.11%	63.94%	61.84%	60.34%	57.72%
SVM	96.28%	97.34%	96.66%	95.40%	92.83%	89.29%	85.12%	79.83%	75.85%	72.45%	69.78%	66.61%	64.29%	61.63%
Original BYEC	96.27%	96.41%	95.98%	95.62%	94.49%	93.46%	91.57%	89.31%	86.09%	83.97%	80.80%	78.91%	77.41%	75.12%
BYEC	96.30%	96.17%	95.95%	95.87%	95.55%	95.32%	95.00%	94.81%	94.17%	93.57%	92.79%	92.42%	91.97%	91.27%

Defect samples with different standard deviations

Figure 5. The accuracy of different classifiers on the defect samples with standard deviations $(0-13)$

6.3. *Number of the sub SVM classifiers*

The number of the SVM sub-classifiers is a key parameter for our BYEC classifier which is defined as k, it decides the particle size that our system can be adjusted. The purpose of this section is to examine how the k affects the accuracy.

We trained five BYEC classifier with $k(5,15,25,35,45)$, these classifiers were trained as described in Section 6.2 with original defect set, adjusted with 10% data and test with 90% data. The experimental result is presented in Fig. 6, the highest accuracy on the original defect set is acquired by the classifier with $k=5$, but it reduces fast compared with the other classifiers that perform lowest accuracy on the defect set with highest noise. The classifier, that parameter k equals 25, presents *3th* higher accuracy on the original data set but acquires the highest accuracy on the defect set with highest bias to original data set. This figure suggests that the BYEC is more adaptive with larger k value but lower accuracy will be presented with the BYEC classifier on the original data set.

As the result indicated in Fig. 6, k should be set with the real production environment. The BYEC classifier should be set with larger k value when higher adaptive performance was needed. However, to avoid information loss, we should set the BYEC with a smaller k value when the changed model has a small bias with the old one.

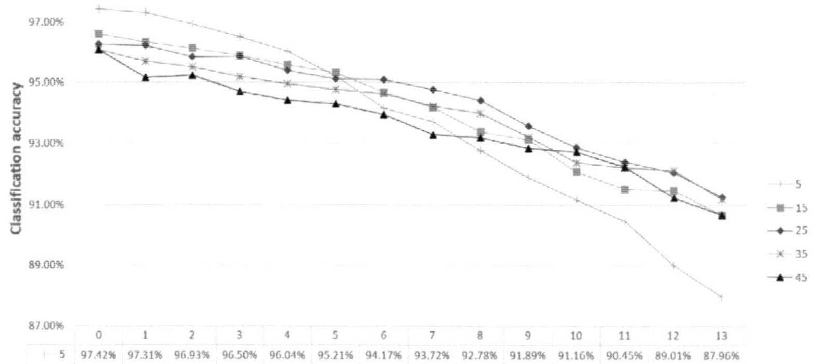

	0	1	2	3	4	5	6	7	8	9	10	11	12	13
5	97.42%	97.31%	96.93%	96.50%	96.04%	95.21%	94.17%	93.72%	92.78%	91.89%	91.16%	90.45%	89.01%	87.96%
15	96.60%	96.34%	96.13%	95.90%	95.60%	95.34%	94.67%	94.18%	93.37%	93.12%	92.08%	91.50%	91.46%	90.66%
25	96.27%	96.23%	95.85%	95.87%	95.41%	95.13%	95.10%	94.76%	94.43%	93.58%	92.87%	92.39%	92.04%	91.25%
35	96.10%	95.72%	95.52%	95.19%	94.95%	94.78%	94.63%	94.24%	94.00%	93.23%	92.37%	92.20%	92.13%	91.19%
45	96.08%	95.17%	95.24%	94.70%	94.42%	94.31%	93.94%	93.29%	93.20%	92.84%	92.72%	92.23%	91.22%	90.67%

Defect samples with different standard deviations

Figure 6. The accuracy of BYEC classifiers with different $k(5,15,25,35,45)$ on the defect samples with standard deviations $(0-13)$

6.4. Set size of the sub SVM classifiers

An important characteristic of classifiers is the size of sample set that used for train. The more samples supplied, more information the classifier machine can learn. To our BYEC classifier, the accuracy of the base classifiers are evaluated more precise with more samples.

The Fig. 7 depicts the accuracy of BYEC classifiers updated by different sizes of samples. All the classifier were performed on the data set described at Section 6.2. We can see that the classifiers trained by 10% reach a lowest accuracy and has least adaptive, but the classifiers trained by 30% reached fairly good performance that the classifiers with more samples have little advantage over this classifier. Even in the defect set with standard deviation 13, the gap between the BYEC classifier trained by 10% and 50% is 1.01%. This illustrated that our BYEC converged very fast and the low requirement for the sample size.

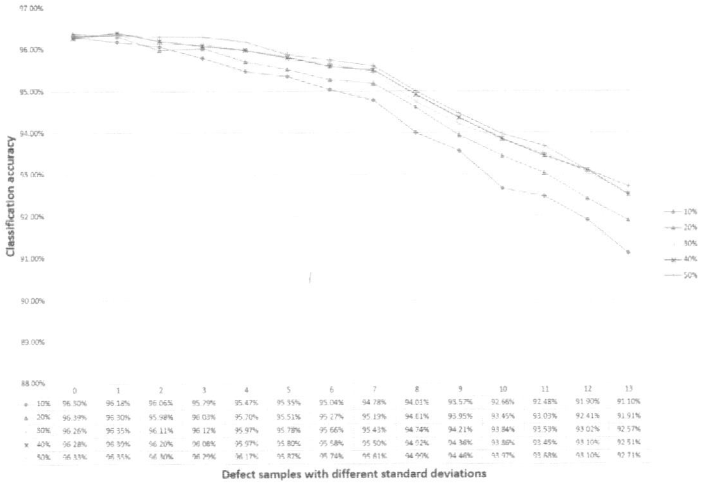

Figure 7. The accuracy of BYEC classifiers adjusted by different

7. Conclusion

Since the accuracy deterioration issue in steel surface inspection system, this research proposes an evolutionary method that can be adjusted with small sample. Firstly, to overcome the information loss, we proposed five kinds of features that cover texture, color and shape respectively. Secondly, random subspace SVM classifiers are proposed to conquer the over-fitting problem. Then, we introduced a Naive Bayes to fuse the results from base SVMs. Finally, we introduced a method to evlvoe the classifier. The experimental result in Section 6.2 indicates that the BYEC algorithm is adaptive with changed steel surface defect data set compared with other algorithms. Research at Section 6.3 suggests that the adaptive of the classifier is highly relative to parameter k, with the growth of the k, the BYEC shows a higher adaptive, but with some accuracy loss. Small sample set requirement was showed in Section 6.4.

We can apply a original BYEC algorithm without any labeled samples to changed model. A new classifier can be trained based on old one with a small sample set and achieve accuracy improvement. The future work is to increase the accuracy both on large sample set and changed production model.

References

[1] N. Neogi, D.K. Mohanta, P.K. Dutta: EURASIP Journal on Image and Video Processing 2014(2014), 1.

[2] Boukouvalas, C., Kittler, J., Marik, R., Mirmehdi, M., & Petrou, M: Proceedings of AMPT95, (1995), 390.

[3] Boukouvalas C, De Natale F, De Toni G, et al., Conference on Quality Control by Artificial

Vision, 97(1997), 49.
[4] JP Yun, C Park, H Bae, H Hwang, S Choi, International Conference on Control, Automation and Systems, 1(2010), 342.
[5] Y Liu, Y Hsu, Y Sun, S Tsai, C Ho, C Chen, The fifth IEEE conference on Industrial Electronics and Applications (ICIEA), (2010), 1667.
[6] JM Blackledge, DA Dubovitskiy, international society for advanced science and technology. J Intell Syst, 3(2008), 76.
[7] SH Choi, JP Yun, B Seo, Y Park, SW Kim, World academy of science, Eng Technol 25,(2007), 66.
[8] F. Dupont, C. Odet, M. Cartont: NDT & E International 30(1997), 3.
[9] C. Unsalan, A. Erci: Recent Advances in Mechatronics. Springer-Verlag Ltd, Singapore, (1999), 468.
[10] S. Ghorai, A. Mukherjee, M. Gangadaran, P.K. Dutta: Instrumentation and Measurement, IEEE Transactions, 62(2013), 612
[11] X.y. Wu, K. Xu, J.w. Xu: CISP'08, Image and Signal Processing, (2008), 528.
[12] K. Song, Y. Yan: Applied Surface Science, 285(2013), 858.
[13] G. Wu, H. Kwak, S. Jang, K. Xu, J. Xu: ICAL 2008 IEEE International Conference, Automation and Logistics, (2008), 2291.
[14] Y.j. Liu, J.y. Kong, X.d. Wang, F.z. Jiang: ICACTE 2010 3rd International Conference, Advanced Computer Theory and Engineering, (2010), 376.
[15] M. Muehlemann, Illumination Technologies, (2000), 1.
[16] T. Ojala, M. Pietik¨ainen, T. M¨aenp¨a¨a: Pattern Analysis and Machine Intelligence, IEEE Transactions, 7(2002), 971.
[17] R.M. Haralick, K. Shanmugam, I.H. Dinstein: Systems, Man and Cybernetics, IEEE Transactions, 6(1973), 610.
[18] V.N. Vapnik, V. Vapnik: Statistical learning theory, 1(1998), 1.
[19] M. Wo´zniak, M. Gra˜na, E. Corchado: Information Fusion, 16(2014), 3.
[20] J.Z. Kolter, M. Maloof, et al: Data Mining, 2003. ICDM 2003. Third IEEE International Conference on, (2003), 123.
[21] N. Dalal, B. Triggs: Computer Vision and Pattern Recognition, 2005. CVPR 2005.IEEE Computer Society Conference on, 1 (2005), 886.
[22] U.H.G. Kreßel: Advances in kernel methods, (1999), 255.
[23] C.W. Hsu, C.J. Lin: Neural Networks, IEEE Transactions on, 13 (2002), 415.
[24] M. G¨oksedef, S¸. G¨und¨uz-O˘g¨ud¨uc¨u: Expert Systems with Applications, 37(2010), 2911.
[25] C. Porcel, A. Tejeda-Lorente, M. Mart´ınez, E. Herrera-Viedma: Information Sciences 184(2012), 1.
[26] C. Cabral, M. Silveira, P. Figueiredo: Pattern Recognition, 45(2012), 2064.
[27] M. Vangelis: CEAS 2006 Third Conference on Email and AntiSpam (CEAS 2006), Mountain View, California USA, (2006)
[28] Y.J. Jeon, D.c. Choi, J.P. Yun, C. Park, S.W. Kim: ICCAS 2011 11th International Conference, Control, Automation and Systems, (2011), 1274.
[29] M. Yazdchi, M. Yazdi, A.G. Mahyari: 2009 International Conference on (IEEE, 2009), Digital Image Processing, (2009), 346.
[30] J.J. Henriksen: South Denmark University, (2007).
[31] R.C. Gonzalez: Digital image processing, Pearson Education India, (2009).

IOP Conf. Series: Journal of Physics: Conf. Series **787**(2017) 012020 doi:10.1088/1742-6596/787/1/0120

Weighted Discriminative Dictionary Learning based on Low-rank Representation

Heyou Chang and Hao Zheng

Key Laboratory of Trusted Cloud Computing and Big Data Analysis, School of Information Engineering, Nanjing XiaoZhuang University, Nanjing, China

E-mail: cv_hychang@126.com, zhh710@163.com

Abstract. Low-rank representation has been widely used in the field of pattern classification, especially when both training and testing images are corrupted with large noise. Dictionary plays an important role in low-rank representation. With respect to the semantic dictionary, the optimal representation matrix should be block-diagonal. However, traditional low-rank representation based dictionary learning methods cannot effectively exploit the discriminative information between data and dictionary. To address this problem, this paper proposed weighted discriminative dictionary learning based on low-rank representation, where a weighted representation regularization term is constructed. The regularization associates label information of both training samples and dictionary atoms, and encourages to generate a discriminative representation with class-wise block-diagonal structure, which can further improve the classification performance where both training and testing images are corrupted with large noise. Experimental results demonstrate advantages of the proposed method over the state-of-the-art methods.

1. Introduction

Low-rank representation has been widely used in computer vision and pattern recognition due to its strong robustness to the noise of corrupted observation data, such as occlusions, lighting variations and pixel corruptions. The basic idea of low-rank representation is to decompose the observation data into a sparse noise matrix and a low-rank matrix, where the low-rank matrix is often described as the linear combinations of dictionary atoms. The dictionary plays an important role on the discrimination of the representation. In the work of [1], the dictionary is pre-specified as original training set. However, with this simple strategy, the training data cannot faithfully represent the test samples due to the noise, and the discriminative information contained in training data cannot be well discovered. Therefore, how to adaptively learn a dictionary from training data has attracted much attention.

Focusing on this issue, a number of dictionary learning methods for face recognition and image classification have been proposed[2]-[6].To deal with the situation where training data are contaminated with large noises, Ma et al presented a discriminative low-rank dictionary for sparse representation (DLRD_SR)[7] algorithm by minimizing the rank of each sub-dictionary. Based on DLRD_SR, Li et al proposed discriminative dictionary learning with low-rank regularization ($D^2L^2R^2$)[8] for image classification. Obviously, these two methods only explore structural information class by class locally and cannot capture the global structure embedded in all classes. To explore the global structure, Zhang et al proposed a discriminative structured low-rank method (DSLR)[9]. For the purpose of acquiring the structure information in the whole database, including training data and testing data, Li et al proposed a semi-supervised low rank matrix recovery algorithm for robust face recognition by exploiting the class-wise block-diagonal structure (LR-CBDS)[10]. Although improved performance has been reported, they cannot effectively exploit the discriminative information between data and dictionary. For example, [9] constructed an ideal-code regularization term, which potentially means that all samples from the same class have the same representation.The term may bring negative effects to the representations of the data. In [10], representations of training samples and testing samples are learnt simultaneously, which is inefficient when the dataset is large or dynamic updating. Moreover, the dictionary in [10] is fixed as training samples, which also restricts the performance when training samples are heavily corrupted by noise.

In this paper, we propose weighted discriminative dictionary learning based on low-rank representation (WDDL_LRR) method. In WDDL_LRR, a weighted regularization term on the data coefficients is introduced to strength the discriminative ability of the dictionary. The regularization term not only contains global structure information between training data and dictionary, but also encourages that the class-specific dictionary is able to well reconstruct the samples for the same class with bad representation ability for the samples from other classes. Benefitting from the weighted regularization term, the representations learned by WDDL_LRR own class-wise block-diagonal structure, which is critical for the classification problem. Extensive experiments show that the proposed approach has better performance than existing state-of-the-art dictionary learning methods.

2. Weighted discriminative dictionary learning based on low-rank representation

With respect to the semantic dictionary$\boldsymbol{D}$, the optimal coefficient matrix **Z** should have class-wise block-diagonal structure as follows:

$$\boldsymbol{Z}^* = \begin{bmatrix} \boldsymbol{Z}_1^* & 0 & 0 & 0 \\ 0 & \boldsymbol{Z}_2^* & 0 & 0 \\ 0 & 0 & \cdots & 0 \\ 0 & 0 & 0 & \boldsymbol{Z}_C^* \end{bmatrix}$$

The coefficient matrix **Z** should be low-rank and sparse[1]. To capture the class-wise block-diagonal structure, we introduce a discriminative regularization term,$r(\mathbf{Z}) = ||\boldsymbol{W} \odot \boldsymbol{Z}||_F^2$, where $\odot$ means the element-wise multiplication operator and $\boldsymbol{W} = [\boldsymbol{w}_1, \dots, \boldsymbol{w}_N]$ is a weighted matrix. $\boldsymbol{W}$ is defined as

$$\boldsymbol{W}(i,j) = \begin{cases} 0, \text{if } \boldsymbol{d}_i \text{ and } \boldsymbol{x}_j \text{ belong to the same class} \\ 1, \text{otherwise} \end{cases}$$

where $\boldsymbol{d}_i$ is the i-th column of **D**, and $\boldsymbol{x}_j$ is the j-th column of the data matrix **X**. This term encourages that $\boldsymbol{Z}_i^j (i \neq j)$ to be 0s and $\boldsymbol{Z}_i^i$ to be large values, which leads $\sum_{j \neq i} ||\mathbf{D}_j \boldsymbol{Z}_i^j||_F^2$to be 0 and $\boldsymbol{Z}_i \approx \boldsymbol{D}_i \boldsymbol{Z}_i^i$. Namely, the class-specific dictionary should be able to reconstruct the samples from the same class and doesn't have the capacity to represent samples from other classes. This is different from the last term in[7], because the last term in[7] requires $\sum_{j=1,j\neq i}^{C} ||\boldsymbol{D}_i \boldsymbol{Z}_j^i||_F^2$being 0, which cannot lead $\boldsymbol{Z}_i^j (i \neq j)$ to be 0 in reverse. Besides, $\boldsymbol{W}$ has class-wise structure and contains the global structure information of training samples and dictionary atoms, which can improve the discriminative ability of dictionary and representation. Based on the above discussion, the final target function is formulated as follows

$$min_{\boldsymbol{D},\boldsymbol{Z},\boldsymbol{E}} ||\boldsymbol{Z}||_* + \lambda ||\boldsymbol{E}||_1 + \beta ||\boldsymbol{Z}||_1 + \alpha ||\boldsymbol{W} \odot \boldsymbol{Z}||_F^2 \quad \text{s.t.} \boldsymbol{X} = \boldsymbol{D}\boldsymbol{Z} + \boldsymbol{E} \tag{1}$$

where λ, β and α are three parameters.

3. Optimization

To solve optimization problem (1), we introduce two auxiliary variables $\boldsymbol{J}$ and $\boldsymbol{L}$ to make the objective function separable. Problem (1) can be rewritten as：

$$min_{D,Z,E}||J||_* + \lambda||E||_1 + \beta||L||_1 + \alpha||W \odot Z||_F^2 \text{ s.t. } X = DZ + E, Z = J, Z = L \quad (2)$$

The optimization of (2) can be solved by ADMM[11]. The updating scheme is as follows.

Updating J Fix the other variables and solve the following problem

$$J^{t+1} = argmin_J \frac{1}{\mu^t}||J||_* + \frac{1}{2}||J - (Z^t + Y_2^t/\mu^t)||_F^2 = \mathbf{U}\Phi_{1/\mu^t}(\mathbf{\Sigma})V^T \quad (3)$$

where $\mathbf{U\Sigma}V^T$ is the singular value decomposition of the matrix $Z^t + Y_2^t/\mu^t$, Y_2^t is a Lagrange multiplier, $\mu^t > 0$ is a penalty parameter and $\Phi_\varsigma(\cdot)$ is the soft-thresholding operator defined as follows [12]

$$\Phi_\varsigma(x) = \begin{cases} x - \varsigma, \text{if } x > \varsigma \\ x + \varsigma, \text{if } x < -\varsigma \\ 0 \quad, \text{otherwise} \end{cases}$$

Updating Z Fix the other variables and solve the following problem

$$Z^{t+1} = argmin_Z \alpha||W \odot Z||_F^2 + \frac{\mu^t}{2}(||(X - E^t + Y_1^t/\mu^t) - D^tZ||_F^2 + ||Z - (J^{t+1} - Y_2^t/\mu^t)||_F^2 + ||Z - (L^t - Y_3^t/\mu^t)||_F^2) \quad (4)$$

In order to update Z as a whole, we denote $||W \odot Z||_F^2 = ||Z - B^t||_F^2$, where $B^t = Q \odot Z^t, Q = S - W$, $S \in R^{K \times N}$ is a matrix whose elements are all 1s. Thus, Z^{t+1} can be updated by

$$Z^{t+1} = \left(D^{t^T}D^t + (\alpha/\mu^t + 2)I\right)^{-1}(D^{t^T}(X - E^t) + L^t + J^{t+1} + (\alpha B^t + D^{t^T}Y_1^t - Y_2^t - Y_3^t)/\mu^t) \quad (5)$$

Updating L Fix the other variables and solve the following problem

$$L^{t+1} = argmin_L \frac{\beta}{\mu^t}||L||_1 + \frac{1}{2}||L - (Z^{t+1} + Y_3^t/\mu^t)||_F^2 = \Phi_{\beta/\mu^t}(Z^{t+1} + Y_3^t/\mu^t) \quad (6)$$

Updating E Similar with **L**, **E** is updated by

$$E^{t+1} = \Phi_{\lambda/\mu^t}(X - D^tZ^{t+1} + \frac{Y_1^t}{\mu^t}) \quad (7)$$

Updating D The updating scheme of D is

$$D^{t+1} = \gamma D^t + (1 - \gamma)\frac{1}{\mu^t}(Y_1^t + \mu^t(X - E^{t+1}))Z^{t+1^T}(Z^{t+1}Z^{t+1^T})^{-1} \quad (8)$$

where $\gamma > 0$ is a parameter that controls the updating step. It is experimentally set 0.1.

4. Experiments and results

We evaluate our approach on three public databases, in which both training and testing images are corrupted, including illumination changes, occlusions and block noises. Our approach is compared with several latest related works, including FDDL[6], DLRD_SR[7], $D^2L^2R^2$[8], DSLR[9], SC-LCPSR[10] and DODL[13]. The parameters α, λ andβ in the dictionary learning phase are set as 0.5, 0.1 and 0.01 respectively. We repeat each experiment 5 times and report the average accuracy.

4.1. Classification

We use a linear classifier for classification as in [9][10]. The linear classifier is learnt as follows

$$\widehat{W} = argmin_W||H - WZ||_F^2 + \kappa||W||_F^2 \quad (9)$$

where H is class label matrix, κ is the weight of regularization term. Given the test data X_{test}, the representations Z_{test} are calculated by solving (1) with α=0. The label for test sample i is given by

$$l = argmin_l(s = \widehat{W}z_i) \quad (10)$$

where s is the class label vector, z_i is the i-th column vector in Z_{test}.

4.2. AR database

A subset from AR database [14] is used, which consists of 2600 images, corresponding to 100 individuals (50 males and 50 females). For each individual, there are two sessions, including 14 non-occluded images with different facial expressions or illumination variations and 12 occluded images (6 by sunglasses and 6 by scarves). Experiments are performed under three different scenarios.

Sunglasses: Seven non-occluded images and one sunglasses image from session 1 are chosen as training samples, the rest seven non-occluded images and five sunglasses images as testing.

Scarf: Seven non-occluded images and one scarf image from session 1 are chosen as training samples, the rest seven non-occluded images and five scarf images as testing.

Mixed (Sunglasses + Scarf): Seven non-occluded images, one sunglasses image and one scarf image from session 1 are chosen as training samples, the rest are used for testing.

The comparison of different methods on AR database is listed in table 1. Our approach achieves the best results and outperforms the state-of-art method by **0.4%** for the sunglasses scenario, **0.9%** for the scarf scenario, and **0.8%** for the mixed scenario.

Table 1. The classification accuracy (%) on AR database.

	FDDL	DLRD_SR	SDLR	DODL	$D^2L^2R^2$	SCLRDL	**Ours**
Sunglasses	89.8	91.7	86.8	91.6	91.4	89.6	**92.1**
Scarf	90.5	89.4	73.7	90.1	90.4	87.2	**91.4**
Mixed	89.8	88.6	74.4	87.4	89.1	84.1	**90.6**

4.3. Extended YaleB database

The Extended YaleB database[15] contains 2,414 frontal-face images of 38 individuals. We first resize the image size to 48×42, and then randomly select half images for each person as training images and the rest as testing images. Comparison of different methods on Extended YaleB database is listed in table 2. It can be seen that our approach outperforms other competing methods. It achieves **1.9%** improvement over FDDL, and **0.5%** improvement over low-rank based dictionary learning methods.

We also present visualization of some decomposition results of testing image in figure 1(a). The top four images show the original images, while the middle four images and the bottom four images are corresponding to the low-rank recovery term $\boldsymbol{DZ}$ and the sparse error term $\boldsymbol{E}$. It turns out that illumination changes can be removed as sparse noises and the diversity of images are preserved well. Figure 1(b) shows representations of the first five class testing samples corresponding to the first five class reconstruction bases. The learned representation has class-wise block-diagonal structure.

Table 2. The classification accuracy (%) on Extended YaleB database.

FDDL	DLRD_SR	SDLR	DODL	$D^2L^2R^2$	SCLRDL	**Ours**
97.2	98.2	96.8	98.6	95.5	97.8	**99.1**

4.4. ORL database

The ORL database [16] contains 400 images of 40 individuals. A randomly located block of each image is replaced with an unrelated random image. For each individual, half images are randomly chosen as the training samples, and the rest are used as testing samples. Experiments are performed under different levels of block-corruption, which is measured by the area proportion between the block and the whole image. Comparison of different methods on ORL database is listed in table 3. When images are clear, FDDL achieves the best performance. With the corruption area increasing, WDDL_LRR performs better than other methods, and achieves **0.1%**, **0.8%**, **1.0%**, **1.5%** and **5.9%** improvement respectively.

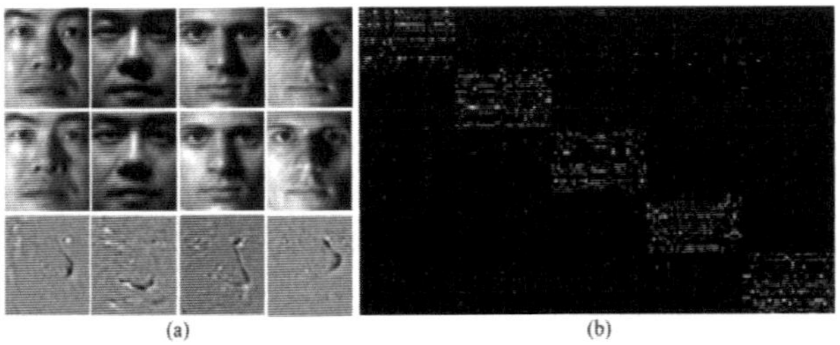

Figure 1.(a) Decomposition results. (b) Representations of testing images.

5. Conclusion

This paper presented weighted discriminative dictionary learning based on low-rank representation, which has a strong discriminative capability for image classification, especially, when the images are corrupted with big noise. By taking the class-wise block-diagonal structure of representation into consideration, a weighted representation term is constructed to enhance the discriminative power of the dictionary. The WDDL_LRR was extensively evaluated on several public databases. The experimental results confirmed that WDDL_LRR outperforms some state-of-the-art methods.

Table 3. The classification accuracy (%) on ORL database.

Corruption	FDDL	DLRD_SR	SDLR	DODL	$D^2L^2R^2$	SCLRDL	**Ours**
0	**96.7**	93.1	91.8	91.0	94.3	95.1	94.3
10	86.8	90.4	84.3	78.0	91.1	91.2	**91.3**
20	74.4	81.2	78.3	59.0	82.6	80.6	**83.4**
30	61.8	76.5	66.7	40.3	77.2	63.7	**78.7**
40	49.1	67.8	56.2	28.7	68.9	50.5	**74.8**
50	36.8	57.2	47.8	22.0	59.8	39.2	**68.3**

References

[1] Guangcan L, Zhouchen L, Shuicheng Y, Ju S, Yong Y and Yi M. 2013.Robust recovery of subspace structures by low-rank representation.*IEEE Transactions on PAMI*,**35(1)**, pp 171–184.

[2] Chih C, Chia W and Yu W. 2012. Low-rankmatrix recovery with structural incoherence for robust facerecognition. *CVPR*.

[3] Alhussein F, Mike D and Pascal F. 2015. Dictionary learning for fast classification based on soft-thresholding. *International Journal of Computer Vision*, **114(2-3)**, pp 306-321.

[4] Meng Y, Dengxin D, Linlin S and Luc G. 2014. Latentdictionary learning for sparse representation based classification.*CVPR*.

[5] Zhuolin J,Zhe L and Larry S D. 2013. Label consistentk-svd: Learning a discriminative dictionary for recognition.*IEEE Transactions on PAMI*,**35(11)**, pp 2651–2664.

[6] Meng Y, Lei Z,Xiangchu F and David Z. 2014. Sparserepresentation based fisher discrimination dictionarylearning for image classification. *IJCV*,**109**, pp 209–232.

[7] Long M, Chunheng W, Baihua Xand Wen Z. 2013. Sparserepresentation for face recognition based on discriminativelow-rank dictionary learning. *CVPR*.

[8] Sheng L,Liangyue L and Yun F. 2014. Low-rank and sparse dictionarylearning. *Image and Vision Computing*, pp61–85.

[9] Yangmuzi Z, Zhuolin J and Larry S D. 2013. Learning structuredlow-rank representations for image classification. *CVPR*.

[10] Yong L, Jing L, Zechao Li, Yangmuzi Z, Hanqing L and Songde Ma. 2014.Learning low-rank representations with class-wise block-diagonalstructure for robust face recognition. *AAAI*.

[11] Stephen B, Neal P, Eric C, Borja P and Jonathan E. 2011. Distributed optimization and

statistical learning via the alternating direction method ofmultipliers, *Foundation and Trends in Machine Learning*,**3** (**1**),pp 1-122

[12] Zhouchen L, Minming C and Yi M. 2010. The augmented lagrangemultiplier method for exact recovery of corruptedlow-rank matrices. *Technical report, UIUC.*

[13] Zhen D, Mingtao P and YundeJ. 2015. Discriminative orthonormal dictionary learning for fast low-rank representation. *ICONIP*.

[14] AleixM and Robert B. 1998. The AR face database.*Technical report, CVC.*

[15] Kuang C L, Jeffrey H and David J K. 2005. Acquiringlinear subspaces for face recognition under variable lighting.*IEEE Transactions on PAMI*,**27**(**5**):684–698.

[16] FerdinandoS and ACHarter. 1994. Parameterisation of astochastic model for human face identification. *WACV*,138 – 142.

Morphology operators construction by adaptive elliptical structuring elements based on nonlinear structure tensor

Chunming Tang, Xinlei Liu, Yanjie Li and Hongbo Zhao

School of Electronics and Information Engineering, Tianjin Polytechnic University, Tianjin 300387, China

E-mail: Tangchunminga@hotmail.com

Abstract. As the linear structure tensor is tending to inaccurately or even wrong estimate the gradient direction of a gray-level image, we present a novel algorithm to construct adaptive elliptical structuring elements via estimating the local anisotropy of an image based on the nonlinear structure tensor. Erosion, dilation, opening, closing and Hit-or-Miss transform are redefined according to the presented structuring elements, which have been applied to some representative images. The processed results and the quantitative analysis show that the novel morphological operators have more advantages in structure adaptation, corner protection, filtering and targets extraction than the others.

1. Introduction

When traditional mathematical morphology is applied to image processing, the shape and size of a structuring element (SE) are fixed, which usually depend on the global feature of a gray-level image. The performance is usually unsatisfactory. Adaptive SEs are then appeared to improve it, whose size and shape can be adjusted depending on the local feature of an image. So far, a variety of methods are proposed. Curic and Landström[1] summarized the following three groups, based solely on connected homogeneous areas in the image, based on weighted combinations of gray-level and spatial information, and structure-based methods. The first one, such as general adaptive neighborhoods, was presented by Debayle and Pinoli[2, 3], which the SEs have no shape and size constraints, the only parameter $m \in Z^+$ controls the homogeneity of the SEs. Hence, this approach has no relationship with the spatial relations of pixels. The second one presents the spatial constraints, taking the geometry of shapes into account, such as the morphological amoebas introduced by Lerrallut et al[4], which applied a weighted gray-level distance to restrict the shape of amoebas. But it is challenging for morphological amoebas to set an explicit weight parameter λ. Amoebas is also sensitive to the small contrast changes that can effect the filtering results. Recently, Curic et al.[5, 6] followed the similar principle and proposed the salience adaptive structuring elements(SASEs) derived from a salience distance map of the original image. The SASEs can cross edges but can not handle the corners. The third one, such as the elliptical adaptive structuring elements (EASEs)[7] introduced by Landström, is based on the Linear Structure Tensor(LST). And the shapes of EASEs vary from lines to disks depending on the rate of anisotropy. However, the LST after Gaussian smoothing, has an important drawback, which is well known, blurring image details, especially the pixels closing to the boundary of two different objects.

Among these approaches, EASEs appear promising because of fully using the anisotropy. Therefore, this paper will present a novel method, which the adaptive elliptical SEs are defined based on the nonlinear structure tensor(NLST)[8] instead of the LST, as the NLST can obtain more accurate information about both local orientation and degree of the anisotropy. Meanwhile, different from reference [7], our novel method will apply the corner and edge strength defined by the eigenvalues of the NLST in order to distinguish the different parts of an image. In addition, it may improve the resolution of the elliptical SEs to make them close to the ideal ellipse, which force us to improve the resolution of the image simultaneously.

The rest of the paper is organized as follows, section 2 introduces the concept of NLST. Section 3 defines the adaptive SEs. Section 4 introduces how to improve the resolution of image and elliptical SEs. Section 5 defines some morphology operators. Experimental results and comparison are given in section 6.

2. Nonlinear structure tensor

It is well-known that Gaussian convolution (GC) is equivalent to linear diffusion. It is natural to address the limitations of GC by using anisotropic diffusion[9,10], that is, smoothing the discontinuities. Let I(x, y) be a 2-D gray-level image. Then the NLST is defined as equation (1):

$$\frac{\partial u_{i,j}}{\partial t} = div(g(\sum_{k,l=1}^{2}\nabla u_{k,l}\nabla u_{k,l}^{\mathrm{T}})\nabla u_{k,l}) \qquad (i,j=1,2) \tag{1}$$

Where, $u_{i,j}$ is a component of the ST. t is an iterative times. $g(\cdot)$[11] is a decreasing diffusion function, which can be rewritten as equation (2):

$$g(\sum_{k,l=1}^{2}\nabla u_{k,l}\nabla u_{k,l}^{\mathrm{T}}) = Vdiag(g_1(\lambda_1), g_2(\lambda_2))V^{\mathrm{T}} = g_1(\lambda_1)\theta_1\theta_1^{\mathrm{T}} + g_2(\lambda_2)\theta_2\theta_2^{\mathrm{T}} \tag{2}$$

Where $\lambda_1 \geq \lambda_2$, which are the two eigenvalues of ST. θ_1 and θ_2 are a pair of orthogonal basis of ST. θ_1 represents the direction of the maximal gradient variation, while θ_2 is orthogonal to θ_1. $g_1(\lambda_1)<g_2(\lambda_2)$ can ensure anisotropic diffusion, rewritten in the (3), (4).

$$g_1(\lambda_1) = \frac{1}{(1+\lambda_1+\lambda_2)^{1.5}} \tag{3}$$

$$g_2(\lambda_2) = \frac{1}{(1+\lambda_1+\lambda_2)^{0.5}} \tag{4}$$

The analysis of the eigenvalues η_1, η_2 from iteration termination of NLST provides local anisotropic features of images. We apply two parameters[12]: corner strength m_C and edge strength m_E, in equations (5) and (6), to describe the regions of interest (ROIs).

$$m_C = \frac{4\eta_1\eta_2}{\eta_1+\eta_2} \tag{5}$$

$$m_E = \eta_1 - \eta_2 \tag{6}$$

Different ROIs can be distinguished by analyzing m_C and m_E. It may be an edge when $m_E >> 0$ and $m_C \approx 0$, or be a homogeneous region when $m_E \approx 0$ and $m_C \approx 0$, or be a corner, when $m_E \approx 0$ and $m_C >> 0$.

3. Adaptive elliptical structuring elements based on nonlinear structure tensor

We define an elliptical SE: $SE(x, y)=[a(x, y), b(x, y), \varphi(x, y)]$, where $a(x, y)$, $b(x, y)$, and $\varphi(x, y)$ are the semi-major axis, the semi-minor axis and the orientation of a SE respectively to a pixel $P(x, y)$. The three parameters values should obey the following three constrains:

1). If pixel (x, y) is at a corner, $a(x, y)$ and $b(x, y)$ should be very small.

2). If pixel (x, y) is at an edge, $a(x, y)$ should be very large, and $b(x, y)$ is very small. The shape of the elliptical SE is close to the line, which direction is perpendicular to the gradient direction.

3). If pixel (x, y) is in a homogeneous region, $a(x, y)$ and $b(x, y)$ should be large. And the shape of the elliptical SE is close to a disk.

We redefine $a(x, y)$and $b(x, y)$ in (7) and (8):

$$a(x,y) = r \cdot \left\{ 1 - \exp\left(-\frac{C_m}{(m_C(x,y)/\beta_1)^m}\right) \right\} \tag{7}$$

$$b(x,y) = a(x,y) \cdot \left\{ 1 - \exp\left(-\frac{C_m}{(m_E(x,y)/\beta_2)^m}\right) \right\} \tag{8}$$

Where r denotes the maximal allowed semi-major axis. β_1 and β_2 are the normalized parameters which are set to be 75% of the maximum of m_C and m_E, respectively. m and C_m are constrained via (9)[12].

$$1 - \exp(-C_m)(1 + mC_m) = 0 \tag{9}$$

Here, we set $m = 1.1$ and $C_m = 0.1877$.

The orientation $\varphi(x, y)$ is defined by the eigenvector[7] of the NLSE in equation (10).

$$\varphi(x,y) = \begin{cases} \arctan\left(\frac{v_{2,y}(x,y)}{v_{2,x}(x,y)}\right), & v_{2,x}(x,y) \neq 0 \\ \pi/2 \ , & v_{2,x}(x,y) = 0 \end{cases} \tag{10}$$

4. Resolution improvement

As the parameters of an elliptical SE being calculated by equations (7), (8) and (10) are continuous, the discretization should be followed. But the discretized SE has undesirable saw-toothed edge. So we have to improve the resolution of the discrete SE and the original image.

Discrete elliptical SE is defined by a set **S** which contains integer coordinates $X(x\_1, x\_2)$, $X \in \mathbf{S}$, all X should satisfy the inequation (11):

$$\left(\frac{x \cdot \cos\varphi - y \cdot \sin\varphi}{a}\right)^2 + \left(\frac{x \cdot \sin\varphi - y \cdot \cos\varphi}{b}\right)^2 \leq 1 \tag{11}$$

In order to improve the resolution of the discrete elliptical SE, we subdivide coordinate $X(x\_1, x\_2)$ into $X'(d/2^k, d/2^k)$. Where d is an integer, and $-r \times 2^k \leq d \leq r \times 2^k$.

A simple and effective method to improve the image resolution is interpolation. But some general methods, such as the bilinear interpolation, are easy to lose the texture feature. As random fractal interpolation[13] can fully express the statistical texture features of the image, we apply it to improve the resolution of images. Random fractal interpolation is an iterative process, which is ended till the desired spatial resolution is required. When the kth iteration, the gray value of the pixel $P'[2^k x-(2^k-1), 2^k y-(2^k-1)]$ in the interpolated image I_H is the value of the pixel $P(x, y)$ in the original image I, the rest pixels' gray values in I_H are obtained by computation[13].

5. Adaptive morphology operators

5.1. Erosion, Dilation, Opening and Closing

The adaptive elliptical SE: $SE(x, y)=[a(x, y), b(x, y), \varphi(x, y)]$ can be calculated by equations (1)-(10) for any pixel $P(x, y)$. Interpolated SE(x, y) is called *ESE*. Only the pixel $P'[2^k x-(2^k-1), 2^k y-(2^k-1)]$ in I_H has its *ESE*, while other pixels in I_H have no corresponding *ESE*. Thus, only pixels P' in I_H will be processed when morphological operators are applied. Our novel adaptive erosion, dilation, opening and closing are defined in (12-15), respectively.

$$IE(x,y)=(\varepsilon(I_H))(i,j)=\underset{(p,q)\in ESE(i,j)}{\wedge} I_H(p,q),\quad (p,q)\in D' \tag{12}$$

$$ID(x,y)=(\delta(I_H))(i,j)=\underset{(p,q)\in ESE^C(i,j)}{\vee} I_H(p,q),\quad (p,q)\in D' \tag{13}$$

$$IO(x,y)=\gamma(I_H)(i,j)=(\delta(\varepsilon))(I_H)(i,j) \tag{14}$$

$$IC(x,y)=\psi(I_H)(i,j)=(\varepsilon(\delta))(I_H)(i,j) \tag{15}$$

Where D' is the domain of I_H, D is the domain of I. $i=2^kx-(2^k-1)$, $j=2^ky-(2^k-1)$, $(x, y) \in D$, $(i,j) \in D'$. $ESE^C(i,j)$ is the reflected neighborhood of $ESE(i,j)$.

5.2. Hit-or-Miss Transform

Hit-or-Miss Transform(HMT)[14] can extract all pixels in an image matched by a given neighboring configuration, which is defined by two disjoint sets, called adaptive SE pair, denoted by (B_{FG}, B_{BG}). B_{FG} defines the set of pixels matching the foreground, while B_{BG}, defines the set of pixels matching the background. However, the traditional HMT uses the fixed SE pair in the global image, and only extracts the object with the same size and shape as B_{FG}. We present an novel adaptive SE pair to extract the objects having a similar geometry shape and different sizes in one image.

In order to extract suborbicular objects with different sizes in an image, the shape of adaptive SE is defined as circle, the size of it is adjusted adaptively according to the results of the chamfer distance transform(CDT) with 5×5 template. The process of calculating the radius of the B_{FG} is as follows.

1). Applying the edge strength m_E (defined in equation (6)) to search the objects' edges in I, according to (16), where *delta* is an empirical threshold.

$$\begin{cases} f(x,y)=m_E(x,y), & if \quad m_E(x,y)>delta \\ f(x,y)=0, & other \end{cases} \tag{16}$$

2). After computing non-maximal value suppression of the $f(x, y)$ in (16), the pixels having the local maximal value are set to 1, having non-maximal value are set to 0. We then obtain an edge image *f_edge* with a single pixel width.

3). *f_edge* is processed by CDT, the result is called $R_{HMT}(x, y)$. The distance of each pixel in R_{HMT} is the radius of B_{FG}, which is defined by (17) [15].

$$B_{FG}(x,y)=\{|(x,y)-(p,q)|<R_{HMT}(x,y)\},(p,q)\in D \tag{17}$$

In addition, B_{FG} and B_{BG} should share an origin, and $B_{FG} \cap B_{BG}=\Phi$. Thus B_{BG} is defined by (18)[15]:

$$B_{BG}(x,y)=\{R_{HMT}(x,y)+u<|(x,y)-(p,q)|<R_{HMT}(x,y)+u+b\},(p,q)\in D \tag{18}$$

Where u is the size of the uncertainty regions[15] and b is the radius of B_{BG}. u and b are fixed for all pixels in the image I.

6. Adaptive morphology operators

6.1. Structure Adaptation

Figure 1(a) is an original image. Figure 1(b-d) are the erosion results by EASEs[7] with r_w=4, M={4, 8, 16}. Figure 1(e-g) are the erosion results by EASEs[7] with r_w=16, M={4, 8, 16}. The erosion results in Figure 1(h-j) are applied by our presented algorithm, we select r={1, 2, 4}. From Figure 1(b-g), we can see the width of the ring with the same gray value is changed in one eroded image. The reason is that r_w needs to keep balanced with M to capture of interested features. Our presented algorithm can make width same of the rings in each erosion operation.

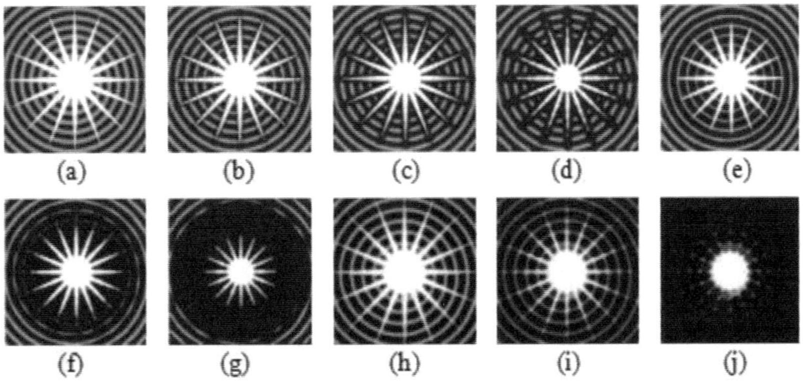

Figure 1. (a) Original image(128×128 pixels); (b-d) Erosion using $r_w = 4$ and $M = 4$, 8 and 16 by EASEs; (f-g) Erosion using $r_w = 16$ and $M = 4$, 8 and 16 by EASEs; (h-j) Erosion by our presented method for $r = 1$, 2 and 4.

6.2. Corner Protection

We concern a synthetic image with fifteen different size, intensity and fuzzy edge diamonds, which is shown in Figure 2(a). Figure 2(b) and (c) are the images after dilation by the SASEs and the EASEs correspondingly. Figure 2(d)-(f) are the erosion, dilation and opening results by our presented method. From the Figure 2(b), (c) and (e), we can see that neither SASEs nor EASEs can protect the diamonds' corners well. But ours perform better than these two. If we define the intersection of a diamond's two adjacent edges as a corner, to the bottom right diamond, the number of missing pixels of one corner in Figure 2(b), (c) and (e) is 13, 7 and 4, respectively, which can prove that our presented method can protect corner best. Additionally, the opening result of our presented method in Figure 2(f) is almost the same as the original image.

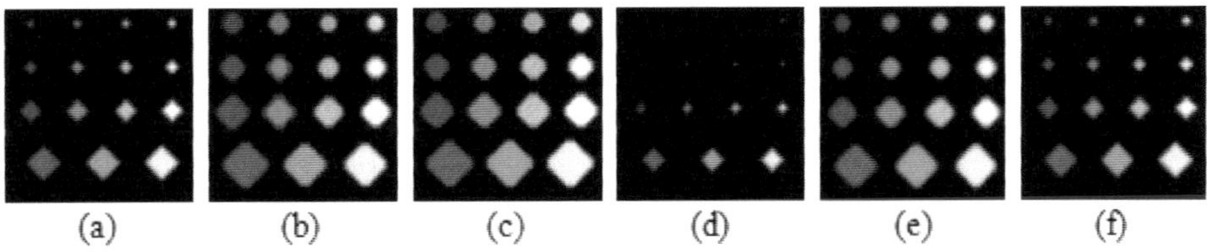

Figure 2. Comparison in corner protection (a) Diamond original image(178×178 pixels); (b) Dilation by SASEs; (c) Dilation by EASEs; (d-f) Erosion, dilation and opening by our presented method.

6.3. Image Filtering

6.3.1. Gaussian Noise Suppression. We consider an original image in Figure 3(a), which is added the Gaussian white noise with $\sigma = 20$ artificially, shown in Figure 3(b). Figure 3(c)-(e) are the images after opening operations by the fixed SE, the EASEs and our presented method respectively. From these results, we can see that Figure 3(c) appears some blocking artifacts, Figure 3(d) blurs the edges of the star, while Figure 3(e) has a much better denoising result compared with them. We have computed the PSNR and RMSE of each filtering result, being listed in Table 1, which can verify our presented opening operation has much better performance than the others in Gaussian noises suppression.

6.3.2. White Artifacts Suppression. Two metallurgic grain boundaries images are shown in Figure 4(a) and(d). The goal is to filter out the white artifacts in the inner part of the grain boundaries. After analysis to the image, we can see that the gradient between the artifacts and the background is the same as the one between the grain boundaries and the background, and the connected region of the white artifacts is much smaller than that of the grain boundaries. Thus, an adaptive maximal allowed

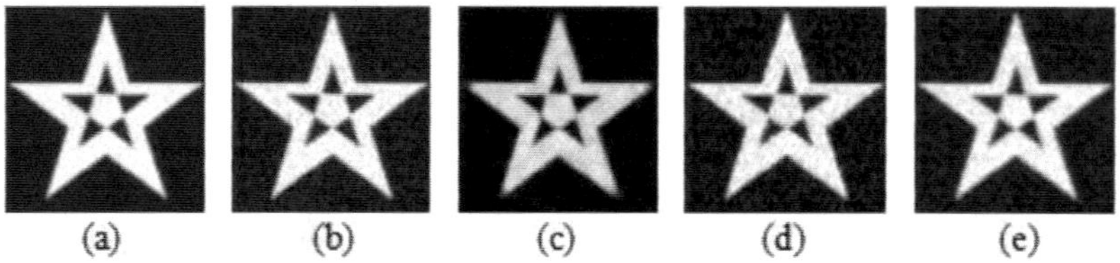

Figure 3. Comparison for fixed SE, EASEs in [7] and our presented method for Gaussian denoising. (a) original image(178×178 pixels); (b) Noise images with standard deviation $\sigma = 20$. (c) opening by fixed disk SE of radius 3. (d) opening by EASEs for $M = 2$. (e) opening by our presented method for $r = 1$.

Table 1. Comparative analysis of the image's performance parameters

Performance parameters	Fixed disk SE	EASEs	Our presented method
RMSE	27.5988	18.0771	10.7628
PSNR	44.4699	52.9323	63.3033

semi-major axis r from the following three steps is proposed: 1)The original image is classified into three feature regions, which are white artifacts, grain boundaries and background. The orientation φ of the elliptical SEs is calculated firstly by equation (10). 2)To any pixel (x_1, y_1) in the grain boundaries or background, r is defined as the number of pixels that belongs to one connected region in $\varphi(x_1, y_1)$ direction. 3)To any pixel (x_2, y_2) in the white artifacts region, r is defined as the maximum of the pixel numbers that belongs to one connected region between in $\varphi(x_2, y_2)$ direction and orthogonal to (x_2, y_2).

Figure 4(b) and (e) are the erosion results of Figure 4(a) and (d) by [7], we can see that the boundaries are blurred and the artifacts are not filtered out completely. The results based on our adaptive r are shown in Figure 4(c) and (f), the grain boundaries are kept better than [7] and the white artifacts are filtered out very completely.

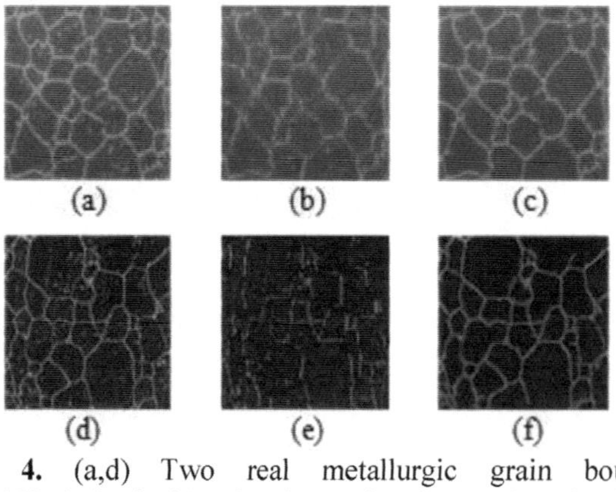

Figure 4. (a,d) Two real metallurgic grain boundaries images(178×178 pixels); (b, e) Erosion by EASEs in [7] with $M=r_w=2$; (c, f) Erosion by our presented method with adaptive radius r.

6.4. Targets Extraction

Figure 5(a) and Figure 6(a) have been conducted to be tested our adaptive HMT. Figure 5(a) is a synthesized image. We compare our proposed HMT results with the classical HMT, shown in Figure 5(b), and the adaptive HMT[15], shown in Figure 5(c). Only the origins of the targets with the same size as the SE have been extracted in (b). The authors of [15] draw a conclusion that this extraction result is relevant to the contrast nor the size, therefore some low contrast disks have not been extracted in (c). Inspiringly, all of the disks have been extracted by our proposed adaptive HMT in (d).

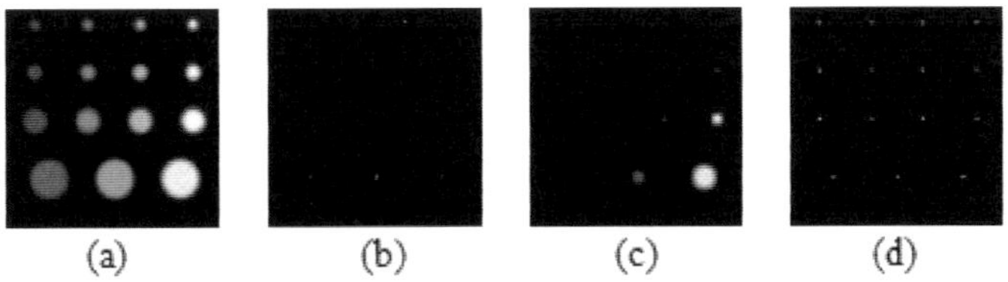

Figure 5. (a,d) Two real metallurgic grain boundaries images(178×178 pixels); (b, e) Erosion by EASEs in [7] with $M=r_w=2$; (c, f) Erosion by our presented method with adaptive radius r.

Figure 6(a) is a graphite cast iron image with complex background. In order to detect the graphite, we let u be different values while keeping $b=1$, b is a radius of B_{BG}. Figure 6(b)-(f) are the HMT results of $u=\{2, 3, 4, 5, 6\}$. These different results are statistic out in Table 2. It is obvious that with the increases of the size of the uncertainly region u, more similar-circle graphite has been extracted. The main reason is when u is small, in Figure 7(a), the dark region denotes the target, and the white point denotes its origin, the B_{BG} can not match the background. When u is big, in Figure 7(b), the B_{BG} can match the background. However, u is not the bigger, the better. When u is set a rather big, the B_{BG} may match an adjacent target, rather than the background, which leads to some targets can not be extracted.

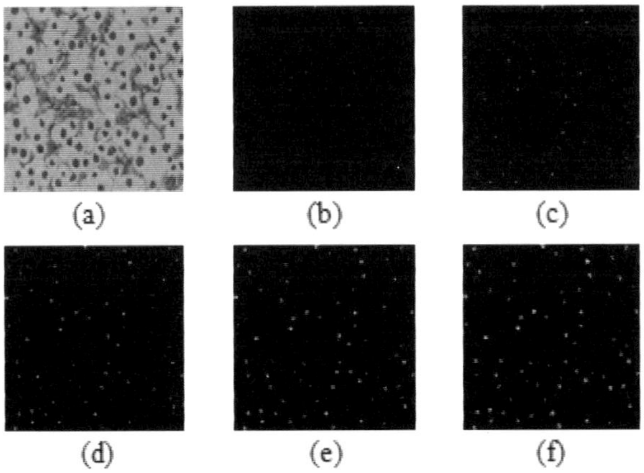

Figure 6. Adaptive HMT for different u and constant $b=1$. (a) original image(305×305 pixels); (b-f) Adaptive HMT with $u = 2, 3, 4, 5, 6$.

Table 2. Statistical results of objects extraction with Figure 6

Values	Rate of extraction
2	29.90%
3	60.82%
4	83.51%
5	86.60%
6	84.54%

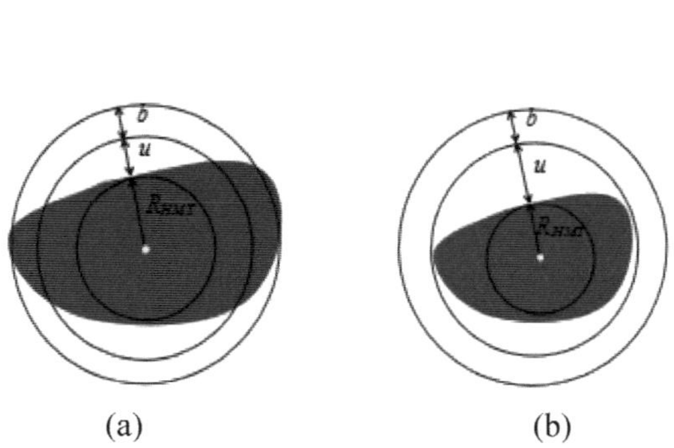

Figure 7. (a) B_{FG} and B_{BG} are used for the extraction of objects with smaller value of u; (b) B_{FG} and B_{BG} are used for the extraction of objects with lager value of u.

7. Conclusion

We present a novel method to construct the adaptive elliptical SEs via estimating the local anisotropy feature of the image based on the NLST. Erosion, dilation, opening, closing and HMT are redefined according to the presented adaptive SEs. After some images are processed by our presented morphology operators, the performance is much better in structure adaption, corner protection, filtering and objects extraction than the other advanced operators.

Reference

[1] Ćurić V, Landström A 2014 Adaptive mathematical morphology : a survey of the field *Pattern Recognition Letters* **47** pp 18-28

[2] Debayle J, Pinoli J 2009 General adaptive neighborhood mathematical morphology *IEEE International Conference on Image Processing* 2249-2252

[3] Debayle J, Pinoli J 2011 Theory and applications of general adaptive neighborhood image processing *Advances in Imaging and Electron Physics* **167** 121-183

[4] Lerallut, Decencière R 2007 Image filtering using morphological amoebas *Image and Vision Computing* **25** pp 395-404

[5] Ćurić V, Luengo Hendriks C L 2012 Salience adaptive structuring elements *IEEE Journal of Selected Topics in Signal Processing* **6** pp 809-819

[6] Ćurić V, Luengo Hendriks C L 2013 Salience-based parabolic structuring functions *11th International Symposium on Mathematical Morphology* **7883** pp 183-194

[7] Landström A, Thurley MJ 2013 Adaptive morphology using tensor-Based elliptical structuring elements *Pattern Recognition Letters* **34** pp 1416-22

[8] Brox T, Weickert J and Burgeth B 2006 Nonlinear structure tensors. *Image and Vision Computing* **24** pp 41-55

[9] Chao SM, Tsai DM 2010 An improved anisotropic diffusion model for detail and edge-preserving smoothing. *Pattern Recognition Letters* **31** pp 2012-23

[10] Tsiotsios C, Petrou M 2013 On the choice of the parameters for anisotropic diffusion in image processing *Pattern Recognition* **46** pp 1369-81

[11] Wang WW, Feng XC 2008 Anisotropic diffusion with nonlinear structure tensor *Siam Journal on Multiscale Modeling and Simulation* **7** pp 963-977

[12] Wu J, Feng Z, Ren Z 2014 Improved structure-adaptive anisotropic filter based on a nonlinear structure tensor *Cybernetics and Information Technologies* **14** pp 112-127

[13] Mandelbrot BB 1982*The Fractal Geometry of Nature* (San Francisco: Freeman).

[14] Murray P, Marshall S 2013 A review of recent advances in the Hit-or-Miss transform. *Advances in Imaging and Electron Physics* Chapter 5 **175** pp 221–282

[15] Ćurić V, Lefèvre S 2015 Adaptive Hit or Miss transform *Springer International Publishing* **9082** pp 741-752

IOP Conf. Series: Journal of Physics: Conf. Series **787** (2017) 012022 doi:10.1088/1742-6596/787/1/012022

Architecture and Key Techniques of Augmented Reality Maintenance Guiding System for Civil Aircrafts

Zhou hong and Lu Wenhua

Room 6408, Building 6, No. 333 Longteng Road, Shanghai China 201620

zhouhongnuaa@163.com

Abstract. Augmented reality technology is introduced into the maintenance related field for strengthened information in real-world scenarios through integration of virtual assistant maintenance information with real-world scenarios. This can lower the difficulty of maintenance, reduce maintenance errors, and improve the maintenance efficiency and quality of civil aviation crews. Architecture of augmented reality virtual maintenance guiding system is proposed on the basis of introducing the definition of augmented reality and analyzing the characteristics of augmented reality virtual maintenance. Key techniques involved, such as standardization and organization of maintenance data, 3D registration, modeling of maintenance guidance information and virtual maintenance man-machine interaction, are elaborated emphatically, and solutions are given.

1. Introduction

Civil aircrafts are highly technology-intensive large-scale electromechanical systems with extremely complex structures. In the traditional aircraft maintenance and repairs process, maintenance personnel often need to consult technical and maintenance manuals or rely on personal skills and experience. Therefore, work efficiency is low, and significant manpower and material resources are consumed. Moreover, because of temporal, conditional and environmental constraints, maintenance cost is difficult to control, and maintenance related errors and maintenance quality problems are highly likely. With the development of digital mockup and virtual reality technologies[1], virtual maintenance technology has offered an opportunity for changing the maintenance mode of aircrafts. Augmented reality (AR) technology is developed on the basis of virtual reality technology, which generates virtual objects that do not exist in the real environment by utilizing computer graphics and visualization technologies and accurately "places" these virtual objects in the real environment via sensor technology. With the help of display device, the AR technology integrates virtual objects into real environment to presents users with a new environment with realistic sensory effects[2-3]. Introduction of AR technology into the maintenance related field can improve the maintenance personnel's perception of information about real-world environment, thereby enhancing maintenance capability and reducing maintenance costs and time[4].

2. Characteristics of AR virtual maintenance

AR maintenance guiding system superimposes various computer-generated virtual assistant maintenance information such as graphics, images, videos and texts in real time to the sight of maintenance personnel through AR technology[5]. Integration of real maintenance scenario with virtual information enhances the scenes observed by the maintenance personnel and generates a rich aircraft maintenance information cognition platform to help the maintenance personnel complete their repair work step by step and achieve "WYSIWYG" maintenance guidance. AR-based virtual maintenance not only avoids the tedious work of building complex real environment, but also has authentic and interactive characteristics[6].

- Authenticity: Different from virtual reality where the users and the actual surrounding environment are isolated[7], in AR, maintenance personnel utilize AR tracking devices such as head-mounted display (HMD)[8], and during the maintenance, real scene images are sent to the server, and detailed aircraft technical and maintenance manuals are displayed in 3D graphics through computation. Virtual information is mapped in real time to the real-world scenarios to achieve scene augmentation or expansion on the maintenance personnel's vision system, in order to strengthen the users' perception and experience, thereby increasing the amount of information and level of understanding of human perception on the real scenarios
- Interactivity: AR maintenance guiding system allows natural interaction between maintenance personnel and virtuality-reality combined scenes[9]. The platform provides natural, humane feedbacks to maintenance personnel's various behaviours according to the relevant data of participants such as gesture, voice and eye movements. It also delivers valuable information to maintenance personnel through real-time interaction. When the user's location in the real world changes, the virtual information also changes accordingly.

3. Architecture of AR virtual maintenance system

The architecture of AR maintenance guiding system created for aircraft maintenance guidance and simulation is shown in Figure 1.

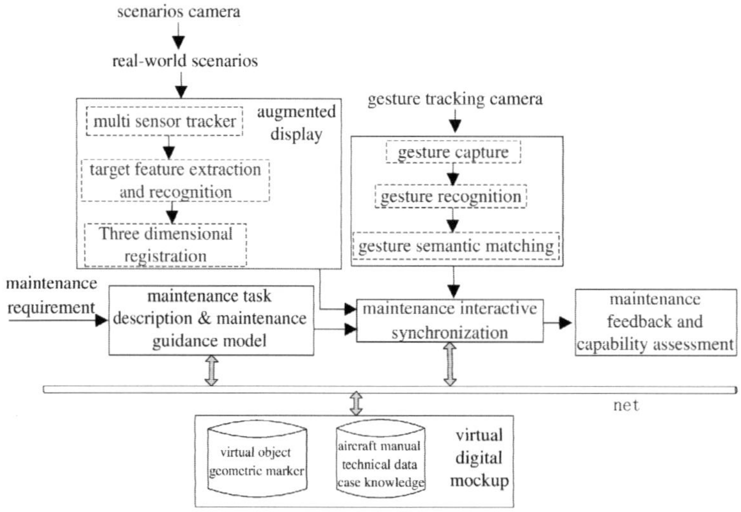

Figure 1. Architecture of AR aircraft maintenance guiding system

The maintenance guiding system consists mainly of five parts.

- Maintenance-oriented virtual digital mockup is the digital product model built using computer technology, which includes computer models for aircraft maintenance object, maintenance tool, maintenance environment, etc. To improve the modeling efficiency and the quality of models, geometry and assembly data of aircrafts can be acquired on the basis of aircraft design department's CAD, PDM models, which are converted or secondarily developed to the information model available for the AR maintenance guiding system. Meanwhile, on the basis of aircraft reliability system, data like aircraft maintenance resources and technical maintenance requirements are acquired to form spatial, temporal and DOF constraints that can support the entire virtual maintenance process.

- To allow the maintenance personnel to perceive changes in the status of surrounding scenarios and maintenance objects, the augmented display module automatically tracks targets such as maintenance objects, maintenance tools and personnel gestures in real-time with tracking devices like camera. It also computes the position and angle information between the tracking device and targets; calculates the precise position of targets in the real scene; determines relevant digital maintenance guidance information; and seamlessly synthesizes the guidance information with real-world scenarios and displays them in real time on the optical see-through HMD based on the line-of-sight and voice commands using 3D registration display technology to present maintenance personnel with a new environment with realistic sensory effects.
- Maintenance task & guidance information module analyzes the maintenance task demands and reorganizes the virtual maintenance assistant information. Maintenance task description model within the module abstractly describes the actual maintenance job process based on the job card and other maintenance requirements. Maintenance task is hierarchically decomposed from top to bottom to give a number of control levels and flow chart of interrelated, mutually restraint maintenance actions varying in level of details. Maintenance guidance model primarily considers how to reorganize and synchronize videos, audios, images, texts, logos and other maintenance assistant information based on the maintenance personnel's maintenance operation process, so that they can be reproduced in accordance with the characteristic virtual-real combination, real-time interaction and 3D registration requirements of the AR system.
- Man-machine interaction module constructs a sight, voice and gesture multi-mode guiding platform, where two interaction-related issues need to be addressed. On one hand, implementation of human interaction with virtual and real scenes not only emphasizes the visual sensory interaction, but also includes behavioral interactions such as language, posture or gesture interactions, so that the maintenance assistant information in the virtual scene are fed back in real time based on the human interactions with real maintenance scenes, state changes, etc. On the other hand, synergy between the two interactions needs to be resolved. The entire system can be maintained by responding to the input information and managing the output information.
- Maintenance feedback & evaluation module provides an objective, accurate evaluation system, which assesses the maintenance capability of maintenance personnel according to the actual operations. Besides, the module analyzes the barriers and difficulty of maintenance and finds defects in the maintenance plans based on the qualitative and quantitative data obtained from the maintenance simulations, which provides maintenance personnel with workable solutions for trying and comparing multiple complex maintenance actions and support plans to find better resource allocation option and the best maintenance scheme.

Based on the above five modules, maintenance guiding system can provide maintenance personnel with an environment for maintenance guidance and assessment where virtual information guidance and real object operation are synchronously integrated. To implement this prototype system, many techniques need to be broken through[10] of which 3D registration, modeling of maintenance guidance information and virtual maintenance man-machine interaction are the key ones to implementing AR virtual maintenance platform.

4. Key techniques of AR maintenance guiding system

4.1. 3D registration technique

AR virtual maintenance guiding system tries to achieve "seamless connection" between the maintenance scenes and virtual information as far as possible. The process of implementing "alignment" and spatial unification between real-world scenes and virtual objects is referred to as the 3D registration display. Generally, two types of registrations are completed with HMD device: 1) registration of wearer's head, i.e. identification of the spatial location of wearer's head and gazing direction; 2) registration of virtual objects, i.e. location of virtual objects in the real space.

In the aircraft maintenance guiding application field, maintenance personnel must be able to see the real world. Besides, human eyes' field of view should be large and intuitive perception is required to be high. Matching precision between virtual objects and real scenes should at least be within the millimeter level, as large registration error will undermine maintenance personnel's correct perception of their surroundings. According to the maintenance and inspection characteristics of aircraft system, natural feature-based registration tracking technique is adopted for implementation of virtual-real fusion, thereby providing maintenance personnel with a maintenance guidance environment with synchronously integrated virtual information guidance and real object operation.

The natural feature-based registration tracking technique is not limited by scene changes. Its specific method is as shown in Figure 2. Firstly, 3 DOF head attitude parameters of maintenance personnel are detected in real time by the sensor, and their gazing direction is calculated. Meanwhile, real scene images are captured through the camera, and natural feature points (such as points and lines that can reflect the shape of objects and pixels that can reflect the texture of objects) are extracted using the machine vision technology. Feature information acquired by the sensor and visual tracking are fused and filtered, thereby estimating the camera pose information to obtain the target objects needing augmented information display. Based on the priori knowledge on scenes stored previously in the database, virtual object database is queried according to the presently identified target information to obtain virtual augmented information (including texts, images, 3D prototype disassembly & assembly processes, etc.) that match with real scenes. Afterwards, spatial coordinates of the virtual camera are adjusted, so that they coincide with the spatial coordinates of camera captured real scenes. Finally, the virtual-real synthesized images consistent in direction with the maintenance personnel's line of sight are output to the optical see-through HMD to achieve the AR effect.

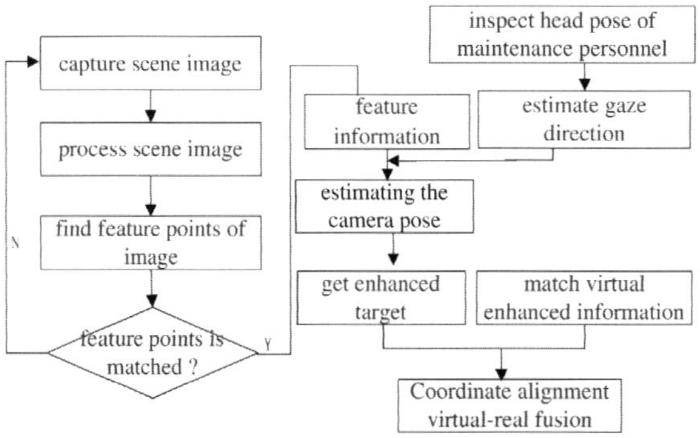

Figure 2. Schematic diagram of natural feature-based registration tracking technique

4.2. Maintenance data standardization technique

AR maintenance guiding information are from aircraft maintenance manuals, catalogs, design department's existing prototype data, etc., which are characterized by different sources and diverse types. To display these information in the augmented scenarios, research on standardization and storage of maintenance information is needed.

Maintenance information standardization primarily studies the extraction of maintenance material from existing aviation maintenance technical manuals (AMM, IPC, EBOM etc.) or other relevant documents; conversion of file formats; and addition of necessary information for generating information data in the forms of audios, videos, images and texts that conform to the requirements of maintenance guiding system, in order to ensure the correct understanding and application of guidance model.

Maintenance information storage technique constructs the data description and storage-retrieval relationship utilizable by the entire system. Through the maintenance relational database, various types of maintenance assistance information are linked together. For example, to organize disassembly and assembly information for typical parts, specific part removal procedure is obtained from the aircraft maintenance manual (AMM) firstly, then corresponding part numbers and their assembly relationship information are acquired via the EBOM. Next, corresponding virtual prototype data are acquired through the part numbers, and removal procedures for corresponding parts are obtained via the illustrated parts catalog (IPC) to form a complete set of disassembly information storage mechanism within the computer, in order to support the implementation of AR maintenance guiding system.

Maintenance storage needs to consider the expandability, scalability and flexibility of the system. Data organizational structures are relatively independent of each other, with weak coupling. When part of the information data is updated, the maintenance guidance information can be granularly changed.

4.3. Maintenance guidance information modeling technique

Maintenance guidance is a process of maintenance information reproduction and interpretive execution. As shown in Figure 3, maintenance guidance information modeling requires creation of model to solve the maintenance task analysis and maintenance job guidance problems[11].

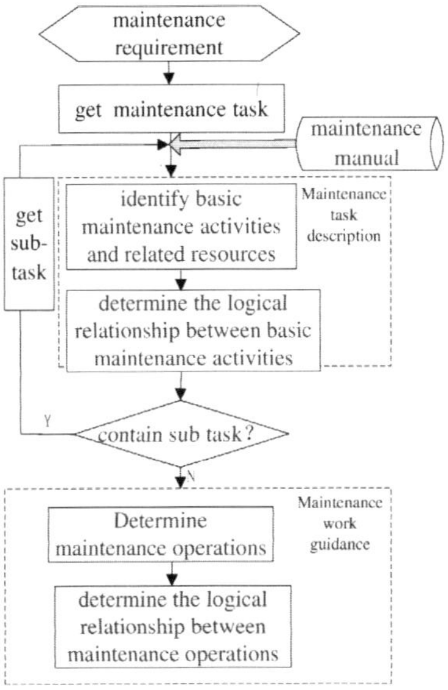

Figure 3. Maintenance guidance modelling procedure

Maintenance task description model focuses on the relationship between maintenance actions, as well as analysis of sequence. A maintenance task can be regarded to comprise a number of maintenance actions, such as preparation, diagnosis, replacement (removal, replacement, installation), adjustment & calibration, maintenance, inspection and original repair. From the perspective of job cards and other maintenance requirements, maintenance tasks are decomposed layer by layer into several actions from top to bottom by consulting the maintenance manuals. Interactive relationship between the maintenance actions is analyzed, as well as resource changes. Maintenance action objects are interconnected into a sequential process. For relatively complex maintenance actions, maintenance work is described using a hierarchical model. Each sub-task is analyzed to determine their respective attributes, behaviors and links to other sub-tasks.

Maintenance work guidance model, on the other hand, emphasizes the relationship between maintenance operations and analysis of sequence. Each maintenance task or action can be completed by specific sequential maintenance operations. Specific maintenance operation is the most essential behavioral unit that cannot be further decomposed. For maintenance actions already decomposed in the maintenance tasks, Maintenance work guidance models are established to expand specific maintenance operation details and therblig unit decomposition. Meanwhile, the models analyze the interactive relationship between maintenance operations and changes in resources then interconnect the maintenance operations into a sequential process. To present the complex operational process, model with conflicting, cyclic and parallel structure is needed, in order to resolve the conflicts between and synchronization of various media information.

4.4. AR virtual maintenance human-computer interaction technique

Real-time natural human-computer interaction can ensure the smoothness of maintenance guidance process. To guide the maintenance "invisibly" and reduce the cognitive load of maintenance personnel, two interaction related issues, i.e. scene interaction and interaction coordination, need to be addressed. Human interaction with virtual + real scenes by sight, voice and gesture multi-modes can achieve good human-computer interaction effect. Among them, gesture interaction is the most natural and intuitive mode. Techniques like machine learning are applied to train large amounts of humanoid image data. In the actual maintenance, human postures or gestures are recognized and tracked. Human postures are recognized by estimating changes in the position of human skeleton and joint points. Semantics of these postures are preset by the user to achieve interaction by controlling virtual objects with simple gestures and body movements.

Interaction coordination is responsible for coordinated control over the maintenance guidance process, which links the specific maintenance actions, maintenance guidance information and AR virtual prototype together, while handling the possible conflicts between various modules. the maintenance guidance procedure is as follows: on the basis of maintenance task description and maintenance work guidance models, personnel –centred maintenance instructions are presented one by one progressively according to the information needs of maintenance personnel with status information as the "drive". Fixation point coordinates are obtained based on gestures, voice commands or eye gaze tracking system, while pose information of objects is acquired by extracting the features of operational objectives in the present scene images combined with digital prototype identification. For the identified targets, appropriate augmentation information is matched from the virtual prototype database for conversion and rendering. Fused display of virtual and real information is completed using AR registration algorithm. Synchronous integration of virtual information guidance with real object operation is achieved.

5. Conclusion

AR maintenance guiding system can complete the assistant maintenance implementation, maintenance guidance and assessment by utilizing 3D registration, image recognition, guide information modeling and human-computer interaction technologies to achieve image acquisition, stereoscopic display and virtual-real fusion functions. AR has obvious advantages in lowering the maintenance difficulty, reducing maintenance errors, and solving the problems of insufficient maintenance staff and massive technical data. With the deepening of research and development of technology, application of AR in civil aircraft maintenance is bound to gradually move towards practicality, which has great significance for optimizing the existing aircraft maintenance mode and bringing new development to the traditional aircraft maintenance industry.

References

[1] Perry T S 2016 Virtual reality goes social *IEEE Spectrum,* **53** pp56-57

[2] Billinghurst M, Clark A and Lee G 2015 A survey of augmented reality *Presence Teleoperators & Virtual Environments* **8** pp 73-272

[3] Nee AYC, Ong SK, Chryssolouris G and Mourtzis D 2012 Augmented reality applications in design and manufacturing. *CIRP Annals - Manufacturing Technology,* **61** pp 657-679

[4] Crescenzio F D, Fantini M, Persiani F,Stefano L D, Azzari P and Salti S 2011 Augmented reality for aircraft maintenance training and operations support *IEEE Computer Graphics & Applications* **31** pp 96-101
[5] Zhao S, Ma D, Zhang Y, Ma S and Yu M 2014 Progressing of augmented reality to assistance maintenances technology *Journal of Graphics* **35** pp 648-654
[6] Neumann U and Majoros A 1998 Cognitive, Performance, and Systems Issues for Augmented Reality Applications in Manufacturing and Maintenance *Virtual Reality International Symposium Proceedings IEEE* pp 4-11
[7] Shen B, Wang T, Zhao J, and Tang Z 2012 Application of virtual reality to simulation in equipment maintenance *International Conference on Measurement, Information and Control* **1** pp 524-528
[8] Gao Y, Liu Y, Cheng D, Wang Y and Optoelectronics S O 2016 A review on development of head mounted display *Journal of Computer-Aided Design & Computer Graphics* **28** pp 896-904
[9] Pier Paolo Valentini 2012 Natural interface in augmented reality interactive simulations *Virtual & Physical Prototyping* **7** pp 137-151
[10] Ma L 2013 *Augmented reality*(BeiJing: China Machine Press)
[11] Zhao X and Zuo H. 2008 Research on interaction technique of augmented reality maintenance guiding system *China Mechanical Engineering* **19** pp 1312-1316

Counterfeit Electronics Detection Using Image Processing and Machine Learning

Navid Asadizanjani, Mark Tehranipoor and Domenic Forte

Electrical and Computer Engineering Department, University of Florida
Gainesville, USA.

E-mail: nasadi@ece.ufl.edu

Abstract. Counterfeiting is an increasing concern for businesses and governments as greater numbers of counterfeit integrated circuits (IC) infiltrate the global market. There is an ongoing effort in experimental and national labs inside the United States to detect and prevent such counterfeits in the most efficient time period. However, there is still a missing piece to automatically detect and properly keep record of detected counterfeit ICs. Here, we introduce a web application database that allows users to share previous examples of counterfeits through an online database and to obtain statistics regarding the prevalence of known defects. We also investigate automated techniques based on image processing and machine learning to detect different physical defects and to determine whether or not an IC is counterfeit.

1. Introduction

Counterfeit ICs are a growing issue in the global market, with some counterfeit ICs infiltrating high-risk applications such as those in the military or medical sectors [1], [2]. Counterfeit ICs may feature a reduced lifespan, fail under conditions mandated by the manufacturer's specification, leak secure data from the system to a remote location (i.e., hardware Trojans [3]), and more. In some cases, the adversary copies the design and labels counterfeit ICs with his own brand name or with the original equipment manufacturer's name using reverse-engineering techniques [4], [5], [6]. Since the adversary does not incur the engineering research and development costs, his products are priced lower, thereby resulting in a loss for authentic products.

The taxonomy of counterfeit component types and defects in electronic components have been discussed in [7]–[8], where the two main classes of defects are classified as physical and electrical defects. Here we will study the physical defects and try to expedite and facilitate the detection and recordkeeping of such physical defects. With the help of high-precision microscopes, most of the physical defects can be identified by a subject matter expert (SME). The SME has to carefully test the entire IC for each defect. This is a labor-intensive and timely process. For a modern electronics manufacturer, it is not feasible for an SME to analyze all the chips and components manually because a single US company can import tens of millions of chips annually. Some examples of these defects focused in this proposal are presented in Fig. 1 [9].

The US Chamber of Commerce has identified 11 issues in the current state of the global electronic supply chain in regard to counterfeiting [10]. In this paper we begin to address the need for stricter testing protocols and quality control practices. We will develop preliminary algorithms that analyze optical images and automatically extract defects using artificial neural networks and image processing algorithms.

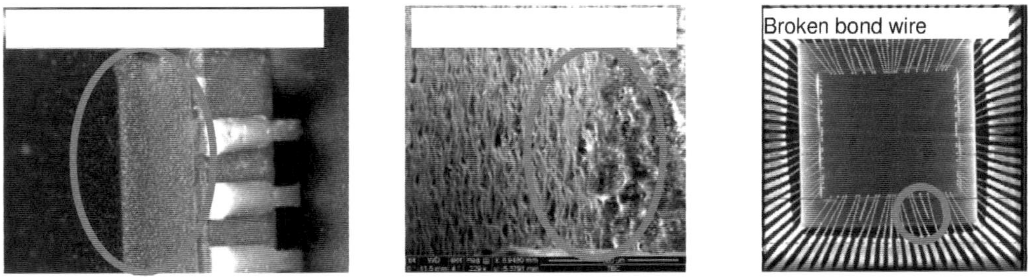

Figure 1. Physical defects seen on electronic components.

2. Automated Defect Detection

We investigate two techniques for detecting defects on ICs are presented: 1) image processing and 2) artificial neural network (ANN)–based algorithms.

2.1. Defect Detection Using Image Processing

There are many image processing algorithms and filters used to detect features on different images. These algorithms are designed in a way to highlight one specific property of a feature based on its histogram. These features can be defects on ICs such as scratches, color variations, ghost marking, etc. It is very important to select the right algorithm and right threshold in order to better highlight the defect. Figure 2a shows top view of an IC with scratches. We have applied image processing techniques on this image including Hough transform, an edge-enhanced maximally stable filter, and Sobel filter to detect these scratches [11]-[13] as shown in Fig. 2b.

Although filters are widely used for detection, the results are not accurate enough to make a decision on the authenticity of a chip. Many defects cannot be identified through this method alone. For example, many counterfeits display variations in color that can be identified based on the intensity of grayscale pixels rather than the placement of binary pixels within the image. To determine whether color variations exist on a given IC, an entirely different image processing method is needed in addition to the scratch detection algorithm.

(a) (b)

Figure 2. The counterfeit IC with scratch defect a) original image b) after image processing.

Such cases increase the complexity of developing a human independent platform to determine the authenticity of an IC based on filters. Ideally, one single algorithm should be used to identify a wide variety of defects under various imaging conditions. This will not be possible without getting help from machine learning. Artificial neural networks are one of the most popular machine learning techniques which we have identified as a promising candidate to achieve this goal. Considering the fact that such an algorithm can train itself and become smarter as more samples are tested, this is a huge advantage to other algorithms and filters.

2.2. Defect Detection Using ANN

ANN is a machine-learning technique that is modeled similarly to the structure of the human brain. A typical ANN consists of neurons in different layers, such as input, hidden, and output layers. The information is stored in the interconnections between neurons across the adjacent layers [14]. The number of hidden layers should be at least one, and the number can be increased depending on the application. The output of a neuron is calculated by using an appropriate activation function for the input value. Step function, *tanh*, and *arctan* are some of the examples of activation functions that will be used in this proposal.

In the structure of a typical ANN each neuron from the input layer is connected to each neuron in the first hidden layer. The input value at a node of a hidden layer is the weighted summation of all the input nodes. The factor by which each node contributes to the value of the input of the next node will decide the weight matrix between the two layers. Thus, the weight matrix stores crucial information which is unique for each class of inputs. The output of the neuron is computed using the activation function. This weighted summation of adjacent layers continues until the output layer. During the training phase, the expected output is compared with the output obtained by the computations. The error is then sent back to the network layers to update the weights.

The backpropagation algorithm (i.e., propagating the error in the backward direction of the network) is a popular artificial neural network algorithm used for training the network. The basic idea behind backpropagation algorithm is to reduce the error energy of a neuron in the neural network. In general, backpropagation method requires a known or desired output for each input value, therefore calculating the error will help to quantify this step. Differentiating the error energy functions with respect to the weights gives the condition for minimum error energy. These error values will be used to adjust the weights associated with input values for each node. This learning phase will be iterated until the error meets an accepted value. The neural network is then trained for more iterations for the same image until the error level is less than the accepted tolerance level.

The complexity of the operations in the ANN depends on factors such as the number of nodes in each layer, the learning rate, the number of iterations done for training, and the number of hidden layers. The artificial neural network is an extremely useful tool for pattern recognition applications. The ANN has been studied extensively for applications such as character recognition, face recognition, etc. We have used backpropagation algorithm to Train ANN. In the next section, we will discuss our implementation of ANN using backpropagation on a chip with scratches.

2.2.1. ANN Implementation. The implementation of the counterfeit IC detection using artificial neural network can be divided into two parts: 1) **training phase**; and 2) **testing phase**. In our experiments, we divide the available set of images into 2 samples for training and 4 for testing. In the future, this proportion can be varied to check the efficiency of the system. In addition, while we focus on scratches only, we can train for multiple defects at once.

Training Phase: We utilized backpropagation algorithm discussed earlier to train the artificial neural network, using tanh, the hyperbolic tangent function, as the activation function. Only one hidden layer was used to reduce the complexity of the ANN, but two and three layers will be investigated in this work to optimize the results.

Figure 3 shows the flowchart for the counterfeit IC detection procedure. The input image is read using the MATLAB function imread, which gives a matrix consisting all pixels of the image. The image is then converted to a column matrix by arranging the columns of the input image matrix one below the other. If the input image is of size M × N, the new column matrix will have a size of MN × 1. This can be reduced by resizing the image using various preprocessing techniques such as binning. However, resizing will decrease the quality of the image and resolution of the features to be detected.

Other preprocessing techniques such as filters to remove pixels over or below a threshold, edge detection, normalizing the image, and converting RGB images to gray scale can be used for preprocessing images and make images ready for ANN analysis. The input image is normalized by dividing pixel values by 255 for grayscale images. The techniques applied for the processing of the image are the same for training and testing till this step. However, during training the output for a particular input image is known.

The weight matrices between input, hidden, and output layers are initialized with random values between −0.5 and +0.5. After preprocessing the images, expected output layer is initialized with its corresponding class. The linear combination of the initial column matrix and the first weight matrix gives the input for the hidden layer. Using the activation function tanh, we get the output of hidden layer. Since the linear combination of input with weight matrix can be very expensive computationally, activation function helps keeping them in the range of −1 to 1. This output will then be multiplied by weight matrix between the hidden and output layer to calculate the input.

Similar to nodes in the hidden layer, the output of output layer is calculated using the activation function. The error at the output is calculated by comparing activation function with the expected output given at the start of training. The error is propagated back to the network. This loop can be continued until we reach a point where the error is within an expected range.

Testing Phase: During testing, we do not use backpropagation algorithm, as the ANN is already trained with error corrections for the weights. In fact, we do not know the output value prior to the computations in ANN.

ANN will then determine if the IC is defected or not depending on its output. We apply the same preprocessing methods that were employed during the training phase for the images. In order to test this technique, we compare the output of the ANN during testing with the expected output and calculate the efficiency of the system. The flow of training and testing of the ANN is explained in Figures 3 and 4.

The output classes considered for this experiment were 0 and 1. The output value 0 represents no defect condition, and the output value 1 indicates that the IC under consideration is defected.

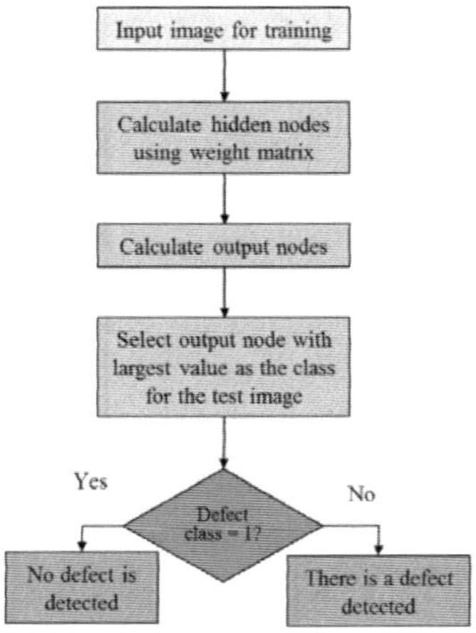

Figure 3. Logical flowchart for artificial neural network testing.

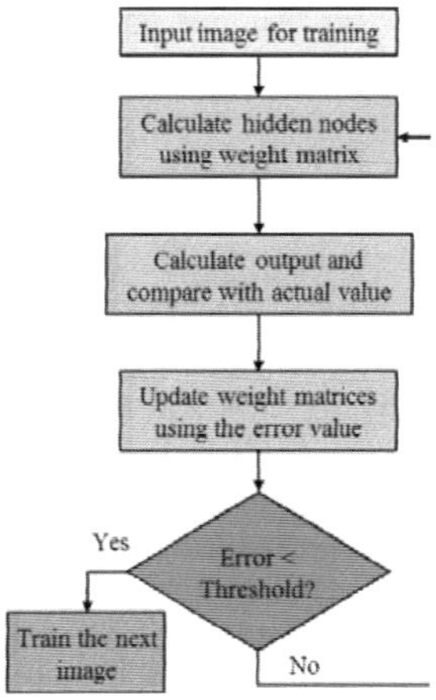

Figure 4. Logical flowchart for artificial neural network training.

2.2.2. Preliminary Detection Results Using ANN. The backpropagation ANN was implemented using MATLAB platform to detect scratches on chips. Standard images were taken from 6 different WSI-PSD813F1 chips with and without scratches. The training was done using two images with and without defects for 10 iterations each. Figure 5 shows the images used for training and ANN was tested for detecting counterfeit ICs. The original image had more than 4 million pixels. This means we need 4 million input nodes, where we have only two output nodes. This was resolved by reducing the size of the images to a resolution of 512×512 during preprocessing stage. The number of hidden nodes were kept as a variable to be adjusted depending on the results. One hundred hidden nodes gave us good results. As we increase the number of nodes, the computational time rises. For our experiments, we set the number of hidden nodes as 100. The number of IC images used for testing were four, out of which two were counterfeit. The weight matrices were initialized using random functions available in MATLAB between −0.5 to 0.5. The images used for testing is shown in Fig. 6. ICs (a) and (c) have scratch defects and are correctly detected by the ANN.

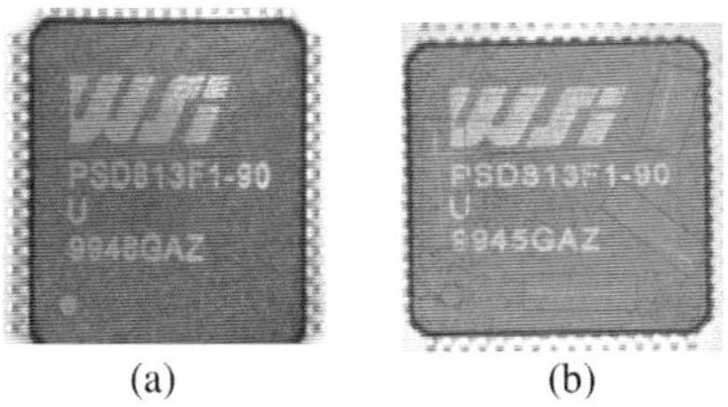

Figure 5. Images used for training a) IC without defect b) IC with highlighted scratches.

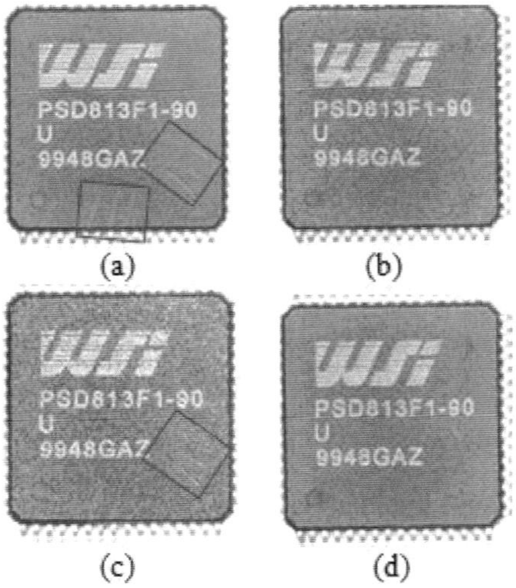

Figure 6. IC images with scratch defect used for testing.

3. Conclusion and Future Work

In this paper, we described a counterfeit database that has been developed to improve recordkeeping of counterfeit incidents, support education and increase awareness about counterfeit ICs/defects, and provide resources for research in automated counterfeit IC and defect detection. As an exemplary instance, data was used to investigate image processing and neural network algorithms for scratch detection. Image processing can successfully detect ICs with scratches, but fails when the scratches are not highly visible or their color is not significantly different from that of the IC surface. Another current limitation is that the exact same algorithm will not apply for other defects. However, the artificial neural network should be more amenable to other types of defects in the future. In future work, we will extend the existing ANN technique to detect other physical defects and to data from additional sources (X-ray, SEM, etc.).

Acknowledgments

The authors would like to acknowledge support from National Science Foundation (grants CCF-1423282, CCF-1559772, and CNS-1558516) for this work.

References

[1] U. Guin D. DiMase and M. Tehranipoor, "Counterfeit Integrated Circuits: Detection, Avoidance, and the Challenges Ahead," Journal of Electronic Testing: Theory and Applications (JETTA), vol. 30, no. 1, pp. 9-23, 2014.

[2] U. Guin, K. Huang, D. DiMase, J. Carulli, M. Tehranipoor, and Y. Makris, "Counterfeit Integrated Circuits: A Rising Threat in the Global Semiconductor Supply Chain," Proceedings of the IEEE, vol. 102, no. 8, pp. 1207–1228, 2014.

[3] U. Guin, D. Forte and M. Tehranipoor, "Anti-counterfeit techniques: From design to resign", Proc. 14th Int. Workshop Microprocessor Test Verification (MTV), pp. 89-96.

[4] K. M. Buddhiraju and I. A. Rizvi, "Comparison of CBF, ANN and SVM Classifiers for Object-based Classification of High Resolution Satellite Images", Proc. IGARSS, pp. 40-43, 2010.

[5] X. Zhang, K. Xiao, and M. Tehranipoor, "Path-Delay Fingerprinting for Identification of Recovered ICs" in Proc. IEEE International Symposium on Defect and Fault Tolerance in VLSI and Nanotechnology Systems (DFTS), 2012.

[6] N. Asadizanjani, "3D Imaging and Investigation of Failure and Deformation in Thermal Barrier Coatings Using Computed X-ray Tomography." (doctoral dissertation) , 2014.

[7] S. Quadir, J. Chen, D. Forte, N. Asadizanjani, S. Shahbazmohamadi, L. Wang, J. Chandy, and M. Tehranipoor. "A survey on chip to system reverse engineering." *ACM Journal on Emerging Technologies in Computing Systems (JETC)* 13, no. 1 (2016): 6.

[8] N. Asadizanjani, S. Shahbazmohamadi, M. Tehranipoor, and D. Forte. "Non-destructive PCB Reverse Engineering Using X-ray Micro Computed Tomography." In *41st International Symposium for Testing and Failure Analysis (November 1–5, 2015)*. Asm, 2015.

[9] www.counterfeit-ic.org

[10] Measuring the Magnitude of Global Counterfeiting available online http://www.theglobalipcenter.com/wp-content/themes/gipc/map-index/assets/pdf/2016/GlobalCounterfeiting_Report.pdf

[11] R. Karri, J. Rajendran, K. Rosenfeld and M. Tehranipoor, "Trustworthy Hardware: Identifying and Classifying Hardware Trojans", Computer, vol. 43, no. 10, pp. 39-46, 2010.

[12] S. Haykin, S. Simon, "Neural Networks and Learning Machines", Vol. 3. Upper Saddle River, NJ, USA: Pearson, 2009.

[13] E. Ardizzone, H. Dindo, and G. Mazzola, "Multidirectional Scratch Detection and Restoration in Digitized Old Images," Eurasip Journal on Image and Video Processing, vol. 2010, no. June 2015, 2010.

[14] U. Guin D. Forte and M. Tehranipoor, "A Comprehensive Framework for Counterfeit Defect Coverage Analysis and Detection Assessment", Journal of Electronic Testing (JETTA), Volume 30, Issue 1, pp 25-40, February 2014.

Noncoherent detection of DQPSK in OFDM systems using predictive VA

Vineel K Veludandi and K Vasudevan

Department of Electrical Engineering, IIT Kanpur, 208016, Kanpur, India

Email: vineelkv@iitk.ac.in, vasu@iitk.ac.in .

Abstract. Noncoherent detection of differential quaternary phase shift keying (DQPSK) signals in OFDM systems is efficiently implemented using a predictive Viterbi algorithm (VA) operating on a trellis with just $S_T = M^{P-1}$ states instead of M^P states, where M denotes an M-ary PSK constellation and P denotes the order of the prediction filter. The prediction filter coefficients are generated based on the channel DFT alone making a high SNR approximation, since the estimation of the noise-variance using training symbols results in loss of throughput.

1. Introduction

Orthogonal Frequency Division Multiplexing (OFDM) is a spectrally efficient, discrete-time implementation of the multicarrier communication system [1]. Data detection in OFDM systems can be performed in three ways. They are

- Coherent detection [2, 3] which require the estimate of the channel impulse response.
- Suboptimal ML detection [4] which require the statistics of the channel fade process.
- Noncoherent detection [5] which does not require any knowledge of the channel fade process nor its statistics.

Channel fade process and its statistics are estimated using training symbols embedded into the OFDM frame [6]. Channel and statistics estimation using training symbols results in loss of throughput and hence the need for noncoherent detection. In [5], predictive VA is used for detecting 8-PSK in Rayleigh flat-fading channel, whereas in this paper, predictive VA is used for detecting DQPSK in OFDM systems. Perfect carrier and timing synchronization is assumed.

A high degree of correlation in the channel frequency response is obtained at the output of FFT, when the channel memory is much less than FFT length. The proposed predictive VA-based receiver exploits this correlation using a linear prediction filter for decorrelating the channel frequency response. There are M^P ways in which a prediction filter of order P can be populated. However, in [5], the complexity of the trellis is reduced to M^{P-1} states, using the concept of isometry.

2. Notation

In this paper, all lower-case and upper-case letters without a tilde e.g., g_k represent real-valued scalars. Letters with a tilde e.g., $\tilde{h}_k$, denote complex quantities. However, complex data symbols are denoted by S_k (without a tilde). Boldface letters represent vectors or matrices. The $(.)^{\mathcal{H}}$ denotes conjugate transpose, $(.)^T$ denotes transpose and $E[.]$ denotes the expectation operation.

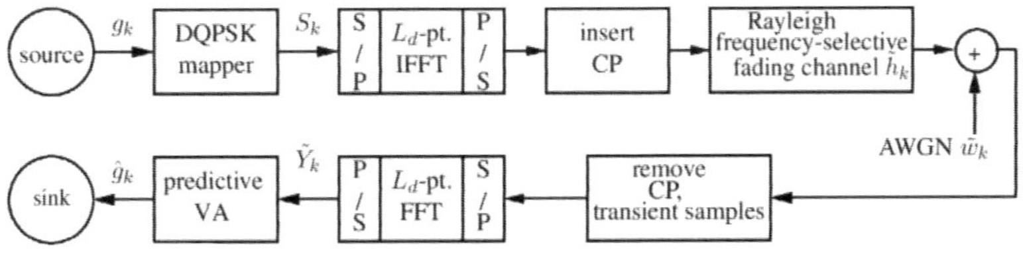

Figure 1: System model. Two consecutive bits of $\boldsymbol{g_k}$ are mapped to QPSK using the differential encoding rules in Table 1.

3. System Model

The block diagram of the system is given in Figure 1. The binary input data g_k $(0 \leq k \leq 2L_d - 1)$ is mapped to differential quadrature shift keying (DQPSK) to get S_k $(0 \leq k \leq L_d - 1)$ according to the differential encoding rules given in Table 1 [1].

Table 1: Differential encoding rules

Dibit $(g_{k-1}g_k)$	Phase change (in radians)
00	0
01	$\pi/2$
10	$3\pi/2$
11	π

The symbol stream S_k is modulated onto the orthogonal subcarriers by an L_d-point IFFT operation. A cyclic prefix (CP) of length $L_h - 1$ is inserted in the time domain to convert linear convolution with the Rayleigh frequency selective fading channel $\tilde{h}_k$ $(0 \leq k \leq L_h - 1)$ into a circular convolution [1].

The channel coefficients are $\tilde{h}_k \sim \mathcal{CN}(0, 2\sigma_f^2)$ are independent over time k in each OFDM frame, that is,

$$0.5 \times E\left[\tilde{h}_k \tilde{h}_{k-m}^*\right] = \sigma_f^2 \delta_K(m) \tag{1}$$

where $\delta_K(.)$denotes the Kronecker delta function and also varies independently from frame to frame, that is, quasistatic. The AWGN samples $\widetilde{w}_k \sim \mathcal{CN}(0, 2\sigma_w^2)$ $(0 \leq k \leq L_d - 1)$ are independent over time k in each OFDM frame, that is,

$$0.5 \times E[\widetilde{w}_k \widetilde{w}_{k-m}^*] = \sigma_w^2 \delta_K(m) \tag{2}$$

and also varies independently from frame to frame.

4. Receiver

The received samples at the FFT output are given by (for $0 \leq k \leq L_d - 1$)

$$\tilde{Y}_k = \widetilde{H}_k S_k + \widetilde{W}_k, \text{ where } \widetilde{H}_k = \textstyle\sum_{i=0}^{L_h-1} \tilde{h}_i e^{-j2\pi ik/L_d}, \widetilde{W}_k = \textstyle\sum_{i=0}^{L_d-1} \widetilde{w}_i e^{-j2\pi ik/L_d}. \tag{3}$$

The autocorrelation of the channel DFT is given by [3]

$$\tilde{R}_{\tilde{H}\tilde{H},m} \triangleq 0.5 \times E\left[\widetilde{H}_k \widetilde{H}_{k-m}^*\right] = \sigma_f^2 \textstyle\sum_{n=0}^{L_h-1} e^{-j2\pi nm/L_d}. \tag{4}$$

Now consider $\tilde{X}_k = \tilde{Y}_k / S_k \approx \widetilde{H}_k$ (at high SNR). The autocorrelation of the $\tilde{X}_k$ is given by

$$\tilde{R}_{\tilde{X}\tilde{X},m} = \tilde{R}_{\tilde{H}\tilde{H},m} + (\sigma_w^2 L_d \delta_K(m) / |S_k|^2). \tag{5}$$

In practice, the prediction filter coefficients are generated using the autocorrelation in (5) which is estimated with the help of training symbols. However, in this paper, the prediction filter coefficients are generated from (4).

Note that the detection rule for the ideal coherent receiver is

$$\hat{S}_k = \min_{S_k} \left|\tilde{Y}_k - \widetilde{H}_k S_k\right|^2. \tag{6}$$

4.1 Noncoherent Detection Principle [5]

In this subsection we give the principle of the operation for the proposed predictive VA-based receiver. The received samples can be represented in matrix form as

$$\widetilde{\mathbf{Y}} = \mathbf{S}^{(q)}\widetilde{\mathbf{H}} + \widetilde{\mathbf{W}},\ 0 \le q \le M^{L_d} - 1 \tag{7}$$

where $\widetilde{\mathbf{Y}}$ is an $L_d \times 1$ matrix of received samples, $\mathbf{S}^{(q)}$ is an $L_d \times L_d$ matrix with elements containing the qth possible transmitted DQPSK symbol sequence, $\widetilde{\mathbf{H}}$ is an $L_d \times 1$ matrix of the channel DFT and $\widetilde{\mathbf{W}}$ an $L_d \times 1$ matrix containing the DFT of the AWGN samples $\widetilde{w}_k$.

The noncoherent ML detector decides in favour of $\mathbf{S}^{(q)}$ that maximizes the conditional probability density function $p(.)$ as

$$\max_q p\left(\widetilde{\mathbf{Y}}\ |\mathbf{S}^{(q)}\right) \Rightarrow \max_q \exp\left(-\tfrac{1}{2}\widetilde{\mathbf{Y}}^{\mathcal{H}}\left(\widetilde{\boldsymbol{R}}^{(q)}\right)^{-1}\widetilde{\mathbf{Y}}\right), \text{ where}$$

$$\widetilde{\boldsymbol{R}}^{(q)} \triangleq \tfrac{1}{2}E\left[\widetilde{\mathbf{Y}}\,\widetilde{\mathbf{Y}}\,|\mathbf{S}^{(q)}\right] = \tfrac{1}{2}\mathbf{S}^{(q)}E\left[\widetilde{\mathbf{H}}\,\widetilde{\mathbf{H}}^{\mathcal{H}}\right]\mathbf{S}^{(q)^{\mathcal{H}}} + \sigma_w^2 L_d\mathbf{I} \approx \tfrac{1}{2}\mathbf{S}^{(q)}E\left[\widetilde{\mathbf{H}}\,\widetilde{\mathbf{H}}^{\mathcal{H}}\right]\mathbf{S}^{(q)^{\mathcal{H}}} = \tfrac{1}{2}\mathbf{S}^{(q)}\boldsymbol{\Phi}\mathbf{S}^{(q)^{\mathcal{H}}} \tag{8}$$

Note that $\widetilde{\boldsymbol{R}}^{(q)}$ is not a diagonal matrix, since channel DFT is highly correlated. Using Cholesky factorization of the autocovariance matrix, it can be shown that $\boldsymbol{\Phi}^{-1} = \widetilde{\mathbf{B}}^{\mathcal{H}}\mathbf{D}^{-1}\widetilde{\mathbf{B}}$, where

$$\widetilde{\mathbf{B}} \triangleq \begin{bmatrix} 1 & 0 & \dots & 0 \\ \tilde{a}_{1,1} & 1 & \dots & 0 \\ \vdots & \vdots & \vdots & \vdots \\ \tilde{a}_{L_d-1,L_d-1} & \tilde{a}_{L_d-1,L_d-2} & \dots & 1 \end{bmatrix}, \quad \mathbf{D} \triangleq \begin{bmatrix} \sigma_0^2 & \dots & 0 \\ \vdots & \ddots & \vdots \\ 0 & \dots & \sigma_{L_d-1}^2 \end{bmatrix}. \tag{9}$$

$\widetilde{\mathbf{B}}$ is the $L_d \times L_d$ matrix of predictor coefficients with $\tilde{a}_{i,\tau}$ being the τth coefficient of the optimum i^{th} –order predictor and $\mathbf{D}$ is the $L_d \times L_d$ matrix where σ_k^2 is the 1-D prediction error variance of the kth –order predictor (minimum mean square-error sense). The prediction coefficients are generated using Levinson-Durbin algorithm based on the autocorrelation of the channel DFT given in (4).

The maximization rule in (8) can be simplified as

$$\min_q \widetilde{\mathbf{Y}}^{\mathcal{H}}\left(\left(\mathbf{S}^{(q)}\right)^{\mathcal{H}}\right)^{-1}\widetilde{\mathbf{B}}^{\mathcal{H}}\mathbf{D}^{-1}\widetilde{\mathbf{B}}\left(\mathbf{S}^{(q)}\right)^{-1}\widetilde{\mathbf{Y}} \Rightarrow \min_q\left(\sum_{k=0}^{P-1}\frac{\left|\tilde{z}_k^{(q)}\right|^2}{2\sigma_k^2} + \sum_{k=P}^{L_d-1}\frac{\left|\tilde{z}_k^{(q)}\right|^2}{2\sigma_k^2}\right) \tag{10}$$

where
$$\left[\tilde{z}_0^{(q)}\ \tilde{z}_1^{(q)} \dots \tilde{z}_{L_d-1}^{(q)}\right]^T \triangleq \tilde{\mathbf{z}}^{(q)} = \widetilde{\mathbf{B}}\left(\mathbf{S}^{(q)}\right)^{-1}\widetilde{\mathbf{Y}}. \tag{11}$$

The first part of the summation in (10) represents the transient-part and the second-part represents the steady-state. In this paper, we only consider the steady-state part. Assuming that a Pth –order predictor completely decorrelates the fade process, the final decision rule can be given as

$$\min_q \sum_{k=P}^{L_d-1}\left|\tilde{z}_k^{(q)}\right|^2, \text{ where } \tilde{z}_k^{(q)} = \sum_{i=0}^{P}\tilde{a}_{P,i}\widetilde{Y}_{k-i}/S_{k-i}^{(q)}. \tag{12}$$

The complexity in (12) increases exponentially with message length L_d. However (12) can be efficiently implemented using a predictive VA algorithm.

4.2 Trellis Construction [1]

In this section, we assume that the DQPSK mapper in Figure 1 is not present. In section 4.3, we motivate the need for the DQPSK mapper. Assume that the *n* bits from the source are mapped to an *M*-ary ($M = 2^n$) constellation according to the mapping $S = \mathcal{M}(.)$. Here, $S_{j,0}$ ($0 \le S_{j,0} \le 2^n - 1$) is the decimal equivalent of the *n* bits $g_k g_{k-1} g_{k-2} \cdots g_{k-n+1}$ in Figure 1, and is referred to as the input digit. The number of states in the trellis would be $S_T = M^P$, since there are M^P ways in which a prediction filter of order *P* can be populated.

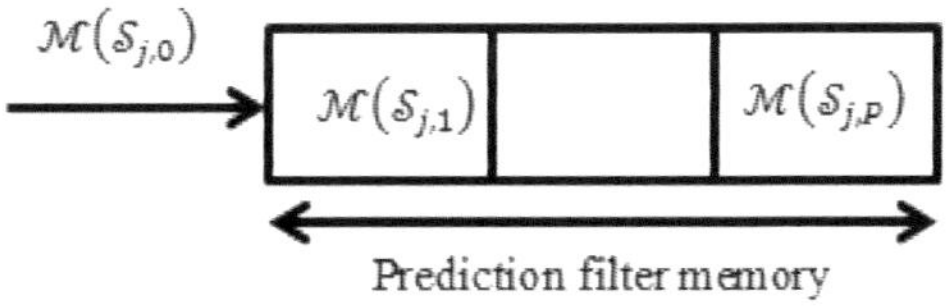

Figure 2: Procedure for constructing trellis.

Note that $\{\mathcal{M}(\mathcal{S}_{j,0}), \mathcal{M}(\mathcal{S}_{j,1}), \ldots, \mathcal{M}(\mathcal{S}_{j,P})\}$ in Figure 2 refers to one of the paths in the trellis. The jth trellis state can be represented by a M-ary P tuple as $\mathcal{S}_j$: $\{\mathcal{S}_{j,1} \ldots \mathcal{S}_{j,P}\}$ for $0 \le j \le M^P - 1$ where $\mathcal{S}_{j,k} \in \{0, \ldots, M-1\}$.

Given the present state $\mathcal{S}_j$ and input $\mathcal{S}_{j,0}$ ($\mathcal{S}_{j,0} \in \{0, \ldots, M-1\}$), the next state is given by $\mathcal{S}_i$: $\{\mathcal{S}_{j,0}\, \mathcal{S}_{j,1} \ldots \mathcal{S}_{j,P-1}\}$, that is, right-shift operation.

4.3 Complexity reduction of the trellis using isometry [5]

Replacing $\mathbf{S}^{(q)}$ by $\mathbf{S}^{(q)}e^{j\phi}$ (here ϕ refers to constant phase) in (8) will yield the same metric in (12) (isometry problem) and this problem can be solved by differential encoding.

Consider the error signal in (12), which is,

$$\tilde{z}_k^{(q)} = \sum_{i=0}^{P} \tilde{a}_{P,i}\tilde{Y}_{k-i}/S_{k-i}^{(q)} = 1/S_k^{(q)} \sum_{i=0}^{P} \tilde{a}_{P,i}\tilde{Y}_{k-i}S_k^{(q)}/S_{k-i}^{(q)}. \tag{13}$$

It can be observed that $\left|\tilde{z}_k^{(q)}\right|^2$ is independent of $S_k^{(q)}$ (isometry) and is dependent only on the phase changes between $S_k^{(q)}$ and $S_{k-1}^{(q)} \ldots S_{k-P}^{(q)}$. That is, $S_k^{(q)}/S_{k-P}^{(q)} = f(\mathcal{S}_{j,0}\mathcal{S}_{j,1} \ldots \mathcal{S}_{j,P-1})$ where $f(.)$ is some function of input digits $\mathcal{S}_{j,0}\mathcal{S}_{j,1} \ldots \mathcal{S}_{j,P-1}$ depending on the differential encoding rules given in Table 1. This suggests that the contents of the prediction filter can always be represented with an unity as its first element. Now, the number of states in the normalized trellis (that is, with differential encoding) required for a Pth-order predictor would be only $S_T = M^{P-1}$.

4.4 Predictive VA algorithm [1]

The predictive VA operates on the normalized trellis. Let $\mathcal{C}_n$ denote the states that converge to the state n ($0 \le n \le S_T - 1$). Let $\mu_{i,n}$ denote the path metric at time instant i ($0 \le i \le L_d$) and state n. Let $v_{i,m,n}$ denote the branch metric at time instant i corresponding to the transition from state m to state n and is given as

$$v_{i,m,n} = \left|\sum_{j=0}^{P} \tilde{a}_{P,j}\tilde{Y}_{i-j}/S_{j,m,n}\right|^2 \tag{14}$$

where $S_{0,m,n}$ denote the input symbol corresponding to the transition from state m to n and the data $S_{j,m,n}$ are the contents of the prediction filter of state m. The algorithm is as follows:

(i) Now set initial values as $i = 0$ and $\mu_{0,n} = 0$ (for $0 \le n \le S_T - 1$), since the receiver does not know the starting state.

(ii) Increase time i by 1

(a) Compute the path metrics at each state n as $\mu_{i,n} = \min_{m \in \mathcal{C}_n} \{v_{i,m,n} + \mu_{i-1,m}\}$.

(b) Store the survivor for each state n.

(c) Having the minimum path metric, trace back along the survivor path and release a symbol corresponding to time $i - \mathcal{D}_V$, where $\mathcal{D}_V$ is the decoding delay of VA.

(iii) Repeat step 4.4.2 until time $i = L_d$.

5. Results

5.1 SNR parameters

Since the overall rate is 2, the SNR per bit is defined as [3]:

$$\text{SNR per bit} = 0.5 \times E\left[\left|\tilde{H}_k S_k\right|^2\right] / E\left[\left|\tilde{W}_k\right|^2\right] = 0.5 \times \left(L_h \times 2\sigma_f^2\right) \times E[|S_k|^2] / (L_d \times 2\sigma_w^2). \quad (15)$$

In Figure 3, we compare the proposed predictive VA-based receiver for different values of prediction orders. In Figure 4, we give the performance of the ideal predictive VA-based detector for different values of P. Note that in Figure 4, the prediction coefficients are generated based on both the channel DFT and the noise, that is using (5) assuming that the receiver has the perfect knowledge of the noise-variance.

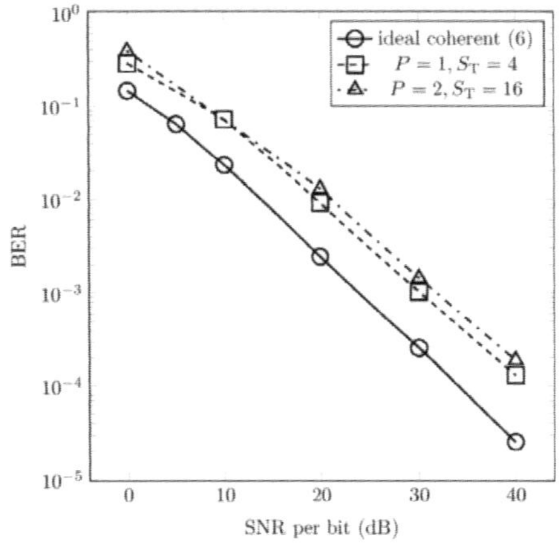

Figure 3: BER performance of the proposed predictive VA-based receiver for $L_d = 1024$, $\sigma_f^2 = 0.5$ and $L_h = 10$ using the predictor coefficients obtained from $\tilde{R}_{\tilde{H}\tilde{H},m}$ in (4).

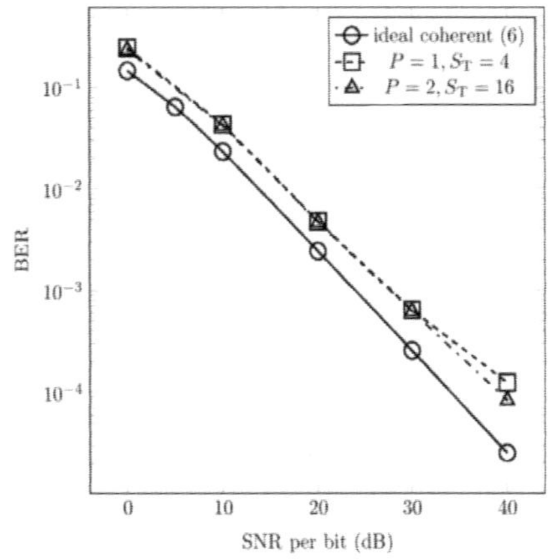

Figure 4: BER performance of the ideal predictive VA-based receiver for $L_d = 1024$, $\sigma_f^2 = 0.5$ and $L_h = 10$ using the predictor coefficients obtained from $\tilde{R}_{\tilde{H}\tilde{H},m}$ in (5)

6. Conclusion

Simulation results show that the performance of the proposed method is as good as the ideal predictive VA-based receiver. The proposed approach is perfectly suitable for applications which demand high throughput, since the proposed detection method is without channel estimation which requires training symbols. Future work can be focussed on incorporating error correcting codes to reduce the transmit power.

References

[1] Vasudevan K 2007 *Digital Communications and Signal Processing* (Hyderabad: Universities Press)

[2] Vasudevan K 2015 Coherent detection of turbo-coded ofdm signals transmitted through frequency selective fading channels with receiver diversity and increased throughput *Wireless Personal Communications* **82** 1623-42

[3] Vasudevan K 2013 *IEEE Int. Conf. on Signal Processing Computing and Control* (Shimla) pp 26-28

[4] Sanzi F, Jelting S and Speidel J 2003 A comparative study of iterative channel estimators for mobile ofdm systems *IEEE Transactions on Wireless Communications* **2** 849-59

[5] Vasudevan K, Giridhar K and Ramamurthi B 2001 Efficient suboptimum detectors based on linear prediction in Rayleigh flat-fading channels *Signal Processing Journal* **81** 819-28

[6] Coleri S, Ergen M, Puri A and Bahai A 2002 Channel estimation techniques based on pilot arrangement in ofdm systems *IEEE Transactions on Broadcasting* **48** 223-29.

IOP Conf. Series: Journal of Physics: Conf. Series **787** (2017) 012025 doi:10.1088/1742-6596/787/1/012025

Adaptive Beamforming with Inadequate Snapshots

Jing YU and Yaan LI

School of Marine Science and Technology, Northwestern Polytechnical University Xi'an, China

E-mail: yujing@nwpu.edu.cn

Abstract. In array signal processing, the covariance matrix used to calculate the adaptive weights is often poor estimated when the snapshot number is inadequate. The prior environmental knowledge can be used to make the estimation more accuracy. In this paper, an alternative knowledge-aided adaptive beamforming approach that is robust to low sample support environment is proposed. In this algorithm the covariance matrix used to calculate the optimum weights is constructed by blending a sample covariance matrix and a priori structured covariance matrix. Numerical simulations demonstrate the proposed algorithm has the potential for substantial performance improvement.

1. Introduction

In recent decades, adaptive beamforming has been widely applied to wireless communications, radar, sonar, speech signal processing, and many other fields [1-3]. With the use of adaptive beamforming, the radiation pattern of arrays can be adaptively changed to reject interferences and can provide significant performance advantages. Capon beamformer [3] is one of the most popular criteria for beamforming, which maximize the output signal-to-interference-plus-noise ratio (SINR) adaptively. However the formulation of the Capon beamformer requires explicit knowledge of the interfering environment, namely the covariance matrix and the steering vector of the interested signal. Uncertainties in these quantities may lead performance degradation. Until recently, many works have been put forward dealing with the steering vector mismatch and improving the robustness of the adaptive beamformers . Most of the existing algorithms use diagonal loading approach to alleviate the aforementioned problems, but the loading levels are difficult to choose.

In this paper, we focus our mind on low sample support (inadequate snapshots) caused covariance matrix uncertainty problem, which may badly affects the performance of beamforming . And a novel knowledge-aided adaptive beamforming algorithm was proposed using the concept of knowledge-aided signal processing techniques. In this algorithm we make full use of the priori covariance matrix and mixing it with the estimated covariance matrix. Use the newly constructed matrix to calculate the optimum adaptive weights can bring exciting performance improvement.

2. System Model

The output of a narrowband adaptive beamformer containing M sensors is given by

$$\mathbf{y}(k) = \mathbf{w}^{\mathrm{H}}\mathbf{x}(k) \tag{1}$$

where k is the time index, $\mathbf{x}(k)=[\mathrm{x}_1(k),\cdots,\mathrm{x}_\mathrm{M}(k)]^\mathrm{T}\in C^{\mathrm{M}\times 1}$ is the kth output snapshot, $\mathbf{w}=[w_1,\cdots,w_M]^\mathrm{T}\in C^{\mathrm{M}\times 1}$ is the complex vector of adaptive weights, and $(\cdot)^\mathrm{T}$ and $(\cdot)^\mathrm{H}$ stand for the transpose and Hermitian transpose, respectively.

The observation vector is given by

$$\mathbf{x}(k)=\mathbf{x}_s(k)+\mathbf{i}(k)+\mathbf{n}(k)=\mathrm{s}(k)\mathbf{a}+\mathbf{i}(k)+\mathbf{n}(k) \tag{2}$$

where $\mathbf{x}_s(k)$, $\mathbf{i}(k)$ and $\mathbf{n}(k)$ are the desired signal, interference and noise components, which are statistically independent. $\mathrm{s}(k)$ is the signal waveform, and $\mathbf{a}$ is the corresponding actual steering vector of the desired signal. The adaptive beamformer weights $\mathbf{w}$ can be obtained by maximizing the signal-to-interference-plus-noise ratio (SINR).

$$SINR=\frac{\sigma_s^2\left|\mathbf{w}^\mathrm{H}\mathbf{a}\right|^2}{\mathbf{w}^\mathrm{H}\mathbf{R}_{i+n}\mathbf{w}} \tag{3}$$

where

$$\mathbf{R}_{i+n}=\mathrm{E}\left\{(\mathbf{i}(k)+\mathbf{n}(k))(\mathbf{i}(k)+\mathbf{n}(k))^\mathrm{H}\right\} \tag{4}$$

is the $M\times M$ interference-plus-noise covariance matrix and $\sigma_s^{\ 2}$ is the signal power.

The optimization problem of maximizing array output SINR can be equivalently written as

$$\begin{cases}\min\limits_{\mathbf{w}} \mathbf{w}^\mathrm{H}\mathbf{R}_{i+n}\mathbf{w}\\ \text{s.t.}\quad \mathbf{w}^\mathrm{H}\mathbf{a}=1\end{cases} \tag{5}$$

which solution is the well-known Capon beamformer.

$$\mathbf{w}_{opt}=\frac{\mathbf{R}_{i+n}^{-1}\mathbf{a}}{\mathbf{a}^\mathrm{H}\mathbf{R}_{i+n}^{-1}\mathbf{a}} \tag{6}$$

Until recently some remedies have been made to Capon algorithm, so in this paper we refer it to the standard Capon beamformer.

In practical applications, the exact interference-plus-noise covariance matrix $\mathbf{R}_{i+n}$ is not known. Therefore the sample covariance matrix $\hat{\mathbf{R}}$ which is estimated using some independent identically distributed (i.i.d.) neighboring snapshots is often used instead of $\mathbf{R}_{i+n}$.

$$\hat{\mathbf{R}}=\frac{1}{N}\sum_{n=1}^{N}\mathbf{x}(n)\mathbf{x}^\mathrm{H}(n) \tag{7}$$

where N denotes the number of i.i.d. snapshots.

Reed, Mallett, and Brennan [4] have shown that the average signal-to-interference plus noise ratio (SINR) loss will be about -3dB if $N\geq 2M$, which is often called RMB rule. However, in real environment, this is no longer true if few i.i.d. snapshots are available to estimate the covariance matrix. For example N is comparable with or even smaller than M, $\hat{\mathbf{R}}$ usually is a poor estimation of $\mathbf{R}_{i+n}$. So the adaptive weights obtained from equation (6) may lead distorted mainbeam and high sidelobes. Whereas in real world, sample heterogeneity frequently occurs, which means many sample snapshots do not share the same spectral properties, the estimation of $\mathbf{R}_{i+n}$ may also be inaccurate. So in recent years, knowledge-aided signal processing techniques are put forwarded to deal with this effect. In these techniques, all kinds of available knowledge about the environments, the array, and the interferences are used to improve the performance.

Until recently two paradigms for knowledge-aided processing methods are presented [5-8]. One paradigm is an indirect approach where a priori knowledge is used for selecting the training data and the other one is a direct approach where a priori knowledge is used to prewhitening the samples. In the latter section, a new direct knowledge-aided adaptive beamforming algorithm will be proposed.

3. Knowledge-aided adaptive beamforming approach

In this part, we consider the eigen-decomposition of the interference-plus-noise covariance matrix first.

It can be expressed

$$\mathbf{R}_{i+n} = \mathbf{E}\{\mathbf{x}(n) * \mathbf{x}(n)^{\mathrm{H}}\} = \mathbf{R}_i + \mathbf{R}_n = \sum_{j=1}^{p} \lambda_j \mathbf{V}_j \mathbf{V}_j^{\mathrm{H}} + \sum_{j=p+1}^{M} \lambda_j \mathbf{V}_j \mathbf{V}_j^{\mathrm{H}} \quad (8)$$

where $\mathbf{R}_i$ and $\mathbf{R}_n$ represent the covariance matrix of interference and noise respectively, $\lambda_1 \geq \lambda_2 \geq \cdots \geq \lambda_p >> \lambda_{p+1} = \lambda_{p+2} = \cdots = \lambda_M = \sigma_n^2$ are the eigenvalues of the covariance matrix. Meanwhile, the eigenvalues of the interferences are much larger than that of noises. $\mathbf{V}_1, \mathbf{V}_2, \cdots, \mathbf{V}_M$ are the corresponding eigen vectors, P denotes the number of the interferences and σ_n^2 is the noise power.

In practice, in a sensor array the power of interferences is always much stronger than that of noise. Therefore, after eigen-decomposition of the covariance matrix, there exist several large eigenvalues and some small eigenvalues. According to the subspace theory, the number of large eigenvalues corresponding to the number of interferences and the number of small eigenvalues denotes the degree of freedom of the thermal noise. Consider an array which consists of 20 sensors with 2 interferences, and the interference-to-noise ratios (INR) are 30dB and 20dB respectively. Fig.1. shows the eigenspectra of the estimated covariance matrix with adequate snapshots. Obviously there existing 2 big eigenvalues and 18 small eigenvalues which correspond to the number of interferences and the noise's degree of freedom respectively. It can also be seen from Fig.1 that the small eigenvalues have the similar magnitudes which are close to the noise level.

Theoretically if the snapshots are limited, for the aforementioned case choose the snapshots less than 40, $\hat{\mathbf{R}}$ may be a poor estimation and near singular. The eigenspectra of $\hat{\mathbf{R}}$ can be shown in Fig.2. The difference between the two cases is in the spread of the eigenvalues which correspond to the random noise. With few snapshots there is inadequate estimation of the noise and a large noise eigenvalue spread occurs. As more samples are used, the estimation improves and the eigenvalues converge on the expected value of the noise. The spread of the small eigenvalue may leads sidelode elevation and mainbeam distortion, thus further affects the beamforming capability.

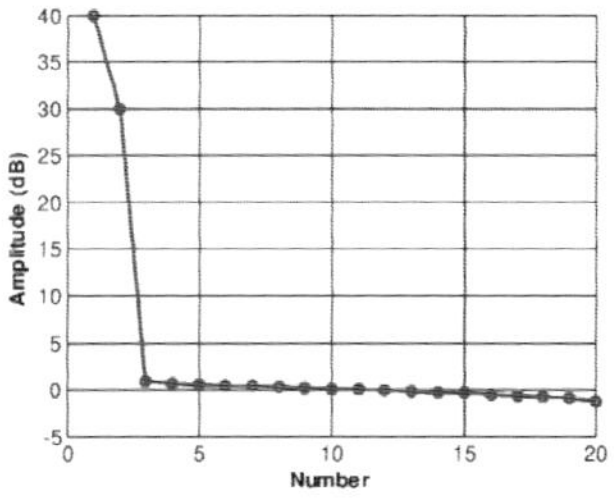

Figure 1. Eigenspectra of covariance matrix with adequate snapshots

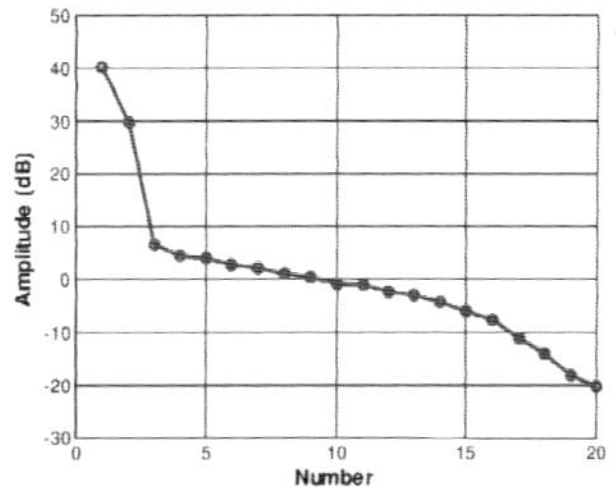

Figure 2. Eigenspectra of covariance matrix with inadequate snapshots

Supposing that a priori structured covariance matrix obtained by previous experiment is available, a reconstructed covariance matrix acquired by blending the estimated covariance matrix with a priori covariance matrix according to the characteristic of eigenspectra can be used to enhance the robustness of beamforming.

Next, we will show how to build the new covariance matrix and the reason why it can be used to improve the robustness of the performance.

The eigen-decomposition of the estimated covariance $\hat{\mathbf{R}}$ is written below:

$$\hat{\mathbf{R}} = \sum_{i=1}^{p} \gamma_i \mathbf{U}_i \mathbf{U}_i^{\mathrm{H}} + \sum_{i=p+1}^{M} \gamma_i \mathbf{U}_i \mathbf{U}_i^{\mathrm{H}} \quad (9)$$

where γ_i and $\mathbf{U}_i$ are eigenvalue and eigenvector, respectively. p denotes the number of interferences which equals the number of large eigenvalues.

The prior covariance matrix in ideal environment with known array structure and interferences can be expressed

$$\mathbf{R}_{prior} = \sum_{j=1}^{p} \alpha_j \mathbf{C}_j \mathbf{C}_j^{\mathrm{H}} + \sum_{j=p+1}^{M} \alpha_j \mathbf{C}_j \mathbf{C}_j^{\mathrm{H}} \tag{10}$$

where α_j and $\mathbf{C}_j$ denote the eigenvalue and eigenvector respectively. p also indicate the number of interferences.

Build a new matrix $\mathbf{R}_{const}$ using the noise subspace of the prior covariance matrix $\mathbf{R}_{prior}$ and the interference subspace of the estimated covariance $\hat{\mathbf{R}}$.

$$\mathbf{R}_{const} = \sum_{i=1}^{p} \gamma_i \mathbf{U}_i \mathbf{U}_i^{\mathrm{H}} + \sum_{i=p+1}^{M} \alpha_i \mathbf{C}_i \mathbf{C}_i^{H} = \mathbf{QGQ}^{\mathrm{H}} \tag{11}$$

$$\mathbf{Q} = \left[\mathbf{U}_1, \cdots, \mathbf{U}_p, \mathbf{C}_{p+1}, \cdots, \mathbf{C}_M\right] \quad , \quad \mathbf{G} = \begin{bmatrix} \gamma_1 & 0 & 0 & 0 & 0 & 0 \\ 0 & \ddots & 0 & 0 & 0 & 0 \\ 0 & 0 & \gamma_p & 0 & 0 & 0 \\ 0 & 0 & 0 & \alpha_{p+1} & 0 & 0 \\ 0 & 0 & 0 & 0 & \ddots & 0 \\ 0 & 0 & 0 & 0 & 0 & \alpha_M \end{bmatrix} \tag{12}$$

Using $\mathbf{R}_{const}$ instead of $\mathbf{R}_{i+n}$ in equation (6) to acquire the adaptive weights may generate stable beam patterns. That is because the reconstructed matrix $\mathbf{R}_{const}$ overcomes the spread of the small eigenvalues.

In equation (11), p should be a known quantity. However, in practical applications, we may not exactly know the number of interferences. Fortunately, we can obtain it by analyzing the characteristic of the eigenspectra of the estimated covariance matrix and the number of the large eigenvalue corresponds to the number of interferences. So, to use this algorithm one must first observe the eigenspectra of the estimated covariance matrix $\hat{\mathbf{R}}$ and distinguish the large and small eigenvalues.

Generally in real systems, interferences are much stronger than thermal noise, so the large eigenvalues are easy to separate from their magnitude. The new approach is easy to be implemented and the simulations in the next section validate the superiority over the Capon beamformer.

4. Simulation results

In our simulations, a uniform linear array with $M = 20$ omnidirectional sensors which has half-wavelength inter-element spacing is considered. Through all simulations, two interfering sources are assumed with plane wave fronts and the directions of arrival are $50°$ and $70°$, respectively. The additive noise is modeled as a complex Gaussian zero-mean spatially and temporally white process. The INR in a single sensor is equal to 20dB. The desired signal is assumed to be from the direction $\theta_s = 0°$. For each scenario, 200 Monte-Carlo runs are used to obtain each simulated point. The proposed knowledge-aided algorithm is compared to Capon beamformer.

The output SINR versus the number of i.i.d. samples is illustrated in Fig.3. In this simulation the signal-to-noise ratio SNR=2dB. It is clear from the picture that the proposed knowledge-based algorithm has better performance compared to the Capon beamformer. Especially, when the i.i.d. samples are limited, the knowledge-based method still has stable capability while the Capon beamformer behaves invalid. As the number of i.i.d. samples becomes larger the performances of two algorithms tend to be the same.

Fig.4 shows the output SINR versus SNR for the fixed training sample size N=200. It can be seen from the figure that the knowledge-based method still has advantages over Capon beamformer when the training samples are plenteous.

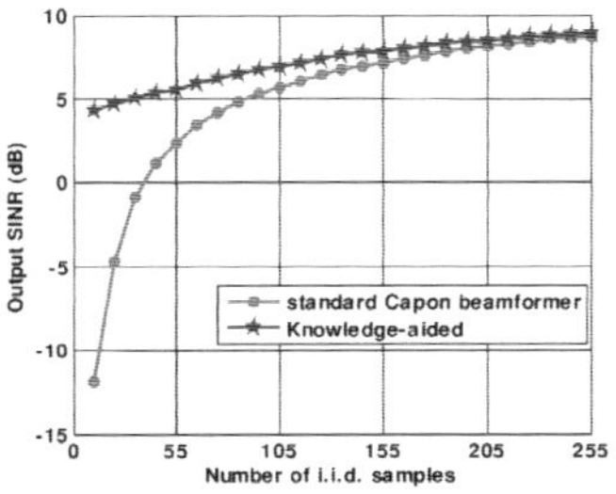

Figure 3. Output SINR against number of i.i.d. samples

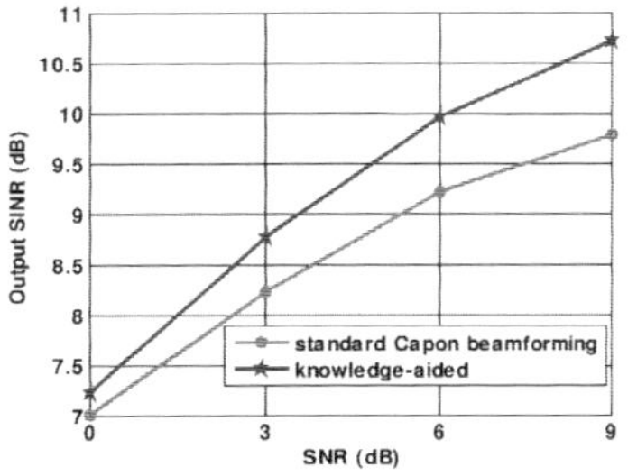

Figure 4. Output SINR against SNR

5. Conclusions

To overcome the covariance matrix uncertainty in low sample support environment a novel knowledge-aided beamforming process is proposed. Unlike the existing diagonal loading based methods in which the loading level should be known, in this new approach we made full use of the prior information of the environment. The covariance matrix used to calculate the optimum weights is constructed by blending a sample covariance matrix and a priori structured covariance matrix. The approach is easy to be implemented and the experiment results demonstrate that in low sample support environment it has superior performance than the Capon beamformer.

Acknowledgment

The work described in this paper is supported by the National Natural Science Foundation of China (Grant No.51409214) and Natural Science Foundation research project of Shaanxi Province (Grant No.2016JQ6071). The authors would appreciate the anonymous reviewers for their valuable comments and suggestions for improving this paper.

References

[1] Jian Li;PetreStoica;Zhisong Wang."On robust capon beamforming and diagonal loading", IEEE Transactions on Signal Processing, vol.51, no.7, July 2003, pp.1702-1715.

[2] Besson O.; Bidon S. "Adaptive Processing with Signal Contaminated Training Samples", IEEE Transactions on Signal Processing,vol. 61, no. 17, Sept. 2013, pp. 4318-4329.

[3] Capon J. "High resolution frequency-wavenumber spectrum analysis", Proceedings of IEEE, vol.57, no.8,Aug.1969, pp.1408-1418.

[4] Reed. I.S.; Mallett. J.D.; Brennan L.E. "Rapid convergence rate in adaptive arrays". IEEE Transactions on Aerospace and Electronic Systems, vol.10, no.6, Nov. 1974, pp.853-863.

[5] Huang Lei, et al. "Robust adaptive beamforming with a novel interfernce-plus-noise covariance matrix reconstruction method", IEEE Transactions on Signal Processing vol.63, no.7, 2015,pp.1643-1650.

[6] Wang Yasen, Qinglong Bao, et al. "Robust adaptive beamforming using IAA-based interference-plus-noise covariance matrix reconstruction", Electronics Letters, vol.52, no.13, 2016,pp.1185-1186.

[7] Zhang Zhenyu, et al. "Interference-plus-noise covariance matrix reconstruction via spatial power spectrum sampling for robust adaptive beamforming'. IEEE Signal Processing Letters, vol.23, no.1, 2016, pp.121-125.

[8] Melvin W.L.; Wicks M, Antonik P. "Knowledge-based space-time adaptive processing for airborne earlywarning radar", IEEE Aerospace and Electronic Systems Magazine, vol.13, no.4, Apr.1998,pp.37-42.

DOA estimation based on multiple beamspace measurements sparse reconstruction for manoeuvring towed array

J Yuan[1], H Xiao[2], Z M Cai[1] and C Xi[1]

[1] Naval University of Engineering, Wuhan, Hubei, China
[2] Air Force Early Warning Academy, Wuhan, Hubei, China

E-mail: yjun_hg@163.com

Abstract. The port-starboard ambiguity in the conventional single towed linear array sonar is one of the most deceiving obstacles which exist in the way of development of spatial spectrum estimation. In order to improve the performance of target detection and Direction of Arrival (DOA) estimation, this paper proposes a novel spatial spectrum sparse reconstruction method based on multiple beamspace measurements (MBM-SR). An array sparse signal model for manoeuvring towed array is established. Then the Mutual Incoherent Property (MIP) is analyzed to ensure the proposed algorithm possessing better spatial spectrum reconstruction property. Simulation results demonstrate that, compared with Conventional Beam Forming (CBF) algorithm, the proposed algorithm has evident advantage in ambiguity suppression ratio (ASR) and estimation performance.

1. Introduction

Towed hydrophone arrays are useful for detecting submerged submarines and monitoring vocalizing marine mammals [1]. The conventional towed arrays comprise multiple omni-directional hydrophone elements drawn behind a vessel. When towing an in-line hydrophone array from a vessel sailing in a straight line, it is impossible to determine which side of the vessel an underwater sound source originates from, because the signal time difference of arrival (TDOA) is same from either side, referred to as 'port-starboard ambiguity'. Port-starboard ambiguity can be resolved by changing the heading of the vessel, but in many applications, the array shape is treated as a straight line which can result in large variance in the bearing estimates and influence the stable target tracking during vessel turning. In order to resolve the port-starboard ambiguity effectively, various towed arrays are studied such as twin-line array [2], array with hydrophone triplets [3] and acoustic vector sensor array [4]. But these arrays are not used widely in military and industrial operations because of increased costs, sophisticated technologies and limited deployment possibilities. Therefore, how to develop methods of signal processing to realize DOA estimation without port-starboard ambiguity for towed hydrophone array is still a challenging research topic. In practice, when the vessel manoeuvres, the towed array may not be a true line array. The distorted array shape induced by the turning of vessel is utilized for resolving port-starboard ambiguity [5, 6], because the symmetric feature of TDOA is destroyed by the distorted array. However, the distorted array can result in reduced signal gain in the beamforming process and limited snapshot number can restrict the use of many adaptive covariance based approaches.

 doi:10.1088/1742-6596/787/1/0

Compressive sensing is a new signal sparse reconstruction technique which requires fewer measurements to reconstruct the signal, which is sparse in some basis vectors [7-10]. In this paper, a novel approach to DOA estimation in beamspace for maneuvering towed array is proposed. Utilizing multiple beamspace measurements, the sparse reconstruction algorithm results in high ambiguity suppression ratio and good estimation performance.

The outline of the paper is organized as follows. In section 2, we present the array signal model for manoeuvring towed array. In section 3, the details of spatial spectrum sparse reconstruction algorithm based on multiple beamspace measurements is discussed. Simulation results are presented in section 4. Section 5 concludes the paper.

2. Array sparse signal model for manoeuvring towed array

2.1. Spatial spectrum sparse reconstruction in beamspace

To simplify the subsequent discussion, the coordinate system of maneuvering towed array is illustrated in figure 1. The towed array is made up with M sensors with equal separation d. Narrowband far field source from the direction θ_q impinges on the towed array. The first sensor that is the closest to the towing vessel is located at original point.

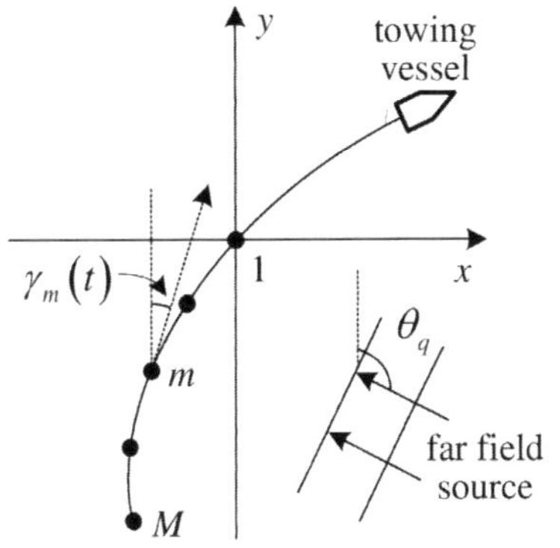

Figure 1. Coordinate system of manoeuvring towed array.

When the towing vessel is turning, the shape of array is distorted. Then with the AR model, the shape of array can be described as [11]

$$\gamma(t)=\Gamma\gamma(t-1)+\mathbf{v}(t)+z(t) \tag{1}$$

$$\gamma(t)=\left[\gamma_2(t),\gamma_3(t),\cdots,\gamma_M(t)\right]^T \tag{2}$$

$$\Gamma=(1-\rho)\mathbf{I}_{M-1}+\rho\begin{bmatrix}0&0&\cdots&0\\1&0&\cdots&0\\0&\ddots&0&0\\0&0&1&0\end{bmatrix}_{(M-1)\times(M-1)} \tag{3}$$

$$\mathbf{v}(t)=\rho\left[\gamma_1(t),0,\cdots,0\right]^T_{(M-1)\times 1} \tag{4}$$

where $\gamma_m(t)$ denotes heading angle of sensor m at time instant t. $\mathbf{v}(t)$ is an update item which is obtained from heading angle of first sensor. ρ is the ratio of current sensor displacement and intersensor spacing. When $\rho = 1$, the heading angle of the current sensor is the same with that of the previous sensor close to it. When $\rho = 0$, the heading angle of sensor is independent from each other. $z(t)$ denotes process white noise.

2.2. *The sparse received signal model*

As illustrated in figure 1, the position of the sensor $m \in \{2,3,\cdots,M\}$ at time instant t can be written in vector form as

$$\boldsymbol{r}_m = \left[-\sum_{i=2}^{m} d \sin\left[\gamma_i(t)\right], -\sum_{i=2}^{m} d \cos\left[\gamma_i(t)\right] \right]^T \tag{5}$$

Consider a source located at θ_q, the direction unit vector can be written as

$$\boldsymbol{u}_q = -\left[\cos(\theta_q), \sin(\theta_q)\right]^T \tag{6}$$

Then we obtain the steering vector as

$$\boldsymbol{a}(\theta_q) = \left[1, \exp\left(-\mathrm{i}2\pi \boldsymbol{r}_2^T \boldsymbol{u}_q / \lambda\right), \cdots, \exp\left(-\mathrm{i}2\pi \boldsymbol{r}_M^T \boldsymbol{u}_q / \lambda\right)\right]^T \tag{7}$$

where λ is wavelength. Based on the theory of sparse reconstruction, the bearing space is uniformly discretized into Q angular directions, and the steering matrix can be written as

$$\mathbf{A}(\Theta) = \left[\boldsymbol{a}(\theta_1), \boldsymbol{a}(\theta_2), \cdots, \boldsymbol{a}(\theta_Q)\right] \tag{8}$$

where $\theta_1, \theta_2, \cdots, \theta_Q$ are all possible directions.

Then we obtain the sparse received signal model of the towed array as

$$\boldsymbol{y} = \mathbf{A}\boldsymbol{x} + \boldsymbol{n} \tag{9}$$

where $\boldsymbol{y}$ is received complex data vector at the M sensors, $\boldsymbol{x}$ is signal vector of all discredited directions, the components of $\boldsymbol{x}$ can be modelled as complex Gaussian with zero mean that is uncorrelated each other, $\boldsymbol{n}$ is independent and identically distributed complex Gaussian noise vector.

3. MBM spatial spectrum sparse reconstruction algorithm

3.1. *Spatial spectrum sparse reconstruction in beamspace*

The beamspace sparse-regularization approach was proposed by Malioutov [9]. The compressive measurement data is transformed from sensor space to beamspace using steering manifold vectors. Based on the conventional beamforming principle, the beam output power can be written as

$$b(\varphi) = E\left\{\left|\boldsymbol{w}^H(\varphi) \cdot \boldsymbol{y}\right|^2\right\} \tag{10}$$

where $\boldsymbol{w}(\varphi) = \frac{1}{M}\left[1, \exp\left(-\mathrm{i}2\pi \boldsymbol{r}_2^T \boldsymbol{u} / \lambda\right), \ldots, \exp\left(-\mathrm{i}2\pi \boldsymbol{r}_M^T \boldsymbol{u} / \lambda\right)\right]^T$ is beamforming steering weighting vector for angle φ and $\boldsymbol{u} = -\left[\cos(\varphi), \sin(\varphi)\right]^T$. Using equation (9) in equation (10) gives

$$b(\varphi)=E\left\{\left|w^H(\varphi)\cdot(\mathbf{A}x+n)\right|^2\right\}=E\left\{w^H(\varphi)\mathbf{A}\cdot xx^H\cdot\mathbf{A}^H w(\varphi)+w^H(\varphi)\cdot nn^H\cdot w(\varphi)\right\}$$
$$=h(\varphi)\cdot p+\left\|w(\varphi)\right\|^2\cdot\sigma_n^2 \tag{11}$$

where $h(\varphi)=\left[\left|w^H(\varphi)\cdot a(\theta_1)\right|^2,\left|w^H(\varphi)\cdot a(\theta_2)\right|^2,\ \dots\ ,\left|w^H(\varphi)\cdot a(\theta_Q)\right|^2\right]$, Q denotes the number of all potential directions. $p=\left[E\left\{|x_1|^2\right\},E\left\{|x_2|^2\right\},\cdots,E\left\{|x_Q|^2\right\}\right]^T$ is sparse signal power vector, σ_n^2 is sensor noise power. Then the beamspace measurement vector is

$$b(\Phi)=\mathbf{H}\cdot p+\left[\left\|w(\varphi_1)\right\|^2,\cdots,\left\|w(\varphi_K)\right\|^2\right]^T\cdot\sigma_n^2 \tag{12}$$

where $\mathbf{H}=\left[h^T(\varphi_1)\ \ h^T(\varphi_2)\ \ \cdots\ \ h^T(\varphi_K)\right]^T$ is sensing matrix.

The sparse signal power vector can be recovered by the following ℓ_1 minimization problem

$$\hat{p}=\arg\min_{p\in\mathbb{C}^Q}\left\|p\right\|_1 \quad \text{s.t.} \quad \left\|\mathbf{H}p-b\right\|_2\le\zeta \tag{13}$$

where $\zeta=\sigma_n^2 M^{-1/2}$.

3.2. *MIP analysis of sensing matrix for manoeuvring towed array*

To ensure good performance of sparse reconstruction, the sensing matrix needs to satisfy reasonable conditions such as Mutual Incoherent Property (MIP) [12] or Restricted Isometry property (RIP) [13]. The mutual coherent between any two columns of $\mathbf{H}$ is defined as

$$\eta(\mathbf{H})=\max_{k\neq l}\mathbf{G}_{kl} \tag{14}$$

where $\mathbf{G}_{kl}=\left|h_k^H\times h_l\right|\cdot\left(\left\|h_k\right\|_2\times\left\|h_l\right\|_2\right)^{-1}$ is the absolute Gram matrix.

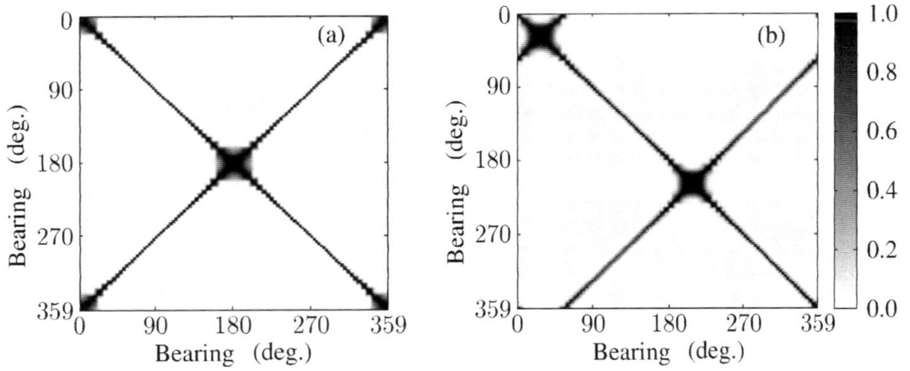

Figure 2. Gram matrix in bearing angle for a manoeuvring towed array. (a) Straight array shape. (b) Distorted array shape.

For example, a 40-sensors towed array ($d/\lambda=0.45$), the matrix $\mathbf{G}$ as a function of bearing angle is presented in figures 2. When the towed array is drawn in a straight line along y axis, the Gram matrix is depicted in figure 2(a). The coherence is high along minor diagonal which is caused by port-starboard ambiguity. When the towing vessel changes course with a steady angular speed, the Gram matrix at some time instant is depicted in figure 2(b), the high coherence moves parallel to the minor diagonal which denotes the ambiguity angle is time-varying during vessel turning.

3.3. MBM spatial spectrum sparse reconstruction

According to the characteristic of port-starboard ambiguity at different time instant, the solution to spatial spectrum estimation of towed array can be imagined by stacking together several error norms of beamspace measurements, each one corresponding to a time-varying array shape. Then equation (13) can be rewritten as

$$\hat{\boldsymbol{p}} = \arg\min_{\boldsymbol{p}\in\mathbb{C}^Q} \|\boldsymbol{p}\|_1 \quad \text{s.t.} \sum_{i=t-N}^{t} \|\mathbf{H}_i\boldsymbol{p} - \boldsymbol{b}_i\|_2 \le \zeta \tag{15}$$

where N is the number of beamspace measurement. The equation (15) is a disciplined convex optimization problem which can be solved efficiently using cvx toolbox.

4. Simulation

In the following simulation, we consider an ULA with M=40 sensors with spacing 1.5 m. The frequency of far field narrowband signal is 450 Hz and the speed of towing vessel is 6 knots. Initially, the towing vessel heads North (0°) and performs a 40° turn toward East.

To evaluate the performance of the algorithm, the following scenario is generated: two signals of interest are placed at 98° and 165° with magnitudes 12 dB and 9 dB respectively. The number of snapshots is L=5, the number of beamspace measurement is N=20 and the number of grid division is Q=360, searching from 0° to 359°. At SNR= -5 dB, the performance of spatial spectrum estimation and ambiguity suppression with MBM-SR and CBF is showed in figure 3. The number of independent Monte Carlo trials is 100 at each SNR.

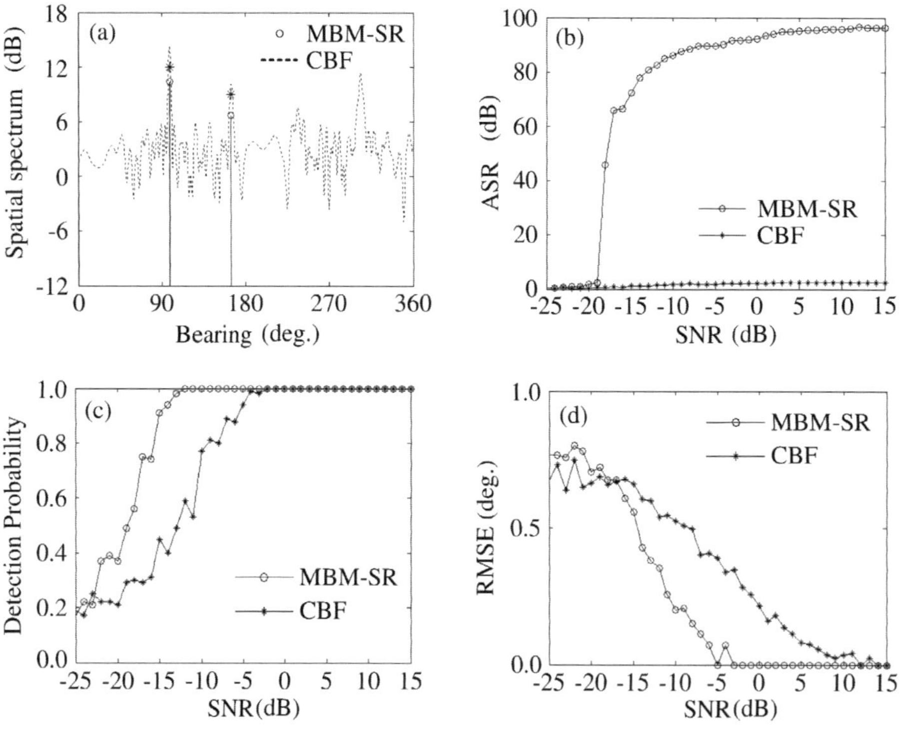

Figure 3. Spatial spectrum of two signals with DOAs of 98° and 165° (SNR= -5 dB).

Figure 3(a) shows that MBM-SR algorithm can suppress two ambiguity targets that located at 237° and 304°. While with CBF, the two ambiguity peaks is still there. The true signals are indicated with asterisks. Figure 3(b) shows that the ambiguity suppression ratio (ASR) of MBM-SR is higher than 90 dB when SNR>0 dB, while the suppression ratio of CBF is low even in high SNR. So MBM-SR is a valid algorithm in ambiguity suppression. Figure 3(c) is the detection probability of two algorithms. It shows that the detection probability of MBM-SR is higher than that of CBF when SNR<-5 dB. Figure 3(d) gives the DOA RMSE of two algorithms. It can be see that the RMSE of MBM-SR is lower than that of CBF, except for very low SNR.

In figure 4, we consider two targets placed at 98° and 100° respectively with SNR is 10 dB. The angle between them is smaller than the Rayleigh limit. Figure 4 shows that MBM-SR can distinguish two adjacent targets and CBF can't provide enough angle resolution to identify them. Therefore, the proposed algorithm possesses higher resolution.

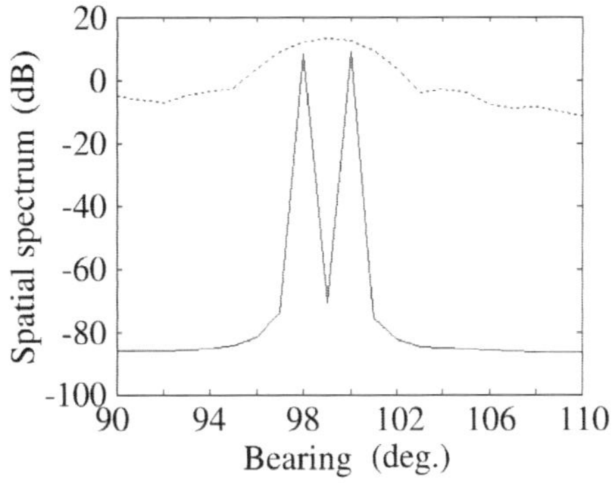

Figure 4. Spatial spectrum of two signals with DOAs of 98° and 100° (SNR= 10 dB).

5. Summary

In this paper, based on multiple beamspace measurements, a novel spatial spectrum sparse reconstruction algorithm for manoeuvring towed array is proposed. The spatial spectrum estimation and ambiguity suppression performance with MBM-SR is then evaluated in a simulation study. It is concluded as following: (1) compared with CBF, MBM-SR can provide evident higher ambiguity suppression ratio; (2) the detection probability and DOA RMSE of MBM-SR is better than that of CBF; (3) the proposed algorithm possesses higher resolution than CBF.

References

[1] Lemon S G 2004 *IEEE Journal Of Oceanic Engineering* **29** 365
[2] Feuillet J P, Allensworth W S and Newhall BK 1995 *J. Acoust.Soc. America* **97** 3292
[3] Groen J, Beerens S P and Doisy Y 2005 *IEEE Journal Of Oceanic Engineering* **30** 348
[4] Nehorai A and Paldi E 1994 *IEEE Transactions on Signal Processing* **42** 2481
[5] He X Y, Jiang X Z, Li Q H and Zhang C H 2004 *ACTA Acustica* **29** 409
[6] Rogers J and Krolik J 2010 *J. Acoust.Soc. America* **128** 3543
[7] Donoho D L 2006 *IEEE Transactions on Information Theory* **52** 1289
[8] Candes E, Romberg J and Tao T 2006 *IEEE Transactions on Information Theory* **52** 489
[9] Malioutov D, Cetin M and Willsky A 2005 *IEEE Transactions on Signal Processing* **53** 3010
[10] Donoho D L, Elad M and Temlyakov V 2006 *IEEE Transactions on Information Theory* **52** 6
[11] Paidoussis M P 1973 *J. Sound Vibration* **29** 365
[12] Elad M 2010 *Sparse and Redundant Representations: From Theory to Applications in Signal and Image Processing* (New York: Springer) p 1–359
[13] Candès E J 2008 *Comptes Rendus Mathematique* 346 589

IOP Conf. Series: Journal of Physics: Conf. Series **787** (2017) 012027 doi:10.1088/1742-6596/787/1/012027

Fast Simulation Method for Ocean Wave Base on Ocean Wave Spectrum and Improved Gerstner Model with GPU

Wenqiao Zhang, Jing ZHANG and Tianchi ZHANG

Harbin Engineering University, Harbin, China

E-mail: hgczwq@163.com

Abstract. For the randomness and complexity of ocean wave, and the simulation of large-scale ocean requires a great amount of computation, but the computational efficiency is low, the real-time ability is poor, a fast method of wave simulation is proposed based on the observation and research results of oceanography, it takes advantage of the grid which combined with the technique of LOD and projection, and use the height map of ocean which is formd by retrieval of ocean wave spectrum and directional spectrum to compute with FFT, and it uses the height map to cyclic mapping for the grid on GPU which combined with the technique of LOD and projection to get the dynamic height data and simulation of ocean. The experimental results show that the method is vivid and it conforms with randomness and complexity of ocean wave, it effectively improves the simulation speed of the wave and satisfied with the real-time ability and fidelity in simulation system of ocean.

1. Introduction

In recent years, with the maturity simulation of ocean wave ,it is applied more and more. Such as reconstruction of Ocean Battlefield Environment, visualization of ocean prediction, training of navigation simulation , special effect production of film, 3D game and so on. Ocean wave is a kind of complicated stochastic process, at present, many scholars have made a lot of achievements for simulation of ocean wave in three areas[1]:The first is based on the physical characteristics of the wave of the sea to simulate, not directly simulates the movement of ocean wave but gets the dynamic state of each inside particle in wave through by calculation of Navier-Stokes equation, such as the method was proposed by Kass is similar to the reality but with a bad real-time effect, there are less application use the method. The second is based on the geometry of shape of ocean wave which constructed by mathematical formula of sines or cosines, it artificially sets wave parameters to simulate a approximately real surface of the sea[2], such as the method was proposed by Laibing Yan is easy to simulate and the real-time ability is better but with a poor authenticity and obviously artificial mechanisms, its suitable range is narrow and often used in the 3Dgame or animation. The third is based on the observational results of sea which use the appropriate ocean wave spectrum to simulate the surface of sea[3], such as the method was proposed by Baolong Wang with better ability of real-time and fidelity, the effect of simulation is well but the peaks of wave is smooth which is relative in realistic ocean in our life.

On the other hand, if each grid node requires calculation in each frame in the simulation, the height of wave in the complexity of ocean wave needs large computational amount, the CPU load is high and it is the most important influencing factors to the real-time ability for simulation of ocean, moreover, large scale simulation of sea needs the technique of LOD[4], but it requires frequent grid computing when the viewpoint in the roaming status,it will affect the rendering efficiency.

This paper presents a method to simulate ocean which aimed at the shortage of simulation in ocean wave spectrum which with peaks of wave is not realistic and the poor ability of real-time in large scale simulation of ocean, it combines with improved Gerstner model and algorithm of FFT to simulate the sea, and it uses the grid which combined with the technique of LOD and projection to cyclic texture mapping to get the dynamic height data, so as to realize the simulation of ocean wave. The experimental results show that the method has the characters of randomness and complexity, it effectively reduces the occupancy rate of CPU and satisfied with the ability of real-time and fidelity for simulation of sea in the real-time simulation system .

2. Gerstner model

Gestner-wave[5] is the first method for Simulation of sea of Computer Graphics(1986 Fourier and Reeves),using the geometric model which is constructed by sines or cosines to simulate ocean is not satisfied with the authenticity of the simulation system, but the Gerstner model can better simulate the growing wave and micro motion in Ocean scene, compared to the sine wave, the peaks of Gerstner model is more sharp when the vertex of wave front move to the peaks, it well makes up for the fault of simulation which is used sine wave. In addition, the model is also proposed the approximate approach to solve the equations of fluid mechanics, it has two obvious advantages:

The normal vector of Gerstner model is easier to solve and it is conducive to rendering and operating to luminous efficiency in simulation of sea.

According to the habit of establishment of three-dimensional coordinates, assume that the static sea level is XoZ plane, Y-axis is vertically upwards, at time t, with the amplitude of A ,the equations(2-1),(2-2) of motion of water particle which in the progress of wave propagation with the amplitude of A as follow[6]:

$$\begin{cases} \vec{x}(\vec{x_0},t) = \vec{x_0} - (\vec{k}/k)A\sin(\vec{k}\cdot\vec{x_0} - \omega t) & (2\text{-}1) \\ y(\vec{x_0},t) = A\cos(\vec{k}\cdot\vec{x_0} - \omega t) & (2\text{-}2) \end{cases}$$

$\vec{x_0}$ is the horizontal maker on the surface of sea when there is have no external forces, $\vec{x_0} = (x_0, z_0)$, $y(\vec{x_0},t)$ is the height of sea surface, $\vec{k}$ is the vector which has same direction with the spread wave,k is the number of waves which wavelength is λ , $k = 2\pi/\lambda$,the image (Figure.1) below shows the Gestner-wave with different parameters :

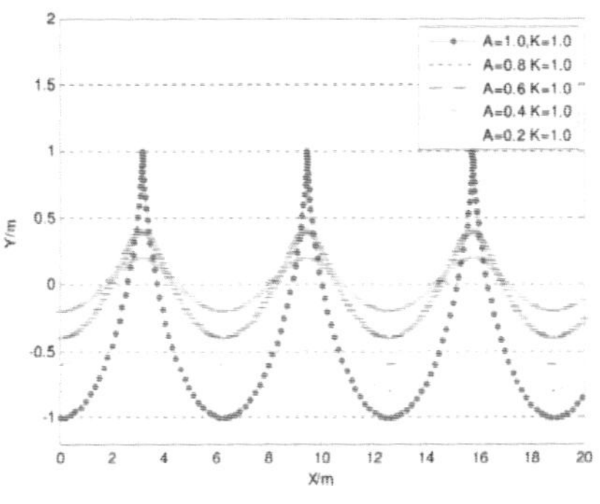

Figure 1. Gestner-wave with different parameters

Moreover, on the basis of the above equations, the wave is regarded as the superposition of N waves of Gerstner, and the equations of motion of the point on the sea level are as follows(2-3),(2-4)[7]:

$$\begin{cases} \vec{x}(\vec{x_0},t) = \vec{x_0} - \sum_{i=1}^{N}(\vec{k_i}/k_i)A_i\sin(\vec{k_i}\cdot\vec{x_0} - \omega t + \phi_i) & (2\text{-}3) \\ y(\vec{x_0},t) = \sum_{i=1}^{N} A_i\cos(\vec{k_i}\cdot\vec{x_0} - \omega t + \phi_i) & (2\text{-}4) \end{cases}$$

3. Ocean wave spectrum

In the early 1940s, Neumann and Pierson are proposed to study ocean wave by using stochastic process, as the so-called method of spectral analysis, this is the beginning of spectral analysis to analyse the waves [8].Ocean wave spectrum is also called energy spectrum which describes the relationship between the energy of the waves and the various components of the waves. And the ocean wave spectrum can be divided into two types: the first is the one-dimensional spectrum which shows the relationship between the wave energy and the frequency of the wave, it also called spectrum of ocean waves. The second is the two-dimensional spectrum which expresses the relationship between the wave energy and the frequency and direction of the waves, it also called directional spectrum of ocean waves.

Our commonly used now one-dimensional spectrum are as: Neumann spectrum, P-M spectrum, Phillips spectrum, Bretschneider spectrum [9], and so on. And the two-dimensional spectrum is generally used the commonly extension function to obtain, such as the function of ITTC(International Towing Tank Conference),ISSC(International Ship Structure Conference),SWOP(Stereo Wave Observation Project),and so forth.

Ocean wave spectrum is the analysis and summary of the observation of regional sea waves, there are lots of researches for the ocean wave spectrum at present. In general, if only consider about the factor of wind, the several ocean wave spectra have been proposed have the following standard form (3-1) when describing the waves:

$$S(\omega) = \frac{A}{\omega_p}\exp[-\frac{B}{\omega^q}] \tag{3-1}$$

A and B are the environmental parameters, in different observation area, wind speed, wind direction, the parameter A, B are rely on these measurements to determine their value, and the value of q, p is mainly considered the weight of high frequency and low frequency.

In addition, the magnitude and direction of the wind are random to led the waves travel in different directions, and even if the wind direction is fixed, there are some deviation about the direction of wave, In other words, the energy of wave is always distributed in different directions. In order to more accurately describe the random wave, introducing the dissipative information of wave in each direction, this is the direction spectrum. Assume θ is the angle which deflected the prevailing wind direction. $D(\theta,\omega)$ is the function of distributed direction, $S(\theta,\omega)$ is the directional spectrum which generally defined as follow(3-2):

$$S(\omega,\theta) = S(\omega)\bullet D(f,\theta) \tag{3-2}$$

And the function of distributed direction should meet the following conditions(3-3):

$$\int_{-\theta_0}^{\theta_0} D(\omega,\theta)\partial\theta = 1 \tag{3-3}$$

$-\theta_0 \sim \theta_0$ is the range of direction which in Energy distribution . ω is the Angular frequency.

4. Simulation method of ocean wave

The speed of simulation of ocean based on ocean wave spectrum is fast and the visual effect of large area of the sea is also better, but the shape of peaks of wave generated by algorithm of FFT is quite smooth, in fact, the shape of peaks is more sharp and the shape of troughs in practical situations even if the weather is calm and flat, and the phenomenon is more obvious in the abominable weather. On the other hand, the Gerstner model has bad performance for simulation of ocean and if the parameters of equations of motion are satisfied with some certain conditions, it could creates the phenomenon of wave interaction as follow (Figure .2)[10],and this situation is not going to happen in the ocean scene. But the shape of peak is actual in the simulation of ocean which based on the method of Gerstner model.

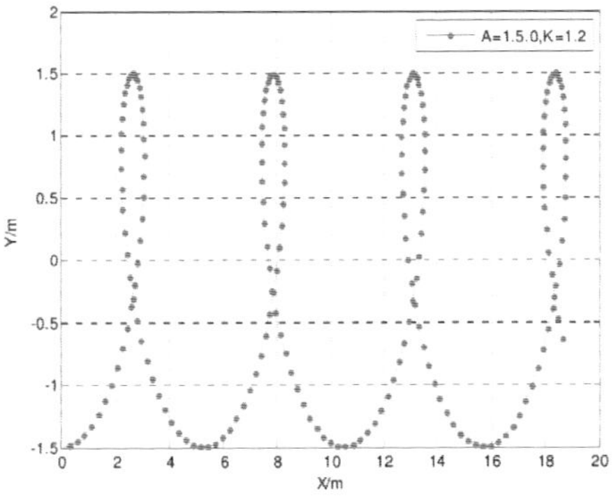

Figure 2. phenomenon of wave interaction

To solve these two problems, this paper combines with advantages of high authenticity of ocean wave spectrum and fidelity of peaks of Gerstner model, considering that the real-time ability of large scale of ocean modeling, it proposes a improved method to simulate the ocean wave .

4.1. Improved gerstner model with controlled factor

In order to prevent the phenomenon of wave interaction which based on simulation of Gerstner model and control the steepness of the peak, this paper add a control factor μ ,the improved formula as follow(4-1,4-2):

$$\begin{cases} \vec{x}(\vec{x_0},t) = \vec{x_0} - \mu(\vec{k}/k)A\sin(\vec{k}\cdot\vec{x_0} - \omega t) & (4\text{-}1) \\ y(\vec{x_0},t) = A\cos(\vec{k}\cdot\vec{x_0} - \omega t) & (4\text{-}2) \end{cases}$$

The simulation of waveforms which uses the above model are shown below (Figure .3)

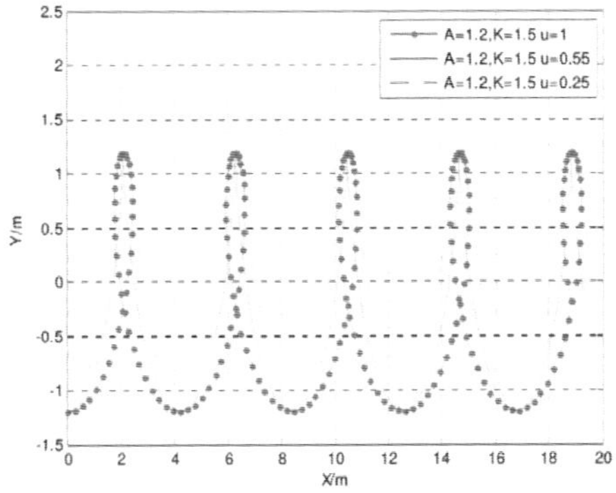

Figure 3. simulation of Gerstner model with Increased

Through repeated experiments, it is found regularity between k ,A, μ is: The condition of simulation without wave interaction is k×A>1,thus,the critical value μ shall conform to the relationship: $\mu \leq 1/(k \times A)$.And in order to generate the complex waveforms in two-dimensional plane, it makes the superposition of N Gerstner models, the equations of motion of the point on the sea level are as follows(4-3,4-4):

$$\begin{cases} \vec{x}(\vec{x_0},t) = \vec{x_0} - \mu \sum_{i=1}^{N} (\vec{k_i} / k_i) A_i \sin(\vec{k_i} \cdot \vec{x_0} - \omega t + \phi_i) & (4\text{-}3) \\ y(\vec{x_0},t) = \sum_{i=1}^{N} A_i \cos(\vec{k_i} \cdot \vec{x_0} - \omega t + \phi_i) & (4\text{-}4) \end{cases}$$

In this way, the profile of waveform generated by the superposition of N waves is shown in the figure(Figure .4) below, the abscissa represents the point on the sea level, the ordinate represents the amplitude about the point, therefore ,using the improved Gerstner with controlled factor can generate the more sharp waveform, this is more in line with the random pattern of ocean waves in the scene.

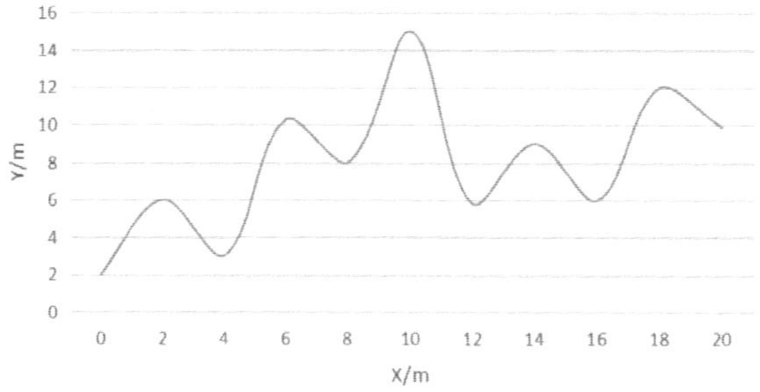

Figure 4. superposition of N waves

4.2. FFT operation of gerstner model and ocean wave spectrum

It has been set up the improved Gerstner with controlled factor in 3.1,Next, it should consider how to get the characteristic value of each wave from the sea wave spectrum, first to establish the grid of sea surface, using the coordinate values of each grid point to calculate the amplitude of wave to realize the simulation of the waves. This paper use the P-M spectrum as the frequency spectrum and use the function of SWOP as the extension function to construct the directional spectrum without loss of generality.

P-M spectrum is based on the Neumann spectrum, its observed data of the North Atlantic were analyzed by the method of spectral analysis,the spectrum shows the full development of the ocean Wave,the formula is shown as the following(4-5) [11] :

$$S_{PM}(\omega) = S_P(\omega)\exp(-\frac{5}{4}(\frac{\omega_P}{\omega})^4) \qquad (4\text{-}5)$$

In the formula, $S_P(\omega) = ag^2/\omega^5$,g is the gravity acceleration, g=9.8m/s2, ω_P is the Peak frequency of frequency spectrum, ω_P =0.855g/U5 ,U5 is the wind speed at 5 m from the sea level, using the international unit m/s, α is the constant of Phillips, α=0.0081.

Due to the P-M spectrum is one-dimensional, and the sea surface is three-dimensional, the above formula can only be described as the distribution of energy with frequency, it needs the function of distributed direction to reflect the distribution of wave energy with frequency in different directions. The method applies function of SWOP as follows(4-6,4-7,4-8)[12]:

$$\begin{cases} D_{SWOP}(\omega,\theta) = \frac{1}{\pi}[1 + p\cos(2\theta) + q\cos(4\theta)], |\theta| \le \pi/2 & (4\text{-}6) \\ p = 0.5 + 0.82\exp[-\frac{1}{2}(\frac{\omega}{\omega_p})^4] & (4\text{-}7) \\ q = 0.32\exp[-\frac{1}{2}(\frac{\omega}{\omega_p})^4] & (4\text{-}8) \end{cases}$$

According to the formula combined with P-M spectrum to get the spectrum of wave number as follows:

$$S_{PM-SWOP}(\omega,\theta) = S_{PM}(\omega)D_{SWOP}(\omega,\theta) \tag{4-9}$$

Then,the $S_{PM-SWOP}$ needs to discretize which means the frequency range and direction range of wave motion are divided,obtaining the interval of angular frequency $\Delta\omega$ and direction $\Delta\theta$,determining the angular frequency $\hat{\omega}_i$ and direction angle, the relationship between wave spectrum and amplitude is as follows[13]:

$$A_{ij} = \sqrt{2S(\hat{\omega}_i,\hat{\theta}_j)\Delta\omega_i\Delta\theta_j} \tag{4-10}$$

By using this formula, it can get the wave amplitude of different frequency and direction angle based on $S_{PM-SWOP}$,at the time, bringing the acquired amplitude A_{ij} of discrete points on the grid of sea level to the improved Gerstner with controlled factor to FFT calculation, and taking the value of calculation to the grid of sea to obtain the ocean wave with high authenticity.

4.3. Texture mapping of grid based on gpu

In the large simulation system of ocean, the CPU resources of computer are not exclusive to the ocean simulation when the system is in a state of multiple tasks, the speed is bound to be affected if the algorithm of ocean simulation is complex. This paper presents a fast method to accelerate the simulation of ocean, firstly, according to the reference, setting up the projection grid in the three-dimensional space, combining with the view rule of human eyes and technique of LOD, replacing the grid in reference to the grid whose center and surrounding is dense and sparse, as figure(Figure. 5) below shows:

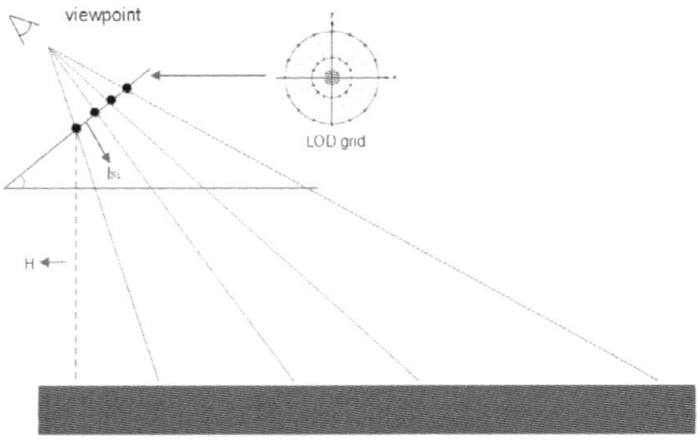

Figure 5. grid combined with the technique of LOD and projection

In the three dimensional world, the grid of concentric circle which uses S as the center, it makes N concentric circles with the radius of increased the distance and M partitions for i-th($1 \le i \le N$) concentric circle, each isotonic point is the vertex of the concentric circle for the grid model. The formula of each radius of the concentric circle as follows(4-11,4-12):

$$\begin{cases} R_i = 2R_{i-1}, i = 1,2,\cdots,N-1 & (4\text{-}11) \\ R_i = R_{\max}, \quad i = N & (4\text{-}12) \end{cases}$$

In the formula, Ri is the radius of i-th concentric circle, Rmax is the max distance for the sea level. And it sets the j-th vertex coordinates(x,y,z) for the i-th concentric circle is(4-13,4-14,4-15):

$$\begin{cases} x = R_i \cos(2\pi / M) & (4\text{-}13) \\ y = l_{Ni} \sin\theta + H, if\ y < (y_{\min} + y_{\max})/2 & (4\text{-}14) \\ z = R_i \sin(2\pi / M) & (4\text{-}15) \end{cases}$$

θ is the angle between the sea level and the projection grid, H is the minimum height from sea level to projection grid, l_{Ni} is the distance between concentric circle of N and i.

The grid combined with the technique of LOD and projection could reduce substantially the load on the CPU, but if the movement of each viewpoint needs to be recalculated the grid of sea, the rendering speed is affected with the frequent calculation of grid. On the other hand, with the emergence and continuous development of mode of GPU programmable, it is necessary to move the rendering and calculating operation of vertex to GPU, it is an effective method to reduce the computational load of CPU. Based on the above two points, a method of texture mapping for vertex on GPU is proposed.

After the grid is established, making normalization for the data of height field which is from the method from 3.2 and mapping them to the number between [0,1] , it saves the form of grayscale in the bitmap file to build the texture map of height, in order to ensure the mapping result is not distorted and do not take up extra storage space, the texture size is calculated by using the following(4-16,4-17):

$$\begin{cases} Amount_{tex} = \max(a,b) & (4\text{-}16) \\ Size = Amount_{tex}^2 & (4\text{-}17) \end{cases}$$

In the formula, a is the number of rings in the radial grid and b is Number of segmentation of rings, Size represents the texture size, a single texture can only show a moment of the sea, in order to realize the dynamic change of wave height, it needs to generate the height maps in a orderly sequence of time, then using the height map to cyclic mapping for the grid on the ocean to realize the dynamic of the ocean wave. The specific approach is: Setting n sampling point of time $f_1 f_2 \ldots f_n$ in time t and using the improved method of this paper to obtain the wave shape. To reduce the number of time sampling and keeping smooth transition for sampling points which is adjacent, using method of interpolation for the the texture map with adjacent time to generate the new texture map. Assuming it needs to interpolate k times between the adjacent point, the calculation formula of the elements of texture is as follows(4-18):

$$h = h_t(1 - i/k) + h_{t+1} \times i/k \qquad (4\text{-}18)$$

h represents the value of the texture element after interpolation, h and ht+1 represent two elements on a height image of texture in adjacent time, respectively. i is the number sequence number of interpolation,$1 \le i \le k$,eventually it will generate a sequence of k×n images of a high degree of texture. The flow chart of this method is as follows (Figure.6):

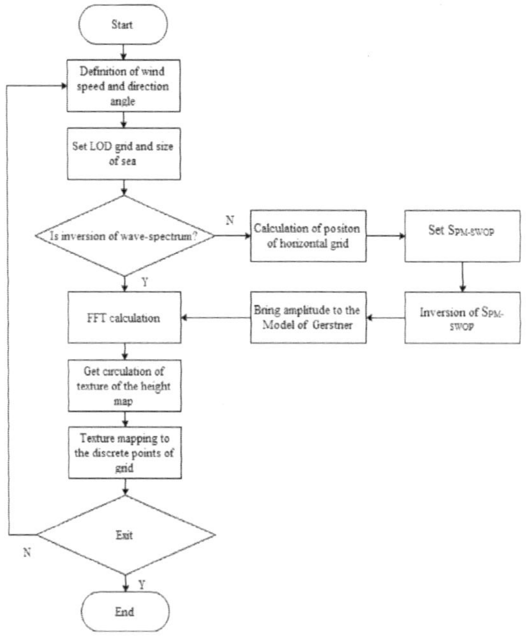

Figure 6. flow chart of improved method

5. Experiment and result analysis

For testing the validity of the proposed method,the paper does the simulation experiments and gives the result of data and graphs.Based on the improved model, using the matlab and osg as development platform to simulate the ocean wave on ordinary PC(I3 3220,3.3GHz, 4G of memory, gtx650 graphics).

According to the method described in the last section,this paper using the simulation software of matlab to simulate the ocean wave which used the method,the simulation results are shown below(Figure.7):

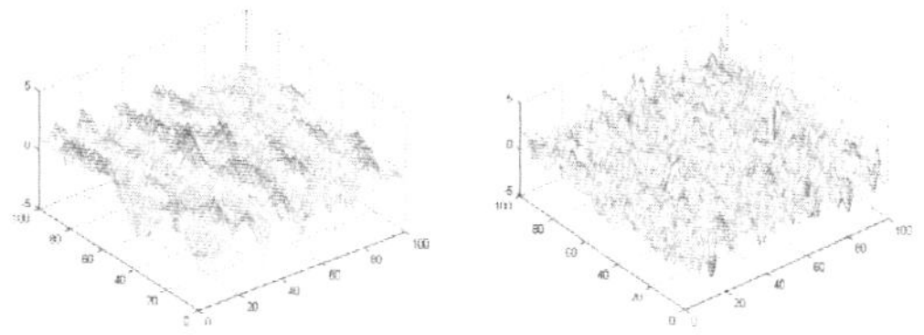

(a)wind speed 10,direction angle 0 **(b)wind speed 20,direction angle 0**

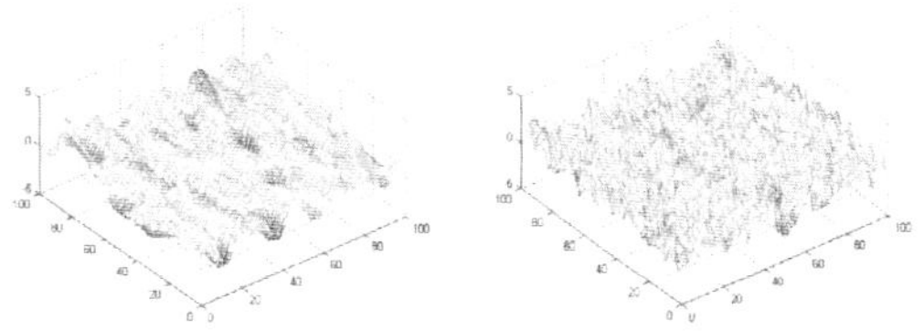

(c)wind speed 10,direction angle 90 **(d)wind speed 20,direction angle 90**

Figure 7. simulation of different wind speed and direction angle

The following figure (Figure.8) is the simulation with wind speed 5 and direction 0 on the OSG platform which using the sequence of rendered image of high texture map based on the technique of texture mapping for GPU vertex.

(a)wind speed 5,direction angle 0 **(b)wind speed 15,direction angle 0**

Figure 8. simulation of the improved method on

To verify the rendering speed of the simulation, this paper designs 8 simulation experiments with different grid resolutions, the grid resolutions in each set of experiments are as follows:

(1) 64 ×48 (2) 128×96 (3) 192×144 (4) 256×192
(5) 320×240 (6) 384×288 (7) 448×336 (8) 512×384

Each experiment adopts the method of circulation of texture and dynamic rendering of LOD to simulate the ocean with different grid resolutions, and it compares the average rendering speed (Figure.9)and average CPU utilization (Figure.10)at once for the experiments.

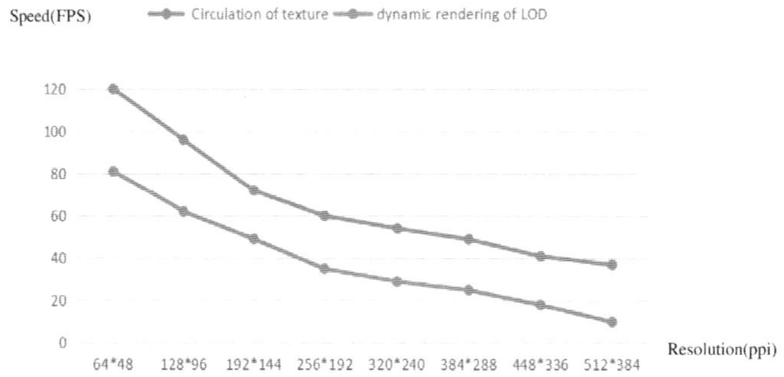

Figure 9. average rendering speed

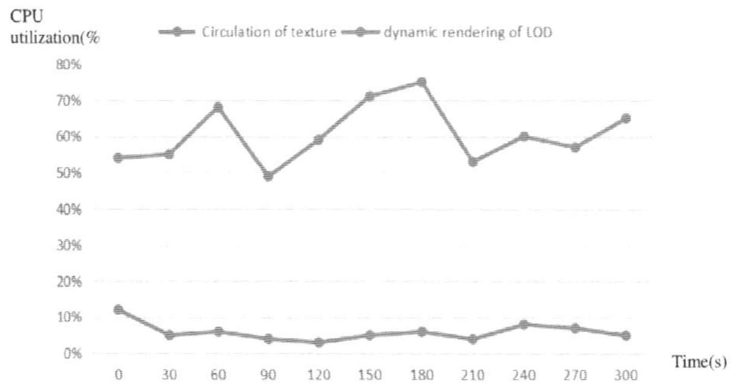

Figure 10. average rendering speed

From the above diagram, it easy to see the result which uses resolution of 512*384 based on the method of circulation still adapt to the real-time ability with FPS 30,but the simulation which uses the dynamic rendering of LOD could not maintain the real-time ability with FPS 30.In the respect of CPU utilization, in the test time of five minutes, the CPU utilization is almost below 10% which uses the method of circulation of texture, conversely, the CPU utilization stays in a high position based on the method of dynamic rendering of LOD. From here, it could use the method of circulation of texture to simulate the ocean which with the high efficiency and real-time ability.

6. Conclusions

This paper goes deep into analyzing the traditional Gerstner model and ocean wave spectrum and on this basis proposed an fast method to simulate the ocean wave on the simulation platform of matlab and OSG,using the height map of discrete point of grid which based on the calculation of FFT of interpolation of SPM-SWOP spectrum and improved Gerstner with controlled factor on the grid combined with the technique of LOD and projection, the obtained simulation of ocean waves with high authenticity. Moreover, it uses the height map to cyclic mapping for the grid on GPU to get the dynamic height data from the grid.The most significant advantage of this method is that it retains the real sense of modeling method of the ocean wave.,freeing the CPU from the heavy burden of calculation and it greatly enhance the speed and efficiency of the simulation.The experimental results show this method not only can keep the real ocean wave effect ,but also could achieve a higher rendering speed, the method is suitable for complex simulation systems of ocean wave modeling.

References

[1] Stefan Jeschke, Hermann Birkholz and Heidrun Schmann, "A Procedural Model for Interactive Animation of Breaking Ocean Waves,"Computer Graphics, 2003.

[2] Jerry Tessendorf, "Simulating Ocean Water," Simulating Nature:Realistic and Interactive Techniques Course Notes, ACM SIGGRAPH 2001.

[3] Zhengyou Li,Xuelong HOU.Ocean Wave Real-time Simulation Based-on Ocean Wave Spectrum and FFT[C]. /Progress in Applied Sciences, Engineering and Technology, Part 4: Selected, peer reviewed papers from the 2014 International Conference on Materials Science and Computational Engineering(ICMSCE 2014), May 20-21, 2014, Qingdao, China.2014:3531-3536.

[4] David Luebke, Martin Reddy, Jonathan D. Cohen, Amitabh Varshney, Benjamin Watson, "level of detail for 3D grphics," ISBN:1558608389,Morgan kaufmann.

[5] A. Fournier and T. Reeves, "A Simple Model of Ocean Waves,"Proceedings of SIGGRAPH 1986, pp.75-84.

[6] Stefan Jeschke, Hermann Birkholz and Heidrun Schmann, "A Procedural Model for Interactive Animation of Breaking Ocean Waves,"Computer Graphics, 2003.

[7] Jensen, L. S. and Goliáš, R., "Deep-water animation and rendering,"Gamasutra article on realtime water, Sept.,
http://www.gamasutra.com/gdce/jensen/jensen_01.htm, 2001.

[8] E. Darles, B. Crespin and D. Ghazanfarpour, "Accelerating andenhancing rendering of realistic ocean scenes," Proceedings of WSCG'2007, Plzen, Czech Republic, January-February 2007.

[9] Omer San,Anne E. Staples.AN EFFICIENT COARSE GRID PROJECTION METHOD FOR QUASIGEOSTROPHIC MODELS OF LARGE-SCALE OCEAN CIRCULATION[J].International journal of
multiscale computational engineering,2013,11(5):463-495.

[10] Kenneth Moreland, Edward Angel, "The FFT on a GPU," Processdings of the ACM SIGGRAPH / EUROGRAPHICS conference on Graphics hardware, pp. 112-119.

[11] William Paul Budgell,Anabela Oliveira,Morten D. Skogen. Scalar advection schemes for ocean modelling on
unstructured triangular grids[J]. Ocean Dynamics,2007,574:.

[12] Stefan Jeschke, Hermann Birkholz and Heidrun Schmann, "A Procedural Model for Interactive Animation of Breaking Ocean Waves,"Computer Graphics, 2003.

[13] Michael Eby,Greg Holloway. Grid transformation for incorporating the Arctic in a global ocean model[J].
Climate Dynamics,1994,104:.

IOP Conf. Series: Journal of Physics: Conf. Series **787** (2017) 012028 doi:10.1088/1742-6596/787/1/012028

Doppler Feature Based Classification of Wind Profiler Data

Swati Sinha[1], T.V. Chandrasekhar Sarma[2] and Mary Lourde. R[1]

[1] BITS Pilani Dubai Campus Dubai

[2] National Atmospheric Research Laboratory Gadanki, India

E-mail: swatisinha2005@rediffmail.com

Abstract. Wind Profilers (WP) are coherent pulsed Doppler radars in UHF and VHF bands. They are used for vertical profiling of wind velocity and direction. This information is very useful for weather modeling, study of climatic patterns and weather prediction. Observations at different height and different wind velocities are possible by changing the operating parameters of WP. A set of *Doppler power spectra* is the standard form of WP data. Wind velocity, direction and wind velocity turbulence at different heights can be derived from it. Modern wind profilers operate for long duration and generate approximately 4 megabytes of data per hour. The radar data stream contains Doppler power spectra from different radar configurations with echoes from different atmospheric targets. In order to facilitate systematic study, this data needs to be segregated according the type of target. A reliable *automated target classification* technique is required to do this job. Classical techniques of radar target identification use pattern matching and minimization of mean squared error, Euclidean distance etc. These techniques are not effective for the classification of WP echoes, as these targets do not have *well-defined signature* in Doppler power spectra. This paper presents an effective target classification technique based on *range-Doppler features*.

1. Introduction

Wind Profiler (WP) radar is a ground based remote probing instrument to study the dynamics of the earth's atmosphere. These are pulsed and coherent radars working in the VHF and UHF bands. These radars perform the Doppler analysis of the received signal and present the data in form of Doppler Power spectra. The spectral parameters like signal power, Doppler shift and spectral width are available from this data. Products like wind velocity, wind direction and turbulence intensity at various heights can be derived from these parameters. Main purpose of the wind profiler is to get wind information at various heights. By setting radar parameters to different values, echoes from different regions of the atmosphere can be obtained. Each region has different atmospheric targets; as an example it is possible to get precipitation in troposphere, meteoric echoes in mesosphere, and magnetic activities in ionosphere. Most of the radars are operated in long sessions and generate a large volume of data. Approximate rate of data generation is 4 Megabytes per hour. For a systematic study, a researcher is required to analyze data volumes of 100 of Gigabytes. For effective use of the data, it must be classified according to the above target types. Traditionally this job is done by human experts. However, in order to process large volume of data, it would be highly desirable to have a fast and robust automated processing method. An algorithm with this functionality has been developed.

A quick overview of historical developments presents early efforts towards the target classification and the rationale behind the work in the field. Initial efforts were directed towards accurate estimation of the spectral moments. In 1970s systematic reporting on the wind profiler echoes was done by Zrnić

[1] and Woodman [2]. Sato [3] presented a method that determines wind velocity by computing first spectral moment and wind turbulence by second central moment of the Doppler spectrum. This led to the development of the methodology of weather modeling from combined observation of Doppler weather radar and wind profiler [4]. The accuracies of the spectral parameters were also estimated in the WP data [5]. A reliable estimation of spectral noise power was established [6]. This was followed by use of appropriate thresholds in weather parameter [7]. These efforts established the relation of spectral parameters wind velocities. However, the determining the atmospheric phenomenon was largely done by human experts.

The Doppler Power Spectra is a representation of the radar echoes on two dimensional range-Doppler plane. Surveillance and imaging radars use techniques like sum of differences, Euclidean distance, and signature correlation for target identification [8]. Signature correlation method works well if the target structure is distinct [9]. In case of strategic radars, the targets are classified into a few known structures. In such cases, techniques of changing radar waveforms followed by seeking consensus are used [10]. However, these techniques are effective for the targets having well defined signatures, e g aircrafts. Echoes from atmospheric targets do not possess even loosely defined range Doppler signatures. This is expected as the received WP signals from atmospheric targets are not wide sense stationary! Therefore, the classification or the identification of the target types can only be done by methods based on statistical criterion or soft computing techniques. One of the early reports on statistical technique was by Silverstein [11]. This technique uses the concept of catalog target representations and match quality distance function. This concept was later formalized as fuzzy logic approach and was used in ground surveillance and acoustic radars [12][13][14]. In such techniques, feature definition for each target class is critical step. Hui-Lin [15] presented feature definition and extraction on time frequency plan in the context of ground penetrating radars.

Systematic work using fuzzy logic and neural network techniques started at the National Center for Atmospheric Research (NCAR), Boulder, Colorado; Cornman [16]. Morse presented improved version of NCAR Improved Moment Algorithm (NIMA) technique capable of identifying the echoes from various wind profiler targets like clear wind, precipitation, clutter and radio frequency interference (RFI) [17]. Ostrovsky [18] presented neural network technique for the classification of the different types of precipitation echoes. The NIMA technique considers each spectral point separately and analyses 5 to 7 features like power spectral density, velocity slope, velocity curvature, gradient etc. The method defines appropriate membership functions like Gaussian, sigmoid, trapezoidal etc. and computes the membership values for each feature. A weighted sum of these values is subjected to a threshold to decide whether the data set belongs to a particular weather phenomenon. NIMA method and associated variations are robust and give precise results. However, the parameters of the membership functions and the thresholds require prolonged observations, statistical analysis and fine tuning. These efforts are often radar specific and computation intensive.

Computationally simple method is required for real time handling of a large data. Also the method must not have any parameters dependent on the radar location or it operating parameters. A classification method was evolved with this motivation. This paper presents a computationally simple technique developed for the purpose of automated classification of large volume of datasets based on prominent atmospheric target. This algorithm is capable of performing near real-time classification of Doppler power spectral data emerging form wind profiler type of radars. This method mathematically formalizes human perception based classification.

Approximately 40,000 data sets of Indian MST radar and 4,000 data sets of Lower Atmospheric Wind Profiler (LAWP) were analyzed to arrive at weather features associated with specific atmospheric phenomenon. As an example, if spectral features indicating Doppler velocities between $4ms^{-1}$ to $12\ ms^{-1}$ and having spectral 5 times the average noise power occur consistently for the range of more than 1050 m, occurring at ranges less than 8 km are present; the data is characterized precipitation echoes. In similar manner, after extensive observations and statistical analysis, characteristic spectral features are determined for different types of atmospheric phenomenon. It may be appreciated that these features are defined in terms of atmospheric parameters like radial wind speed, signal strength, range and the standard deviation of the radial wind speed. The spectral feature definition is independent of the radar related parameters. Therefore, this method can be implemented on any radar or WP type. The classification algorithm sequentially searches for the features of different atmospheric phenomenon.

This paper presents classification of three type of radar targets; namely the Ionospheric activity, Clear Air Turbulence (CAT) and Precipitation Echoes. The flow chart of the algorithm is given in Fig. 1. This structure makes the algorithm modular and any new atmospheric phenomenon could be added by adding one more step in the search sequence. The algorithm was implemented on the Indian Mesosphere–Stratosphere–Troposphere (MST) radar located at Gadanki, India. The operating parameters of Indian MST radar (location 13.4N, 79.17E) are as given in table 1.

Table 1. Operating parameters of MST Radar

Parameter	Parameter value
Transmission Frequency	53 MHz
Average power	40 KW
Operation mode	DBS, 6 Beams, N10,E10,S10,W10, Z_x,Z_y
Pulse width	16 μs; binary coded, of 1 μs.
FFT points	512
Doppler Resolution	0.0305 Hz.
Sampling Start (after Tx)	24 μs
No. of range bins	150
Starting range	3.6 Km
Range Resolution	150 m
Maximum radial velocity	±22.11 ms^{-1}
Radial velocity resolution	0.08636 ms^{-1}

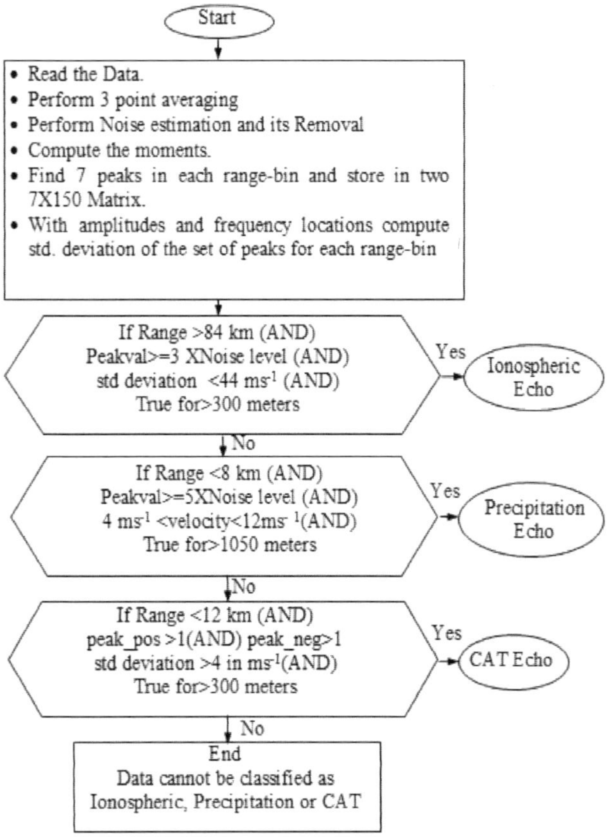

Figure 1. Flow Chart of Classification Algorithm.

2. Classification of different targets from Doppler Power Spectra

The WP radars segregate the echoes based on into different range bins depending on the time of flight of the received signal. The Doppler power spectra are computed at all the range bins presented by stacking them one above other as shown in Fig. 2, 3, 4. The figures are representative examples of ionospheric echoes, precipitation echoes and clear air turbulence (CAT) echoes respectively. In these figures, the abscissa gives the radial wind speed obtained by multiplying the Doppler frequency values by $\lambda/2$. The ordinate is the range. A brief description of the classification criterion and associated mathematical conditions corresponding to each atmospheric target are given in following subsections

2.1. Echoes of ionospheric winds

The Ionosphere is an atmospheric layer that contains the charged particles like electrons ions. It surrounds the earth at altitudes between 80 km to more than 1000 km and is classified into D, E and F layers depending on the types of particles. During the day, the D and E layers become heavily ionized; thereby increasing the RF reflectivity of that layer. In spite of large distance, on the MST radar Doppler power spectra, the ionospheric echoes show good signal strength. They present themselves as prominent peaks clustered together. The spread of the ionospheric echoes is found to be less than $44ms^{-1}$. The Doppler frequency shift is according to the radial velocity of the ionospheric winds. Understanding this fact, following conditions are used to identify the ionospheric echoes.

If {R(i)>80&&peakval(i,j)>4*N(i)&&std_dev(i)<44} (1)

Where, R(i) is the range of ith Range bin, peakval(i,j) is the amplitude for the ith range bin j^{th} peak selected from power spectral component, N(i) is the RMS noise level of ith range bin. The mathematical expressions use the symbol '&&' for logical function 'AND'. If this condition is true for more than 300 meters, more than 2 range-bins in this case, the target is identified as ionospheric winds.

2.2. Precipitation echoes

Rain or precipitation is a tropospheric phenomenon and the echoes are observed at the ranges up to 6 to 8 km. The falling rain is the target-with-velocity approaching the radar. These echoes would result in positive Doppler frequency and appear on the right half of the range-Doppler plane. The terminal velocities of falling rain drops are generally between $8ms^{-1}$ to $12\ ms^{-1}$. Following conditions are put to identify the precipitation echoes.

if {R(i) <8 && peakval(i,j)>5*N(i) && 8<velocity(i,j)<12 (2)

where, R(i) is the range of ith Range bin, and peakval(i,j) has same significance as in earlier section, N(i) is the is the RMS noise level of ith range bin and velocity(i,j) is the velocity of the falling rain drops in ms^{-1}. If this condition is true for around 1050 meters, more than 7 range bins in our case, the target is classified as precipitation echo.

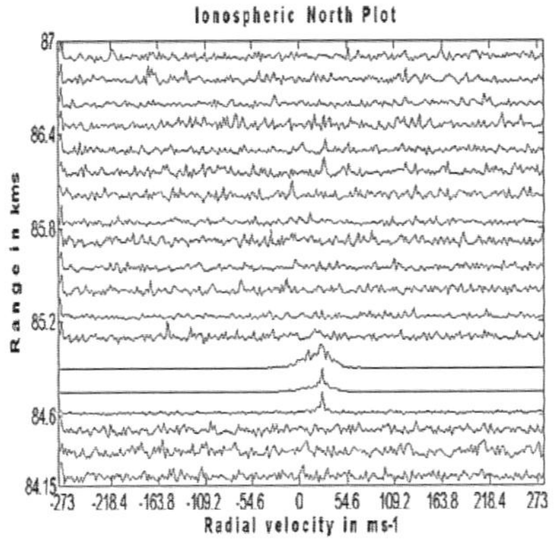

Figure 2. Ionospheric Echo 2D plot

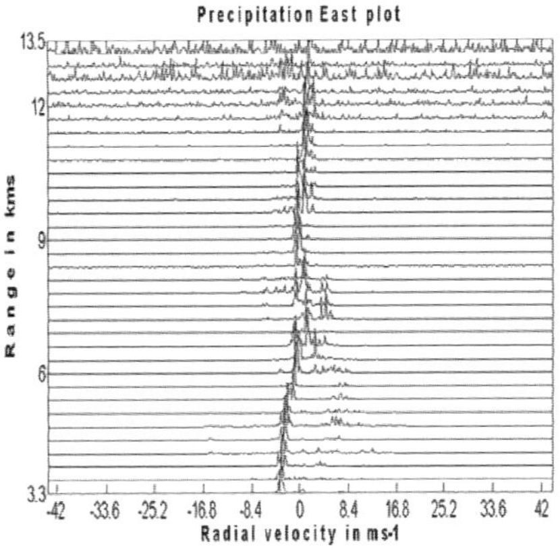

Figure 3. Precipitation Echo 2D plot

IOP Conf. Series: Journal of Physics: Conf. Series **787** (2017) 012028 doi:10.1088/1742-6596/787/1/012028

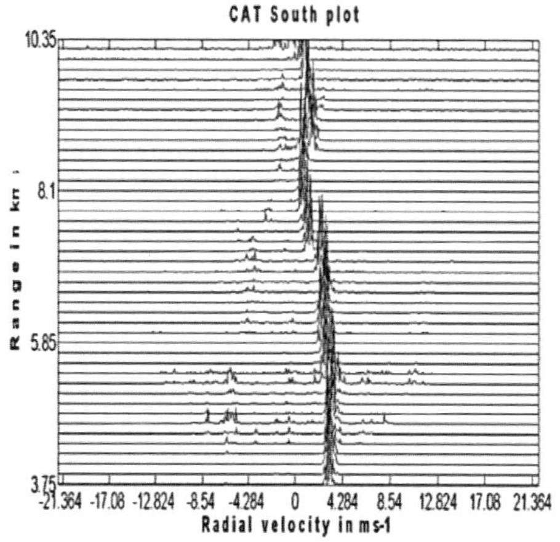

Figure. 4 CAT Echo 2-D plot

2.3. Clear air turbulence echoes (CAT)

Occasionally, high turbulence is observed in atmospheric layers. On range-Doppler plane this presents itself as moderately strong Doppler components spread over complete Doppler range. This patterns spans over a few range-bins depending on the vertical extent of the turbulence. The CAT phenomenon generally occurs in troposphere (ranges below 12 km). The classification programme identifies the7 highest peaks in each range bin.

if {(std_dev(i)>4 && peak_pos(i,j)>1 && peak_neg(i,j)>1)} (3)

Where std_dev(i) is the standard Deviation of radial velocity of selected peaks in the ith range bin. The lower limit of the standard deviation is set to 4(in ms^{-1}), as the CAT echoes are more spread. The peak_pos(i,j) is the positive amplitude of the seven peaks selected for the ith range bin and j^{th} power spectral component and peak_neg(i,j) is the negative amplitude of the seven peaks selected for the ith range bin and j^{th} power spectral component. If this condition occurs for more than 300 meters, 2+ range bins in our case, the target is classified as clear air turbulence.

3. Results and discussions

Previous section presents the search criteria for three target types. The algorithm was tested on 3000 sets of Indian MST radar data. The results are given in table 2. This table mentions the range and velocity limits for individual phenomenon and the percentage of match between the automated classification and the classification done by human expert.

Table 2. Summary of the results on MST radar at Gadanki

Type of Atmospheric Phenomenon	No. of Data sets	Range of Vales of the Critical Parameters			Comments (% match) Auto Vs/Human
		Range (km)	Radial Velocity or Std Dev of radial velocity	Peak power / Noise level	
Ionospheric(D)	500	75-80	<44 ms^{-1}	4	100% match
Ionospheric (E)	500	80 -110	<44 ms^{-1}	4	100% match
Precipitation	1000	0-8	4-20 ms^{-1}	10	90 % match
CAT	1000	0-12	>4 ms^{-1}	4	90 % match

While implementing this algorithm the limit values need to be converted into number of range bins and number of Doppler bins. These conversion computations are done using the radar operational parameters available in the header of the data set. As example, radar operating pulse width of 1μ sec, and inter pulse period of 1 millisecond, number of coherent integration of 512 FFT points, will give range resolution of 0.15 km and radial velocity resolution of 0.0854 ms^{-1}

4. Conclusions

The newly developed algorithm is capable of detecting the presence of 3 different types of atmospheric targets/ phenomenon. This detection is done by searching range-Doppler features specific corresponding to the atmospheric phenomenon. The limit values in the mathematical conditions are finalized after studying multiple data sets. This algorithm was implemented in Matlab using simple mathematical expressions on the components of Doppler Power spectra. There is some percentage of misclassification. However, we expect that in large scale classification, such shortcomings are tolerable. This approach is capable of reliably identifying the atmospheric phenomenon as presented in earlier sections. The performance of this method does not get affected by different radar parameter settings, atmospheric conditions and the time of the day etc. The computational simplicity is the main advantage of this algorithm. It can be implemented for real time classification on any wind profiler.

Acknowledgements
The authors thank the scientists at National Atmospheric Research Laboratory (NARL) Gadanki, India for their help in providing the radar data and guidance for the work done.

References

[1]. Zrnić DS. Estimation of spectral moments for weather echoes. IEEE Transactions on Geoscience Electronics, Oct; 1979, vol. GE-17, No. 4,, 118-128.
[2] Woodman Ronald F. Spectral Moment Estimation in MST Radar. Radio Science; 1985, vol. 20, 1185-1195.
[3] Sato Toru, Woodman RF, Spectral parameter estimation of CAT radar echoes in the presence of fading clutter. Radio Science; 1982, vol. 17, 817-826.
[4] Dombrowsky R. Complementary Use of Modern Doppler Radars and Profilers in the Upper Air Network, Operational Use of Wind Profiler Data in United States Weather Forecasting,, CIM O/OPAG-UPPER-AIR-/ET-RSUAT&T-1 Doc.4.1.(1), World Meteorological Organization (WMO), Geneva, Switzerland, 2005.
[5] Gan T., Yamamoto MK, Hashiguchi H., Okamoto H., Yamamoto M. Error estimation of spectral parameters for high-resolution wind and turbulence measurements by wind profiler radars, Radio Science; 2014, vol. 49, 1214-1231.
[6] Hildebrand Peter H., Sekhon R.S. Objective Determination of Noise Level in Doppler Spectra, Journal of Applied Meteorology; June 1974, vol. 113, 808-811.
[7] Riddle Anthony C, Hartten Leslie M., Carter David A. Johnston Paul E., Williams Christopher R., A Minimum Threshold for Wind Profiler Signal-to-Noise Ratios, Journal of Atmospheric and Oceanic Technology; Dec. 2011, vol. 29, No.7, 889-895.
[8] Allistair Moses, Rutherford Matthew J., Valavanis Kimon P. Radar-Based Detection and Identification for Miniature Air Vehicles, IEEE International Conference on Control Applications (CCA); Sept. 2011, 28-30.
[9] Vespe M., Baker C.J. , Griffiths H.D. Radar target classification using multiple Perspectives, IET Radar Sonar & Navigation; August 2007, vol. 1, No. 4, 300-307.
[10] Sowelam Sameh M., Tewfik Ahmed H. Waveform Selection in Radar Target Classification, IEEE Transactions on Information Theory; May 2000, vol. 46, No. 3, 1014-1029.
[11] Silverstein Paul B., Sands O. Scott, Garber Fred D. Radar Target Classification and Interpretation by Means of Structural Descriptions of Backscatter Signals, IEEE AES Systems Magazine; March 1991, 3-7.

[12] Andrić Milenko, Durovic Zeljko, Zrnić Bojan. Ground Surveillance Radar Target Classification based on Fuzzy Logic Approach, EUROCON, November 2005, 22-24.

[13] Andrić Milenko, Bondžulić Boban, Zrnić Bojan, Kari Aleksandra, Dikić, Goran. Acoustic Experimental Data Analysis of Moving Targets Echoes Observed by Doppler Radars, Journal of Mechanical Engineering; 2012, vol. 58, No 6, 386-393.

[14] Andrić Milenko S., Bondžulić Boban P., Zrnić Bojan M. The Database of Radar Echoes from Various Targets with Spectral Analysis, 10th Symposium on Neural Network Applications in Electrical Engineering; 2010, 187-190.

[15] Zhou Hui-Lin, Tian Mao, Chen Xiao-Li. Time-Frequency Representations for Classification of Ground Penetrating Radar Echo Signal, Proceedings of International Symposium on Intelligent Signal Processing and Communication Systems, Dec. 2005, 597-600.

[16] Cornman LB., Goodrich RK, Morse CS. Ecklund WL. A Fuzzy Logic Method for Improved Moment Estimation from Doppler Spectra, Journal of Atmospheric and Oceanic Technology; 1998, vol. 15, 1287-1305.

[17] Morse CS., Goodrich RK, Cornman LB. The NIMA Method for Improved Moment Estimation from Doppler Spectra, Journal of Atmospheric and Oceanic Technology; Sep. 2001 vol. 19, 274-295.

[18] Ostrovsky YP., Yanovsky FJ. Use of Neural Network for Turbulence and Precipitation Classification Procedure, 11 Int. Conf. on Mathematical Methods in Electromagnetic Theory, Ukrain; June 2006, 161-163.

Rearrangement and Grouping of Data Bits for Efficient Lossless Encoding

Ajitha Shenoy K B[1] , Meghana Ajith[2] and Vinayak M Mantoor[3]

[1, 3]Department of Information and Communication Technology, Manipal Institute of Technology, Manipal University, Manipal, Karnataka, India
[2] Department of Computer Applications, Manipal Institute of Technology, Manipal University, Manipal, Karnataka, India

E-mail: ajith.shenoy@manipal.edu, meghana.ajith@manipal.edu, vinayak.mantoor@manipal.edu

Abstract. This paper describes the efficacy of rearranging and grouping of data bits. Lossless encoding techniques like Huffman Coding, Arithmetic Coding etc., works well on data which contains redundant information. The idea behind these techniques is to encode more frequently occurring symbols with less number of bits and more seldom occurring symbols with more number of bits. Most of the methods fail if there is a non-redundant data. We propose a method to re arrange and group data bits there by making the data redundant and then different lossless encoding techniques can be applied. In this paper we propose three different methods to rearrange the data bits, and efficient way of grouping them. This is first such attempt. We also justify the need of rearranging and grouping data bits for efficient lossless encoding.

1. Introduction

Data compression is important in the field of information theory and signal processing. Good encoding techniques reduce amount of memory needed to store the large file and also make transmission of such file over network faster. Good encoding implies efficient way of utilizing computer memory and network bandwidth. There are many encoding techniques defined in the literature. Compression techniques are classified into three categories. Loss less compression, Lossy compression and hybrid compression. Well known lossless compression techniques are run-length coding, Shannon-Fano Coding [1], Huffman coding [2], LZW encoding [3, 4], arithmetic coding [5, 6], lossless predictive encoding etc.[1]. Lossy compression techniques are discrete cosine transformation, KL transformation and lossy predictive encoding, differential pulse code modulation, delta modulation, wavelet transform etc[1]. Hybrid encoding techniques are combination of lossy and lossless encoding techniques. Examples are JPEG, MPEG, DVI, H.26X etc[1]. M. Kafashan et.al. in their paper proposed new rectangular partitioning methods for lossless binary image compression. They proposed 2D RLE which is efficient than 1D RLE scheme [7]. In this paper our work focuses on lossless compression techniques. Most of the lossless encoding techniques rely on frequency of occurrence of symbols in the given document. If all the symbols in the given document are distinct then all the lossless compression techniques like Huffman, coding, arithmetic coding, and predictive coding fail to encode it using fewer number of bits. We propose a new scheme which transforms the given data into redundant data so that lossless compression can be applied efficiently. In the next section we give definition and concepts which are needed to understand our work described in this paper. The rest of the paper is organized as follows. Section 2 gives basic concepts and definitions used in our work. Section 3 gives motivation for rearranging and grouping, Section 4 describe our proposed work and last section contains conclusion and future work.

2. Preliminaries

Entropy plays important role in lossless compression. Famous scientist Claude E. Shannon of Bell Labs defined entropy of information source as follows:

Definition 2.1 (Entropy) [1, 8, 9]: Entropy η of an information source with alphabets $S = \{ s_1, s_2, \ldots, s_n \}$ is defined as $\eta = \Sigma^n_{i=1} P_i \log_2 (1/P_i)$, where P_i denote probability of occurrence of symbol s_i in given information.

Entropy denotes measure of disorder in a system. More entropy implies more disorder. As far as compression is concerned, entropy denote minimum average number of bits needed to represent each symbol s_i in S. It specifies the lower bound for the average number of bits needed to code each symbol s_i in S. Any lossless compression technique is good if average number of bits need to encode each symbol in the given information is approximately equal or equivalent to the entropy measure η. So entropy is very important measure to compare lossless compression techniques. Most widely used lossless encoding techniques are Huffman coding, Arithmetic coding and Run length coding. All these coding techniques are based on variable length coding. Main key idea in these compression techniques are: Frequently occurring characters are encoded with lower number of bits and seldom occurring characters are coded with higher number of bits [1]. Hence it is very much important that the given information should consist of redundant information so that these techniques can be applied on it efficiently.

Definition 2.2 (Compression Ratio)[1]: Compression ratio is defined as B_0/B_1, where B_0 denote number of bits needed to represent the given data before compression and B_1 denote number of bits needed to represent data after compression.

Note that more compression ratio implies more compression and vice versa. In the next section we describe our method of rearranging the given data so that it contains more redundant information and thus suitable for applying lossless encoding techniques on it efficiently.

3. Motivation

Let us justify the need of rearrangement of data. With the following example we will explain why rearrangement of data bits is important for lossless encoding.

Example 2.1: Let $D = D_0.\ D_1, \ldots, D_{255}$, where $D_i \neq D_j$ for $i \neq j$ and D_i belong to set $S=\{0, 1, \ldots, 255\}$ for all $0 \leq i \leq 255$.That is D is rearrangement of symbols in S in random order. Suppose we want to encode this using loss less compression technique like Huffman or Arithmetic coding, it will take 8 bit to represent each symbol in the information. Shannon's entropy measure also indicates that minimum average number of bits needed to encode each symbol in D requires 8 bits. So we can't do better than this. The only way to do better is rearrangement of data bits and grouping them in such a way that the grouped values are redundant and hence efficiently apply lossless compression techniques on grouped values. Example 2.1 motivates us to find rearrangement of data bits in such a way that, we can apply lossless encoding techniques efficiently on rearranged data.

4. Proposed Method

Suppose that the given data is $D = D_1.\ D_2, \ldots, D_m$, where D_i be n - bit number for all $1 \leq i \leq m$. Now we explain rearrangement scheme for the given data D. Since each D_j is n-bit number for all $0 \leq j \leq m$, we can write bit representation of each D_i in given data D. So the entire data D can be expressed as $m \times n$ bit string. i.e.,

$D_1 = b_{11}, b_{12} \ldots\ldots b_{1n}$ $D_2 = b_{21}\ b_{22} \ldots\ldots b_{2n}$ etc... $D_m = b_{m1}\ b_{m2} \ldots\ldots b_{mn}$

Where, $D_i = b_{i1}\ b_{i2} \ldots\ldots b_{in}$ represent n-bit representation of symbol D_i in D. Now we define the method of rearranging data bits. We arrange data bits in three different types as follows:

Type i: Arrange m x n bit sequence in the given order without any change i.e. b_{11}, b_{12} …….. b_{1n} b_{21} b_{22} …….. b_{2n} ………. b_{m1} b_{m2} …….. b_{mn}

Type ii:. Arrange the *m x n* bit sequence in the following manner.

MSB (D_1. D_2, …, D_m) Next MSB (D_1. D_2, …, D_m)…….. LSB(D_1. D_2, …, D_m)

i.e. b_{11}, b_{21} …….. b_{m1} b_{12} b_{22} …….. b_{m2} ………. b_{1n} b_{2n} …….. b_{mn}

Type iii: Consider each significant bit of data separately as independent bit streams.

Stream 1: MSB (D_1. D_2, …, D_m) = b_{11}, b_{21} …….. b_{m1}

Stream 2: Next MSB (D_1. D_2, …, D_m) = b_{12} b_{22} …….. b_{m2}

……………………………………………. Stream n: LSB(D_1. D_2, …, D_m) = b_{1n} b_{2n} …….. b_{mn}.

For the given input data, our aim is to find best suitable rearrangement and grouping of data bits. We group *(n + k) bits* (where k is varied from –n/2 to n/2) in each type of arrangements and calculate Shannon's entropy value η. Finally we select the rearrangement of bits which takes less entropy value. Less entropy implies less disorder in the given data. Now we formally state the algorithm for grouping bits in each arrangement and to select best rearranged data bits and best value of k, which is efficient for loss less encoding

Algorithm: Finding the Best Rearrangement

```
{   Type = NULL;
    For each of the Type i, Type ii do the following
      {     Best_k = 0;
            Best-Entropy = infinity;
            For k = – n/2, – n/2+1,…, 0, 1, 2, …,n/2 do
                {     Group n + k bits   // (from  m x n  bits we can get [m x n / (n +
                      k) ] values)
                      Convert decimal equivalent of each binary n + k  bits
                      Find Shanon's entropy measure  η for  [m x n / (n + k) ] decimal
                      values
                      If (Best-Entropy > η) {     Best-Entropy = η
                                                  Best_k = k
                                                  Type = type i or ii for which
                                                  Best_k is updated.}}}
    For Type iii do the following
      {     For each Stream i (i = 1, 2, ….n) do the following
            {     For k = – n/2, – n/2+1,…, 0, 1, 2, …,n/2 do
                  {     Group n + k bits   of  Stream i.
                        // (from  m  bits we can get [m / (n + k) ] values)
                        Convert decimal equivalent of each binary n + k  bits
                        Find Shanon's entropy measure  ηi for  [m  / (n + k) ] grouped
                        decimal values //  where ηi denotes entropy of Stream i grouped
                        values}
                  Let N denote number of bits needed to represent entire data D, then
                  N =  ( η1 + η2 + ..... + ηn) m/[n + k]
                  N1= Best_Entropy x  m x n/(n+k)
                  // where N1 denote, number of bits needed to represent entire data using
                  // best Rearrangement (Type) found so far in the previous for loop for
                  // type i & ii
                  If (N  <  N1 ) then  {Type = Type iii ; Best_k = k}}}
```

Algorithm 4. 1: Finding Best Rearranged Data Bits

Note that if *m* and *m x n* is not exactly divisible by (n + k) add trailing zero to the bit string and make it divisible by *(n + k)*. It is trivial to note that time complexity of above algorithm is *O(nm)*. Usually n is either 8 or 16 or 24 bits, hence algorithm takes linear time in m, i.e. O(m). Once we get Best-Entropy, type of rearranged data bits (type i or ii) and Best_k values, we can consider group of *(n + Best_k)* bits from that type and efficiently apply existing lossless encoding techniques on it. Note that decoder should know type of arrangement of data bits, Best_k, m and n values for decoding the data. These additional information should be sent to decoder to decode the data. Since decoder has the information about m, n and Best_k it can remove trailing zeroes(if any). Next we will give one example which shows why rearranging data bits and grouping it into n + k bits needed.

Example 4.2: Let D denote 256 consecutive numbers from 0, 1, 2, …. ,255.
Shanon's entropy measure for the given data D is $\eta = 8$ bits. Hence minimum average number of bits needed to represent each symbol in D is 8 bits. Therefore, Number of bits needed to represent D is at least 256 x 8 = 2048 bits. We consider rearrangement of data bits with respect to Type i, ii and iii. Let k = 0.

Type i: with Type i arrangements of data bits and k = 0, we get same data D as it is. Hence Entropy won't change ($\eta = 8$)

Type ii arrangement of data bits. We get,
$(00.....0)_{128\text{ times}}$ $(11....1)_{128\text{ times}}$ $(00.....0)_{64\text{ times}}$ $(11....1)_{64\text{ times}}$ $(00.....0)_{64\text{ times}}$ $(11....1)_{64\text{ times}}$ $(00.....0)_{32\text{ times}}$ $(11....1)_{32\text{ times}}$ $(00.....0)_{32\text{ times}}$ $(11....1)_{32\text{ times}}$ $(00.....0)_{32\text{ times}}$ $(11....1)_{32\text{ times}}$ $(00.....0)_{32\text{ times}}$ $(11....1)_{32\text{ times}}$ $(00.....0)_{16\text{ times}}$ $(11....1)_{16\text{ times}}$ $(00.....0)_{16\text{ times}}$ $(11....1)_{16\text{ times}}$ $(00.....0)_{16\text{ times}}$ $(11....1)_{16\text{ times}}$ $(00.....0)_{16\text{ times}}$ $(11....1)_{16\text{ times}}$ $(00.....0)_{16\text{ times}}$ $(11....1)_{16\text{ times}}$ $(00.....0)_{16\text{ times}}$ $(11....1)_{16\text{ times}}$ $(00.....0)_{16\text{ times}}$ $(11....1)_{16\text{ times}}$ $(00.....0)_{16\text{ times}}$ $(11....1)_{16\text{ times}}$ $(00.....0)_{8\text{ times}}$ $(11....1)_{8\text{ times}}$ $(00.....0)_{8\text{ times}}$ $(11....1)_{8\text{ times}}$ $(00.....0)_{8\text{ times}}$ $(11....1)_{8\text{ times}}$ $(00.....0)_{8\text{ times}}$ $(11....1)_{8\text{ times}}$ $(00.....0)_{8\text{ times}}$ $(11....1)_{8\text{ times}}$ $(00.....0)_{8\text{ times}}$ $(11....1)_{8\text{ times}}$ $(00.....0)_{8\text{ times}}$ $(11....1)_{8\text{ times}}$ $(00.....0)_{8\text{ times}}$ $(11....1)_{8\text{ times}}$ $(00.....0)_{8\text{ times}}$ $(11....1)_{8\text{ times}}$ $(00.....0)_{8\text{ times}}$ $(11....1)_{8\text{ times}}$ $(00.....0)_{8\text{ times}}$ $(11....1)_{8\text{ times}}$ $(00.....0)_{8\text{ times}}$ $(11....1)_{8\text{ times}}$ $(00.....0)_{8\text{ times}}$ $(11....1)_{8\text{ times}}$ $(00.....0)_{8\text{ times}}$ $(11....1)_{8\text{ times}}$ $(00.....0)_{8\text{ times}}$ $(11....1)_{8\text{ times}}$ $(00.....0)_{8\text{ times}}$ $(11....1)_{8\text{ times}}$ $(00.....0)_{8\text{ times}}$ $(11....1)_{8\text{ times}}$ (00001111)……32 times ……. (00001111) (00110011)……32 times ……. (00110011) (01010101)……32 times ……. (01010101)

Now grouping n + k bits (let us take k = 0 here for simplicity). i.e. grouping 8 bits we get
0, 0, 0, 0, 0, 0, 0, 0, 0, 0, 0, 0, 0, 0, 0, 0, 255, 255, 255, 255, 255, 255, 255, 255, 255, 255, 255, 255, 255, 255, 255, 255, 0, 0, 0, 0, 0, 0, 0, 0, 255, 255, 255, 255, 255, 255, 255, 255, 0, 0, 0, 0, 0, 0, 0, 0, 255, 255, 255, 255, 255, 255, 255, 255, 0, 0, 0, 0, 255, 255, 255, 255, 0, 0, 0, 0, 255, 255, 255, 255, 0, 0, 0, 0, 255, 255, 255, 255, 0, 0, 0, 0, 255, 255, 255, 255, 0, 0, 255, 255, 0, 0, 255, 255, 0, 0, 255, 255, 0, 0, 255, 255, 0, 0, 255, 255, 0, 0, 255, 255, 0, 0, 255, 255, 0, 0, 255, 255, 0, 255, 0, 255, 0, 255, 0, 255, 0, 255, 0, 255, 0, 255, 0, 255, 0, 255, 0, 255, 0, 255, 0, 255, 0, 255, 0, 255, 0, 255, 0, 255, 15, 15, 15, 15, 15, 15, 15, 15, 15, 15, 15, 15, 15, 15, 15, 15, 15, 15, 15, 15, 15, 15, 15, 15, 15, 15, 15, 15, 15, 15, 15, 15, 51, 51, 51, 51, 51, 51, 51, 51, 51, 51, 51, 51, 51, 51, 51, 51, 51, 51, 51, 51, 51, 51, 51, 51, 51, 51, 51, 51, 51, 51, 51, 51, 85, 85, 85, 85, 85, 85, 85, 85, 85, 85, 85, 85, 85, 85, 85, 85, 85, 85, 85, 85, 85, 85, 85, 85, 85, 85, 85, 85, 85, 85, 85, 85.

Now entropy of this new values $\eta_{new} = 1/256\ [32 \log_2 (256/32) + 32 \log_2 (256/32) + 32 \log_2(256/32) + 80 \log_2(256/80) + 80 \log_2(256/80)] = 2.17$ Which is lesser than the previous η.

Type iii : By grouping 8 bits we get

Stream 1: 0, 0, 0, 0, 0, 0, 0, 0, 0, 0, 0, 0, 0, 0, 0, 0, 255, 255, 255, 255, 255, 255, 255, 255, 255, 255, 255, 255, 255, 255, 255, 255. Entropy $\eta_1 = 1$

Stream 2: 0, 0, 0, 0, 0, 0, 0, 0, 255, 255, 255, 255, 255, 255, 255, 255, 0, 0, 0, 0, 0, 0, 0, 0, 255, 255, 255, 255, 255, 255, 255, 255. Entropy $\eta_1 = 1$

Stream 3: 0, 0, 0, 0, 255, 255, 255, 255, 0, 0, 0, 0, 255, 255, 255, 255, 0, 0, 0, 0, 255, 255, 255, 255, 0, 0, 0, 0, 255, 255, 255, 255. Entropy $\eta_1 = 1$

Stream 4: 0, 0, 255, 255, 0, 0, 255, 255, 0, 0, 255, 255, 0, 0, 255, 255, 0, 0, 255, 255, 0, 0, 255, 255, 0, 0, 255, 255, 0, 0, 255, 255. Entropy $\eta_1 = 1$

Stream 5: 0, 255, 0, 255, 0, 255, 0, 255, 0, 255, 0, 255, 0, 255, 0, 255, 0, 255, 0, 255, 0, 255, 0, 255, 0, 255, 0, 255, 0, 255, 0, 255. Entropy $\eta_1 = 1$

Stream 6: 15, 15, 15, 15, 15, 15, 15, 15, 15, 15, 15, 15, 15, 15, 15, 15, 15, 15, 15, 15, 15, 15, 15, 15, 15, 15, 15, 15, 15, 15, 15, 15. Entropy $\eta_1 = 1$

Stream 7: 51, 51, 51, 51, 51, 51, 51, 51, 51, 51, 51, 51, 51, 51, 51, 51, 51, 51, 51, 51, 51, 51, 51, 51, 51, 51, 51, 51, 51, 51, 51, 51. Entropy $\eta_1 = 1$

Stream 8: 85, 85, 85, 85, 85, 85, 85, 85, 85, 85, 85, 85, 85, 85, 85, 85, 85, 85, 85, 85, 85, 85, 85, 85, 85, 85, 85, 85, 85, 85, 85, 85. Entropy $\eta_1 = 1$

Hence total number of bits needed to represent entire data = (1 + 1 + 1+ 1+ 1 + 1 + 1 + 1)256/8 = 256 bits. Which is optimal possible compression with compression ratio = 2048/256 = 8 .

5. Conclusion and Future Work

In this paper we proposed a method for rearranging the data bits and grouping them. This is first such attempt. We defined three ways of rearranging the data bits and then grouping them efficiently. The rearranged and grouped data bits are efficient for lossless compression since it contains less disorder. We justified our claim by taking suitable examples, which clearly shows the benefit of rearranging and grouping of data bits. Note that decoding is straight forward, while sending compressed data we need to send type of rearrangement, best value of k, m and n. The decoder will arrange the data based on this information and get the original data. As a future work it will be interesting to check how this technique work on medical images and X-ray images, since medical records like patient file, X-rays, MRI scan report etc. need lossless compression techniques. It would also be nice to see the loss less JPEG's performance on the given data bits, rearranged and grouped data bits. It is obvious to note that our algorithm can be implemented in parallel, so it would be nice to see the performance of rearranging and grouping on big data.

References

[1] Ze-Nian Li and Mark S. Drew 2004 *Fundamentals of Multimedia,* Pearson Prentice Hall.

[2] D.A. Huffman 1952 "A Methodfor the Construction ofMinimum-Redundancy Codes' *Proceedings Of the IRE* [Institute of Radio Engineers now the IEEE] 40(9): 1098-1101.

[3] J. Ziv and A. Lempel 1977 "A Universal Algorithm for Sequential Data Compression" *IEEE Transactions on Infonnation Theory* 23(3): 337-343.

[4] J. Ziv and A. Lempel 1978 "Compression ofIndividual Sequences Via Variable-Rate Coding" *IEEE Transactions on Infonnation Theory* 24(5): 530-536.

[5] Rissanen and G.G. 1979 Langdon "Arithmetic Coding," *IB1'vi Journal of Research and Development 23(2):* 149-162.

[6] I.H. Witten, RM. Neal, and J.G. Cleary 1987 "Arithmetic Coding for Data Compression" *Communication of the ACM,* 30(6): 520-540.

[7] M. Kafashan, H. Hosseini, S. Beygiharchegani, P. Pad, and F. Marvasti 2010 "New rectangular partitioning methods for lossless binary image compression," IEEE Int. Conf. Signal Processing, pp. 694–697.

[8] C.E. Shannon 1948 "A Mathematical Theory of Communication' *Bell System Technical Journal* 27: 379-423 and 623-656.

[9] C.E. Shannon and W. Weaver 1949 *The Mathematical Theory oj Communication* Champaign IlUniversity of Illinois Press.

Alternative Line Coding Scheme with Fixed Dimming for Visible Light Communication

M T Niaz, F Imdad and H S Kim

Sejong University, Department of Information & Communication Engineering, 209 Neungdong-ro, Gwangjin-gu, Seoul, Republic of Korea, 05006.

E-mail: hyungkim@sejong.edu

Abstract. An alternative line coding scheme called fixed-dimming on/off keying (FD-OOK) is proposed for visible-light communication (VLC). FD-OOK reduces the flickering caused by a VLC transmitter and can maintain a 50% dimming level. Simple encoder and decoder are proposed which generates codes where the number of bits representing one is same as the number of bits representing zero. By keeping the number of ones and zeros equal the change in the brightness of lighting may be minimized and kept constant at 50%, thereby reducing the flickering in VLC. The performance of FD-OOK is analysed with two parameters: the spectral efficiency and power requirement.

1. Introduction

Light emitting diodes (LEDs) have been adopted for lighting mainly owing to their lower energy consumption and longer lifetime. They have provided a way to replace conventional illumination devices, e.g., incandescent, the gas discharged, etc. [1-3]. LED light sources offer a small size, a low forward voltage and drive current, excellent brightness in the visible spectrum, and the option of emission at a single wavelength or a range of wavelengths [4]. These properties of LEDs have enabled the possibility of wireless communication for data transfer. The use of light for transferring data is known as visual-light communication (VLC) and was drafted as the IEEE 802.15.7 standard [5].

In VLC systems, pulse width modulation (PWM) and pulse amplitude modulation (PAM) are usually used to achieve dimming support [6]; in contrast, on/off keying (OOK) and pulse position modulation (PPM) are usually used for data transmission. There are variants of the PPM available in the literature [7]: L-pulsed position modulation (LPPM), inverted PPM (I-PPM), and multipulse PPM (MPPM) [8]. Most of these modulation schemes were originally designed for free-space optical wireless communication (FSOWC) based on infrared (IR) transmission and not for VLC. The core component of VLC is the use of off-the-shelf LED lights. These LED lights exhibit nonlinear behavior when operated under normal circumstances. This nonlinearity is directly related to the dimming and flickering of the LED that is experienced in the VLC system. There are two ways to address the abovementioned issues: 1) to control the input current by introducing some bias and 2) to use the appropriate modulation scheme by fixing the current and voltage at the optimum values. Rather than the first method that requires a change in the circuit of the LED, the efficient way to address the issue is to use a modulation scheme along with line coding schemes that are easier to implement.

In order to resolve the abovementioned issues, an alternative line coding scheme called fixed - dimming OOK (FD-OOK) is proposed. The main objective of FD-OOK is to make the number of ones and zeros equal within each code word, which will make the dimming level equal to 50% and also reduce flickering. The performance of FD-OOK is evaluated in terms of both the normalized power requirement defined by the required optical power and the spectral efficiency measured by the bit rate given at a specific bandwidth.

The main contributions are as follows:

- The FD-OOK scheme is proposed so that a constant 50% dimming level is maintained at all times.
- A new encoder and decoder are proposed to implement the FD-OOK scheme.
- Flickering is controlled to keep the brightness level of the LEDs constant.

The rest of the paper is outlined as follows. Section 2 discusses the VLC system model, and Section 3 describes the proposed FD-OOK scheme in detail. In Section 4, the performance is evaluated through simulation, and Section 5 finally concludes the paper.

2. System Model

The VLC system model considered in this study is shown in Fig. 1. It is assumed that the receiver containing photodiodes (PDs) is located at the desktop surface height, and line-of-sight (LOS) is assumed for the communication scenario as well as illumination.

The flux produced by the LED lamp can be computed as:

$$\phi_i = N \times \phi_{max} \quad (1)$$

where i is the ith LED lamp, N is the dimming factor, and ϕ_{max} is the maximum luminous flux for each LED lamp when the input waveform has a maximum-amplitude direct-current (DC) component. N is computed as

$$N = a/T \quad (2)$$

where a denotes the number of high pulses in a particular code word waveform, and T is the time period of one code word waveform.

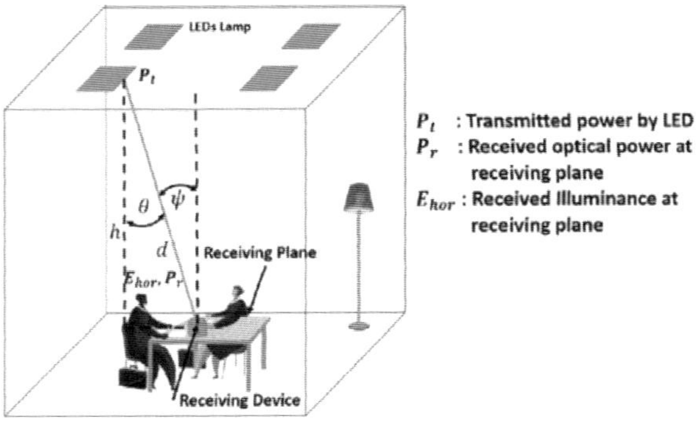

Figure 1. Visible light communication system model

Hence, (1) can be expressed by substituting (2) for N as

$$\phi_i = (a/T) \times \phi_{max} \quad (3)$$

and we can see that the dimming range can go from 0% to 100% by changing the variables a and T.

The illuminance expresses the brightness of the illuminated surface. Let E_j denote the illuminance at some workplace j from all LED lamps, which is calculated as [9]

$$E_j = \sum_{i=1}^{l} \frac{(m+1)\phi_i}{2\pi r^2} \cos^m(\theta) \cos\Psi \tag{4}$$

where m is the Lambertian order, r is the distance between the LED and the PD, and θ and Ψ are the angles of irradiance and incidence, respectively. The recommended illuminance value for homes and offices set by the ISO is 400–800 lux.

The total energy irradiated by the LED lamp is indicated by the total transmitted optical power, which determines the signal strength of optical communication [10]. VLC uses intensity modulation and direct detection (IM/DD). IM is obtained by varying the bias current of the LED lamps. DD is achieved by capturing the photocurrent induced by the PD, which is proportional to the optical power incident on it. In VLC, $x(t)$ represents the optical power from the LED lamps and not the amplitude; thus, it must satisfy

$$x(t) \geq 0 \text{ and } \lim_{T\to\infty} \frac{1}{T} \int_0^T x(t)\,\mathrm{d}t \leq NP \tag{5}$$

here P is the average optical power, and $0 \leq N \leq 1$ is the dimming factor [11]. The brightness depends on the LED lamp dimming factor and the average optical power. As an example, if an LED lamp is at full brightness ($N = 1$) or 50% brightness ($N = 0.5$), the average power of x(t) is P or $0.5P$, respectively.

3. Proposed FD-OOK Scheme

In this study, the above mentioned issues are solved by introducing an alternative encoding and decoding scheme for VLC, which is proposed in our patent [12]. The scheme has two major objectives. First, the proposed scheme maintains a 50% dimming level, and second, flickering can be avoided by maintaining equal numbers of ones and zeros. One additional benefit from this encoder is that it maintains a steady running length of not more than four ones or zeros at any time.

Table 1. Code-word Comparison between different line coding scheme

Decimal	HEX	4-Bit Binary	Existing 4B/5B	Proposed Scheme
1	0	0000	11110	010011
2	1	0001	01001	110001
3	2	0010	10100	110010
4	3	0011	10101	100011
5	4	0100	01010	110100
6	5	0101	01011	100101
7	6	0110	01110	100110
8	7	0111	01111	010110
9	8	1000	10010	011001
10	9	1001	10011	101001
11	A	1010	10110	101010
12	B	1011	10111	001011

Table 2. Simulation Parameters

Parameter	Value
Room dimensions [m]	5 × 5 × 3
Height of receiving plane [m]	0.8
Number of array of LEDs	3600
Center luminous intensity [cd]	0.73
Single LED bulb power [mW]	20
Semiangle at half power [°]	30
Photodiode	0.4

13	C	1100	11010	101100
14	D	1101	11011	001101
15	E	1110	11100	001110
16	F	1111	11101	011100

responsivity	
Field of View at the receiver [°]	70
Reflective index of concentrator	1.5
A_r [cm^2]	1
R_b [bps]	10^6
N_o	10^{-21}
m	1

3.1 Encoder and Decoder Design

The encoder is designed in such a manner that it will keep the number of ones and zeros equal. The first task of the encoder is to create a code word such that it has an even-numbered length. For this, the proposed encoder changes the 4-bit hex number into a 6-bit code word. The 6-bit code word will have the following properties:

- By making the code-word length even numbered, the numbers of one and zeros can be made equal.
- The proposed code word will have a code run equal to four bits.

In VLC, the fixed -dimming index is equal to r/n, where a time slot is n bits, and r bits amongst these n bits are used to represent meaningful data. Because the proposed scheme uses $n = 6$ and $r = 3$, the fixed - dimming index becomes 0.5 or 50%. This means that the dimming level is maintained as a constant at 50%, regardless of the data sent. In other words, the ratio of ones to zeros in a code word is always 1:1. A time slot of $n = 6$ bits results in a total of 64 code words to choose from. Among these 64 code words, there are 20 code words in which the number of ones and zeros are equal. 16 code words out of the 20 are used by the data, the remaining four code words can be used for other tasks such as synchronizing and ideal patterns. This can be understood from the truth table and comparing it with the existing 4B/5B line coding pattern in Table 1.

3.2 Algorithm for the Encoder and Decoder

The encoder is designed in consideration of the number of ones in the code word. This scheme generates a 6-bit code word that has equal numbers of ones and zeros. To ensure that there are equal numbers of ones and zeros, a counting method is applied. This method counts the number of ones in the 4-bit input and generates a 6-bit output that has equal numbers of ones and zeros.

In Table 1, the sum of the ones in a 4-bit binary pattern has five cases: zero, one, two, three, and four. The encoder handles each case individually and encodes the 4-bit pattern in the unique 6-bit pattern with equal numbers of ones and zeros. The method of encoding is described in Algorithm 1 while the decoding algorithm is described in Algorithm 2.

4. Performance Evaluation

To evaluate the performance of the proposed encoding and decoding scheme, a simulation was performed. The indoor environment in Fig. 1 was considered. Four LED lamps were installed at a height of 3 m above the floor with 60 × 60 LEDs in each lamp [13]. The simulation parameters are given in Table 2.

Algorithm 1 Proposed FD-OOK encoder

Algorithm 2 Proposed FD-OOK decoder

```
Input: 4bit pattern x, sum
Output: Encoded Code word
append 0 0 to the most significant position of x
if sum = 0 then
        flip the least significant bits of x
        change the most significant bits to 0 1 of x
else if sum = 1 then
        if the last three least significant bits of x = 0 then
                flip least significant bit of x
                change the most significant bits to 0 1 of x
        else
                change the most significant bits to 1 1 of x
        end
else if sum = 2 then
        change the most significant bits to 10 of x
else if sum = 3 then
        if the last three least significant bits of x = 1 then
                flip least significant bit of x
                change the most significant bits to 0 1 of x
        else
                change the most significant bits to 0 0  of x
        end
else if sum = 4 then
        flip last two least significant bits of
        change the most significant bits to 0 1 of x
end
```

```
Input: Encoder Code word cw
Output: 4 bit binary output
Set v = value of the most two significant bit of cw
Set v1 = value of the least two significant bit of cw
If v = 11 || 10 || 00 then
        remove v from cw
else if v = 01 then
        if v1 = 11 then
                flip v1
                remove v from cw
        else if v1 = 00 then
                flip v1
                remove v from cw
        else if v1 == 01 || 10 then
                flip the least significant bit
                remove v from cw
        end
end
```

The simulation was conducted to check the performance of the scheme on two main parameters: the normalized power requirement and the spectral efficiencies. The proposed FD-OOK scheme is compared with existing modulation schemes such as variable on-off keying (VOOK) and variable pulse position modulation (VPPM), which focus on dimming control. Fig. 2 shows the normalized required power with respect to the dimming factor N for each modulation scheme. FD-OOK exhibits better performance than VPPM and VOOK. VPPM and VOOK have almost the same performance for the same system model. For the region enclosed within the concerned window, it can be seen that FD-OOK can only work for $N = 0.5$ or a 50% dimming index. As compared to other schemes, FD-OOK requires the lowest optical power to operate with $N = 0.5$. In addition, the graph is symmetric along the $N = 0.5$ dimming index. From the simulation, it is seen that the optimal required power is achieved when the ratio of ones and zeros is 1:1 at the dimming index of 0.5. Fig. 3 shows the spectral efficiencies of the modulation schemes. Considering the concerned window again because FD-OOK has a 50% fixed - dimming index, a spike appears at $N = 0.5$. The VPPM scheme has poor spectral efficiency compared to VOOK and FD-OOK. The performance of VOOK is better, as it is spread over a large dimming-index range from 0.2 to 0.8. Because we are only concerned with the value of $N = 0.5$, FD-OOK outperforms VOOK and VPPM.

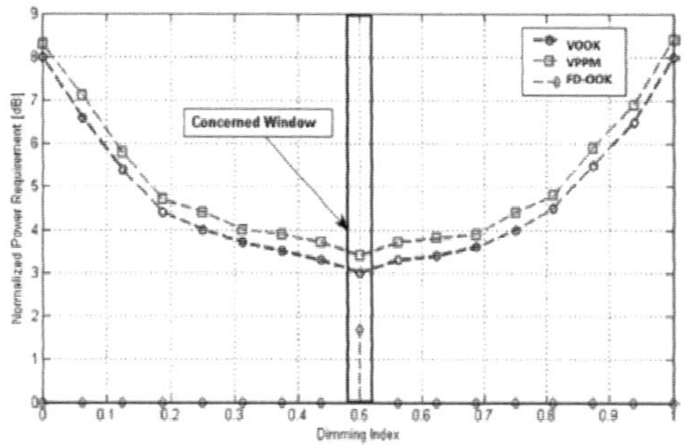

Figure 2. Normalized power requirement.

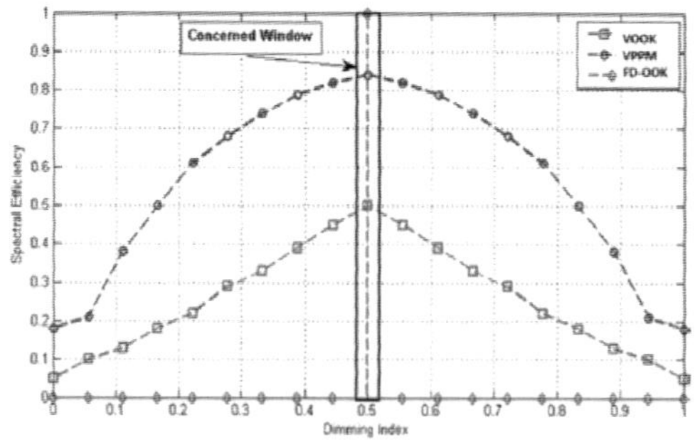

Figure 3. Spectral efficiency.

5. Conclusion

In this paper, a new scheme FD-OOK was proposed to address the problems of dimming and flickering present in VLC, and its encoder and decoder designs were proposed. To overcome the problem of dimming, the FD-OOK encoder scheme maintains a constant 50% dimming level. This, in turn, also reduces flickering by keeping the number of ones and zeros equal within a code word. The performance was evaluated for the normalized power requirement and spectral efficiency. The FD-OOK scheme was compared with other well-known schemes and its performance was better. In future work, the proposed scheme will be implemented as a real-time system, and the results for the data rate will be investigated.

Acknowledgments

This work was supported by the National Research Foundation of Korea (NRF) grant funded by the Korea government (MSIP) (2016R1A2B4008457) and the Strengthening R & D Capability Program of Sejong University.

References

[1] Chi N, Hass H, et al, 2015 *IEEE Wireless Communications* Visible light communications: Demand factors, benefits and opportunities **22** 2 5-7.
[2] International Energy Agency, http://www.iae.org. Accessed Mar 2015.
[3] Halonen L et al., 2010 Guidebook on Energy Efficient Electric Lighting for Buildings. Ph.D. dissertation, Dept. Electron. Lighting Unit, Aalto Univ. School Sci. Technol., Espoo, Finland,
[4] Kavehrad L, 2010 *IEEE Commun. Mag.* Sustainable energy-efficient wireless applications using light. **48** 66-73.
[5] Rajagopal S et al., 2009 5.7 Merged Draft Text ETRI-Samsung-CSUS-Intel, IEEE 802.15-09-0786-01-0007.
[6] Lee C. G., 2011 Visible light communication: In Advanced Trends in Wireless Communication. Paterson, NJ, USA: Mutamed Khalid.
[7] Noh J, et al, 2015 *Optics Communications* A dimming controllable VPPM-based VLC system and its implementation **343** 34-37.
[8] Lee K and Park H, 2011 *IEEE Photon. Technol. Lett.* Modulations for visible light communications with dimming control **23** 1136-1138.
[9] Park H and Barry J. R., 1995 *Proceeding IEEE International Conference Communications (Seattle, WA)* Modulation analysis for wireless infrared communications pp 1182–1186.
[10] Grubor J et al., 2005 *Proc. ITG (Fachbericht)* High-speed wireless indoor communication via visible light pp 1-5.
[11] Komine T and Nakagawa M, 2004 *IEEE Trans. Consum. Electron.* Fundamental analysis for a visible light communication system using led lights **50** 100-107.
[12] Kim H, and Niaz M T, 2016 Method for transmitting and receiving data in visible-light communication system, and device therefor. Patent no. WO/2016/024674.
[13] Kumar N and Lourenco N F, 2010 *J. Eng. Appl. Sci.* Led-based visible light communication: A brief survey and investigation **5** 296-307.

Time Granularity Transformation of Time Series Data for Failure Prediction of Overhead Line

Yan Ma[1], Wenbing Zhu[1], Jinxia Yao[1], Chao Gu[1], Demeng Bai[1] and Kun Wang[2]

1 State Grid Shandong Electric Power Research Institute, Jinan, China 250002
2. Shandong Zhongshi Yitong Group Co. Ltd., Jinan, China 250002

E-mail: yanpony@126.com

Abstract. In this paper, we give an approach of transforming time series data with different time granularities into the same plane, which is the basis of further association analysis. We focus on the application of overhead line tripping. First all the relative state variables with line tripping are collected into our big data platform. We collect line account, line fault, lightning, power load and meteorological data. Second we respectively pre-process the five kinds of data to guarantee the integrality of data and simplicity of analysis. We use a representation way combining the aggregated representation and trend extraction methods, which considers both short term variation and long term trend of time sequence. Last we use extensive experiments to demonstrate that the proposed time granularity transformation approach not only lets multiple variables analysed on the same plane, but also has a high prediction accuracy and low running time no matter for SVM or logistic regression algorithm.

1. Introduction

At present the data generated from grid operation and equipment monitoring exponentially increase and the environment of electric big data forms. Different monitoring indicators are collected into separate business information systems at various intervals. In order to integrate the data from different systems including production management system (PMS), energy management system (EMS), online condition monitoring system and meteorological system [1], we build a big data platform of condition assessment for electric transmission and transformation equipment. The time interval of experimental data in PMS system depends on the period of specific experiments and maintenance plan. The voltage and current are stored every five minutes. The online condition monitoring system has different collecting intervals according to different monitoring devices and in general its data are collected every 24 hours. Meteorological system has two kinds of monitoring indicators: real-time state variables and predictable state variables, which also have different time granularities.

Overhead transmission lines are exposed in the wild all year round and usually suffer severe natural environment and serious disaster attack, which cause degrees of damage for safe operation and lead to the trip of transmission lines. In order to find the rules between line tripping and its relative state variables, a feature representation method of data in different time granularities is required to make conjoint analysis feasible. The relative data on line tripping are the fault description information including phase, incoming switch, outgoing switch, line fault position, line fault reason and whether reclosing successfully, account information including line ID, equipment code, line name, voltage class and maximum allowable electric current, lightning information, real-time meteorological information, one hour power load information before line fault including current, active power and reactive power.

All the above information has different collecting time granularities, where loading data are every five minutes, real-time meteorological data are every 10 minutes and the time granularities of the other data are uncertain. In the paper we combine data characteristics of line tripping, and mainly research the uniform transformation method for time series data with different time granularities, which is fundamental for the future association analysis between line tripping and its relative state variables.

2. Related work

Time series data is a kind of high dimensional data related with time. Mining the valuable information and knowledge from massive time series data is one of the mainstream researches in data mining domain [2]. The studies of time series data mining include similarity measurement, classification, clustering, pattern discovery and association rules of time series data, etc. In the paper, we focus on the data pre-processing step of association rules [3].

Time granularity is regarded as temporary resolution. Time granularity has many levels. A single time granularity such as day, hour, second has absolute time length [4]. Each time interval can be measured by a time granularity with the same time length. If multiple attributes in the same time range have different time granularities, these attributes have different numbers of records. If we try to combine these attributes and analyze them, an approximate representation method of time series data with different time granularities is necessary. Since the structure of time series data is complicated, directly mining on the original data has a low efficiency and affects the accuracy and reliability of mining results [5]. Therefore, finding a proper representation method for multiple status data and meanwhile keeping the main information are the foundation of the further analysis.

At present, common representation methods consist of discrete wavelet transformation [6], piecewise linear representation [7], symbolic representation [8] and etc. Discrete wavelet transformation maps time domain to frequency domain through wavelet transformation and use a small amount of wavelet coefficients to represent original time series data. The method has a high degree of dimension reduction, but the inappropriate wavelet coefficients lose massive information. Piecewise linear representation uses k consecutively connected lines to represent a time sequence with length n. It could reduce data dimensions but the selection of k is critical. Symbolic representation first discretizes time sequence and then map each part of time series waveform to a symbol. It use research results of character string, but it is hard to discretize time series and define the meaning of symbols.

Before the further association analysis, we first preprocess time series data using the proposed transformation method. It combines aggregated representation and trend extraction methods, which absorbs the simplicity of aggregated representation and the refining of trend extraction. We analyze each state variable, which guarantees not to lose lots of information and also reduces the time complexity of data mining. The proposed method realizes the feasibility of jointly analyzing multiple attributes with different time granularities.

3. Time granularity transformation method

We first discuss the time range of different state variables. In order to cover the error of inconsistent clock, the time ranges of lightning and l data are set as 10 minutes before line fault and 2 minutes after line fault. The time range of power load information is set as one hour before line fault. Through data analysis, lightning, meteorological and power load data need doing time granularity transformation to have the same time granularity with line account data and fault data. Line account information and fault description data correspond to a device or a fault. We next separately process different attributes related to line tripping according to their characteristics.

3.1. Lightning data

If there is no lightning data in the defined time range before or after a fault, the attribute value is set to 0. If there are many lightning data before or after a fault, we just record the maximum, minimum and average value, which uses three values in the same time interval to represent the time series information of lightning data.

3.2. *Meteorological data*

Meteorological data are provided by weather monitoring station. A line may cover two or more weather monitoring stations, and the real-time meteorological data are sent 10 minutes at a time. We choose the recent meteorological data from the weather monitoring station near to the fault and record its real-time data including wind direction, wind speed, gust wind direction, gust wind speed, amount of precipitation, relative humidity, temperature, air pressure, visibility, maximum wind direction, maximum wind speed and etc.

3.3. *Power load data*

Power load data include many information, thus we use a relatively complicated method to represent them. Power load data include three attributes: electric current, active power and reactive power. We first respectively compute their average values one hour before fault. Assuming the time sequence of electric current one hour before fault is denoted as $I = \{i_1, i_2, ..., i_n\}$, its average value is computed as $\bar{i} = \sum^{n} i / n$. Meanwhile, we compute the variation factors of electric current, which include head variation factor and tail variation factor. Head variation factor and tail variation factor are represented as $\Delta I_s = i_1 - \bar{i}$ and $\Delta I_e = i_n - \bar{i}$. The average value, head variation factor and tail variation factor of electric current as new state variables are added into line fault records.

In addition, in order to represent the long term trend of time series data, we introduce the conception of local extremums as the key point of trend variation ratio. Local minimum point is defined as follows. If the electric current i_j satisfies one of the following conditions:

1) For $1 < j < n$, there exist the subscripts r, t satisfying $1 \le r < j < t \le n$, and i_j is the minimum value among $i_r, ..., i_t$; 2) For $j = 1$, there exists the subscript r satisfying $j < r \le n$, and i_j is the minimum value among $i_1, ..., i_r$; 3) For $j = n$, there exists the subscript t satisfying $1 \le t < j$, and i_j is the minimum value among $i_t, ..., i_n$. Then the electric current i_j is defined as extreme minimal value.

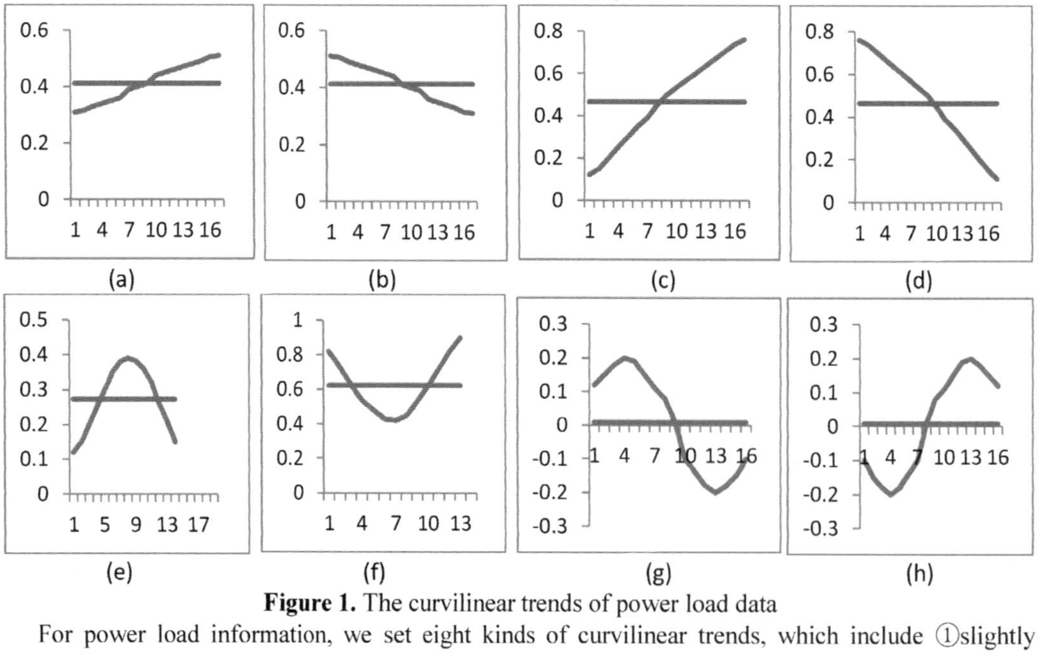

Figure 1. The curvilinear trends of power load data

For power load information, we set eight kinds of curvilinear trends, which include ①slightly ascending, ②slightly descending, ③obviously ascending, ④obviously descending, ⑤first descending then ascending, ⑥first ascending then descending, ⑦generally ascending , ⑧generally descending. Their concrete images are shown as figure 1. Taking current as examples, the unit of Y axis is ampere and the unit of X axis is the number of time intervals.

Their concrete methods of trend judgment are as follows.

1) We first determine whether there exist extremums during the time period. If there is no extremum, we could know that its trend is categorized to trend ①, trend ②, trend ③ or trend ④. If there is a extremum, its trend is trend ⑤or trend ⑥. If the extremum is extreme maximum, it is trend ⑤; else if the extremum is extreme minimum, it is trend ⑥. If there are two extremums, the trend are categorized as trend ⑦ or trend ⑧.

2) If there are more than two extremums, we further divide time series data. After the segmentation, go to step 1)

3) Taking the electric current as an example, we compute its head variation factor and tail variation factor as $\Delta I_s = i_1 - \bar{i}$ and $\Delta I_e = i_n - \bar{i}$. The minus between head variation factor and tail variation factor is $\Delta I = i_1 - i_n$.

4) After determining the range of trend ① to trend ④ in step 1), we further give determination conditions. If $0 < |\Delta I| < \delta$ stands, it is slightly changing trend. If $|\Delta I| \geq \delta$ holds, it is obviously changing trend, where δ is determined as 20 percent of the varying minus during a year for the state variable. If $\Delta I < 0$ holds, we know it is ascending trend; else it is descending trend.

5) After determining the range of trend ⑦ and trend ⑧, we next use ΔI as the judgment condition. If $\Delta I < 0$ stands, it is generally ascending trend; else it is generally descending trend.

After getting the trends of time series data, we use symbols to represent the eight kinds of trends, and add them into our data records.

4. Experiments

In the section, we make extensive experiments to validate the performance of our proposed time granularity transformation method.

We first choose 10000 fault records from unplanned outage lines, which include fault line ID and fault happening time. Combing them with our given time range for each state variable, we could obtain all the relative state variables with line tripping through searching our big data platform. Then we preprocess the original data using the proposed transformation method. So far we get line fault data. Meanwhile, according to the aforementioned line ID, we choose a time interval when the line runs smoothly. In the similar way we preprocess and obtain line normal data.

We use SVM (support vector machine) and logistic regression algorithm to run the above data. The data pattern is organized as table 1, where the new generated variables are input attributes and the line state is goal attribute.

Table 1. Data pattern of line state prediction

Line account variables	Real meteorological variables	Lightning current maximum	Lightning current minimum	Lightning current average value	Current maximum	Current minimum	Current average value	Current head variation factor	Current tail variation factor	Current changing trend	…	active power head variation factor	active power tail variation factor	active power changing trend	Line state

After running SVM and logistic regression algorithm according to the above data pattern, we get the prediction value of line state. Figure 2 and figure 3 shows their predication accuracy and running time. As seen from figure 2 we know the predication accuracy increase with the increasing number of records and SVM has a higher classification accuracy than logistic regression algorithm. In figure 3 their running times increase with the increasing number of data records as well and logistic regression algorithm has a shorter running time.

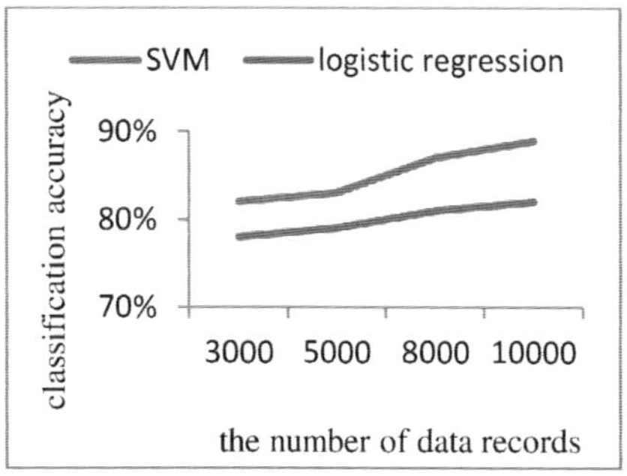

Figure 2. Classification accuracy with the number of data records.

Figure 3. Running time with the number of data records.

In all, no matter SVM or logistic regression algorithm, time granularity transformation method makes jointly analyzing multiple state variables feasible and largely reduces running time to an acceptable level. The most important advance is transformation method does not push down the classification accuracy.

5. Conclusions

In the paper, we propose a new transformation method for time series data with different time granularity. It totally considers the characteristics of overhead line and its concrete state variables. We aim for five kinds of business data: account data, fault data, lightning data, power load data and meteorological data. After the research, we could put all the state variables on the same plane, which lays a foundation for further joint analysis. After running SVM and logistic regression algorithm we validate that the proposed transformation method makes the classification accuracy and running time acceptable enough for failure prediction.

Acknowledgement

This work was supported by the institute-level project in State Grid Shandong Electric Power Research Institute (2016-15). The authors are grateful for the anonymous reviewers who made constructive comments.

References

[1] Peng X, Deng D, Cheng S, Wen J, Li Z and Niu L 2015 *Proceedings of the CSEE*, **35** 503.
[2] He X. Research on Key Issues in Time Series Data Mining 2014 *Ph.D. dissertation.*
[3] Li H. Research on Feature Representation and Similarity Measure Methods in Time Series Data Mining 2012 *Ph.D. dissertation.*
[4] Shao X, Hao W, Jin D, Wang Y 2011 *Journal of Computer Applications* 31.
[5] Xi W, Li P, Guo X, Xu A, Jiang Y, Zhang L and Wu Y 2014, *Power System and Clean Energy* **30**.
[6] Akansu A, Haddad R 1992 *Academic Press*, ISBN 978-0-12-047141-6.
[7] Yan C, Fang J, Wu L and Ma S 2013 *Journal of Information & Computational Science* **10** 2747.
[8] Lin J, Keogh E, Wei L and Lonardi S 2007 *Data Mining & Knowledge Discovery* **15** 107.

Optimizing System Call Latency of ARM Virtual Machines

Song-Woo Sok[1,3], Young-Woo Jung[1,4] and Cheol-Hun Lee[2,5]

[1]High-Performance Computing Dept., Electronics and Telecommunications Research Institute, 218 Gajeong-ro, Yuseong-gu, Daejeon, 34129, Korea
[2]Department of Computer Science & Engineering, Chungnam National University, 99 Daehak-ro, Yuseong-gu, Daejeon 34134, Korea

[3]swsok@etri.re.kr, [4]jungyw@etri.re.kr, [5]clee@cnu.ac.kr

Abstract. This paper introduces ViMo-S, a type 1 hypervisor for ARMv7 and ARMv8-based ARM server systems. It supports full virtualization to run existing operating systems and applications unmodified. It uses ARM hardware virtualization extensions to optimize the performance of virtual machines, especially system call latency. Therefore, its virtual machines' system call latency is near physical machine's, while other hypervisors like Xen and KVM show relatively slower and unstable performances in benchmark tests.

1. Introduction

ARM-based processors are dominant in the mobile computing market because of low power consumption and high performance per watt ratio. As the performance of ARM processors keeps close to x86 processors, ARM servers are becoming popular for its low power and high-density characteristics.

These ARM-based servers give benefits of lower TCO with less power consumption, low heat emission, and smaller space needs. To utilize the server more efficiently, Supporting consolidation based on virtualization technologies is necessary [1].

Traditionally, ARM architectures had no support for virtualization technologies. Since ARMv7 architecture, hardware virtualization extensions are added to support virtualization software [2]. It helps virtualization software to be more lightweight, efficient and able to run existing operating systems and applications unmodified.

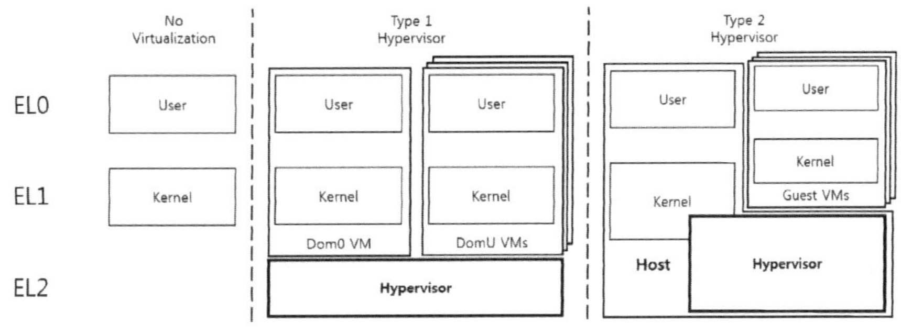

Figure 1. ARM exception levels and software layers of non-virtualized and virtualized systems

Figure 1 shows exception levels of ARM processors which include the virtualization extensions. ARM architectures also support EL3, most privileged exception level. However, it is specially designed to make an isolated, secure world and mostly used for mobile devices. Moreover, some ARM server SoCs do not support EL3 exception level. So, EL3 mode is not considered in this paper.

Without virtualization, EL1 is the highest exception level, and it is used to execute operating system kernel. While user programs run in EL0, the least privileged exception level. To support virtualization and run existing operating systems and applications unmodified, virtualization extensions add more privileged exception level, EL2. In Figure 1, non-virtualized, type 1 virtualized and type 2 virtualized software layers are shown.

Xen is a popular type 1 hypervisor that supports both x86 and ARM architectures. As a type 1 hypervisor, Xen ARM is executed before virtual machine (VM)'s OS boot. Then, it creates a special VM, domain0 which has responsible for controlling hardware devices and executing I/O operations on behalf of other VMs, called domainUs [3].

KVM is a type 2 hypervisor technology, and it runs as a part of host operating systems. It was developed to support full virtualization on x86 architectures. When it was ported to ARM architectures, KVM had to be redesigned to adopt virtualization extensions [4]. The result is called KVM/ARM, and its core component was separated into two parts. One part called as highvisor is executed as a part of the host's operating system, the other part called as lowvisor runs in EL2 to use benefits of ARM virtualization extensions. KVM supports full virtualization, but full virtualized I/O operations have lower performance compared to para-virtualized ones. Therefore, KVM supports VirtIO para-virtual I/O device to increase performance and reduce overheads. VirtIO is a standard para-virtual I/O interface for virtualized computing environment.

This paper presents ViMo-S which is a type 1 hypervisor based on ARM virtualization extensions. It supports full virtualization for CPU and Memory, while it uses VirtIO interface for I/O operations of VMs. If an operating system has VirtIO drivers, it can be executed on ViMo-S' domainUs without modifications.

2. Design and implementation of ViMo-S

ViMo-S is the following project of ViMo (Virtualization for Mobile), a micro virtual machine monitor for ARM mobile systems. It supported full virtualization on ARM architectures that have no hardware virtualization support [5].

ViMo-S has been developed for ARM server applications based on ViMo architecture. Because ViMo-S employs ARM hardware virtualization technologies, it becomes simpler and more efficient compared to ViMo.

2.1. Design of ViMo-S

2.1.1. CPU Virtualization. Because ARM virtualization extensions are more suitable to type 1 hypervisors [1], ViMo-S was planned to be a type 1 hypervisor from the start. So, ViMo-S is executed right after the boot loader in EL2 level; then it initializes the first VM, domain0, and boots its operating system.

All VMs can have one to four virtual CPUs (VCPUs). When a VCPU is scheduled to a physical core, ViMo-S saves the current context of the core to the related VCPU then loads scheduled VCPU's context to the core. Therefore, 2 or more VCPUs can share a core to run.

User applications execute many system calls when they are running. ViMo-S sets system calls to be routed to EL1, not to EL2. Because the VM's kernel directly handles all system calls and there is no intervention of the hypervisor, system call latency of the VMs on ViMo-S is near to a non-virtualized system.

2.1.2. Memory Virtualization. The ARM virtualization extensions support hardware 2-stage memory translation to support memory virtualization for VMs. The first stage's page tables are configured by operating systems of VMs, and its translated address is not a real physical address (PA) but an intermediate physical address (IPA). If the second stage page tables are configured in EL2, then an IPA can be translated to a PA by hardware MMU.

ViMo-S allocates VM's physical memory and configures second stage page tables for each VMs. ViMo-S allocates memory area exclusively to each VMs, so, a VM cannot access to other VM's memory area. However, ViMo-S can create shared memory space by configuring each 2nd stage page tables of VMs to have same physical memory address entries.

2.1.3. I/O virtualization. ViMo-S supports VirtIO interface for para-virtualized I/O operations. If domainU's operating system has VirtIO drivers, it can use a console, disk and network VirtIO virtual devices to execute I/O operations.

Figure 2 shows the processing path of virtual I/O requests. DomainU's VirtIO requests are trapped to ViMo-S and ViMo-S delivers I/O requests to the domain0's VirtIO backend driver. Then, the VirtIO backend driver maps the requests to real hardware I/O operations and calls hardware drivers to process the requests.

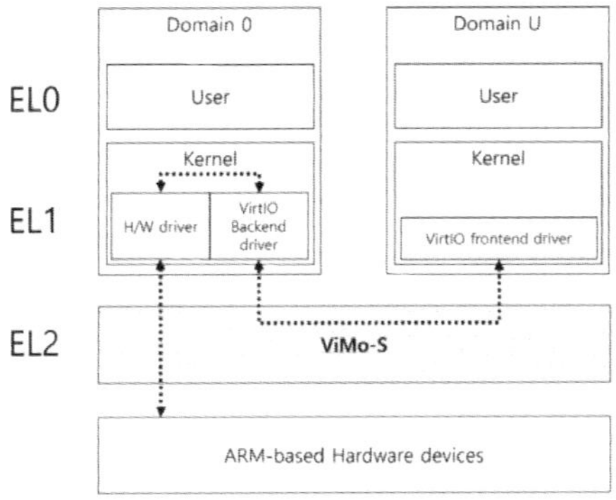

Figure 2. I/O Virtualization of ViMo-S

2.2. Implementation

ViMo-S has been implemented on X-C1 Server Development Platform board developed by AppliedMicro.

The main SoC of the board is AppliedMicro's X-Gene 1, an ARMv8 compliant processor. It has eight cores and runs at 2.4 GHz clock speed. The X-C1 has 16GB of DDR3 RAM, three ports of Gigabit Ethernet and two SATA ports. A Samsung 850 EVO 500GB SSD was used as storage device.

We used Linux 3.15-rc8 kernel for domain0 and Linux 3.18.0 for domainUs. Ubuntu 14.04 ARM64 version was used for both domain0 and domainU's root file systems.

3. Benchmark tests and results

3.1. Benchmark test configuration

To evaluate the performance of ViMo-S, we used lmbench 3.0 to check system call latencies of virtualized and non-virtualized environments on the X-C1 board. The tests included ViMo-S, Xen ARM, and KVM/ARM configurations. The lmbench is a benchmark suite to measure system's micro performance factors like system calls, memory accesses, network, and disk performances [6].

Table 1 shows configurations for each system.

Table 1. Configurations for benchmark tests

	Version	Domain0/Host Core/Memory	Domain0/Host Kernel	DomainU/Host Core/Memory	DomainU Kernel
ViMo-S	-	1core/2GB	Linux-3.15-rc8	1core/512MB	Linux-3.18.0
Xen ARM	4.4.0	1core/2GB	Linux-3.15-rc8	1core/512MB	Linux-3.15-rc8
KVM/ARM	Linux-3.15-rc8	1core/2GB	Linux-3.15-rc8	1core/512MB	Linux-3.18.0

3.2. Raw Linux and Domain0s

First, we conducted lmbench tests of non-virtualized Linux, ViMo-S and Xen's domain0 VMs.

Table 2 shows the results of lmbench system call latency tests of raw Linux and domain0 VMs. For easy comparison, it includes latency ratios based on raw Linux results also.

Table 2. lmbench test results for raw Linux and domain0s (unit=microseconds)

	Simple syscall	Simple read	Simple write	Simple stat	Simple fstat	Simple open/close	Protection fault	Pipe latency
Raw Linux	0.1318	0.1845	0.2312	0.8316	0.1936	2.4462	0.2057	4.7348
ViMo-S	0.1316	0.1833	0.2308	0.8406	0.1951	2.1689	0.1993	4.6733
Xen ARM	0.3834	0.5356	0.6008	2.5131	0.4516	5.9182	0.5072	13.6326
ViMo-S/Linux	99.8%	99.3%	99.8%	101%	100.8%	88.7%	96.9%	98.7%
Xen/Linux	291%	290%	260%	302%	233%	242%	247%	288%

As shown in Table 2, raw Linux and ViMo-S have very similar results, while Xen Domain0 shows relatively slower performances. Xen's system call latencies are 2.3 to 3 times slower than ViMo-S and raw Linux.

3.3. DomainUs and KVM/ARM guest

We did same lmbench tests on KVM/ARM's guest, ViMo-S and Xen's domainU VMs.

Table 3 shows the results of lmbench system call latency tests of KVM guest, domainU VMs. The raw Linux results from Table 2 are also included for comparison.

Table 3. lmbench test results for KVM/ARM guest and domainUs (unit=microseconds)

	Simple syscall	Simple read	Simple write	Simple stat	Simple fstat	Simple open/close	Protection fault	Pipe latency
Raw Linux	0.1318	0.1845	0.2312	0.8316	0.1936	2.4462	0.2057	4.7348
ViMo-S	0.1221	0.2279	0.3104	1.0876	0.1825	3.0000	0.0992	6.8388
Xen ARM	0.3935	0.6091	0.6129	2.3186	0.4395	5.7837	0.3724	14.0291
KVM/ARM	0.1218	0.1963	0.2913	2.7275	0.1964	6.9162	0.2673	108.6665
ViMo-S/Linux	93%	124%	134%	131%	94%	122%	48%	144%
Xen/Linux	299%	330%	265%	279%	227%	236%	181%	296%
KVM/Linux	92%	106%	126%	328%	101%	283%	130%	2295%

ViMo-S's domainU has relatively slower performance than ViMo-S's domain0. However, its latencies did not exceed 1.5 times of raw Linux's.

Xen's domainU shows similar performance with Xen's domain0's. It has 1.8 to 3.3 times slower than raw Linux.

KVM/ARM's guest VM shows interesting results. For some system call tests, its latencies are slightly faster than ViMo-S but significantly slower for simple stat, simple open/close and pipe latency tests.

4. Conclusion and Future Works

This paper proposed ViMo-S, a new type 1 hypervisor for ARM architecture. ViMo-S supports ARM virtualization extensions to optimize VM's performance and minimize virtualization overheads. We showed that ViMo-S has minimized system call latency for VMs and provides similar performance with bare metal machines.

According to our experiment results, ViMo-S has better and more stable system call latency performances compared to Xen and KVM/ARM.

We are trying to optimize I/O performance of ViMo-S and doing more benchmark tests with more complex benchmark tools and real world applications.

Acknowledgments

This work was supported by Institute for Information & communications Technology Promotion (IITP) grant funded by the Korea government (MSIP) [R0101-16-237, Development of General-Purpose OS and Virtualization Technology to Reduce 30% of Energy for High-density Servers based on Low-power Processors]

References

[1] Dall C, Li S W, Lim J T, Nieh J and Koloventzos G. 2016 ARM Virtualization: Performance and Architectural Implications. In *Proceedings of International Symposium on Computer Architecture* ISCA 2016 .

[2] Varanasi P and Heiser G 2011 Hardware-supported virtualization on ARM. In *Proceedings of the Second Asia-Pacific Workshop on Systems* p11. ACM.

[3] Stabellini S and Campbell I. 2012 Xen on arm cortex a15. *Xen Summit North America, 2012.*

[4] Dall C and Nieh J. 2014 KVM/ARM: the design and implementation of the Linux ARM hypervisor. In *ACM SIGPLAN Notices* Vol. **49**, No. 4, pp. 333-348. ACM.

[5] Oh S C, Kim K, Koh K and Ahn C W 2010 ViMo (virtualization for mobile): a virtual machine monitor supporting full virtualization for ARM mobile systems. In *Proceedings of Advanced Cognitive Technologies and Applications, COGNITIVE.*

[6] McVoy L W and Staelin C 1996 lmbench: Portable Tools for Performance Analysis. In *USENIX annual technical conference* pp. 279-294.

Detection of Pigment Networks in Dermoscopy Images

Khalid Eltayef, Yongmin Li and Xiaohui Liu

Computer Science Department, Brunel University London, Uxbridge, UB8 3PH, UK

E-mail: khalid.eltayef@Brunel.ac.uk, Yongmin.Li@Brunel.ac.uk, XiaoHui.Liu@brunel.ac.uk.

Abstract. One of the most important structures in dermoscopy images is the pigment network, which is also one of the most challenging and fundamental task for dermatologists in early detection of melanoma. This paper presents an automatic system to detect pigment network from dermoscopy images. The design of the proposed algorithm consists of four stages. First, a pre-processing algorithm is carried out in order to remove the noise and improve the quality of the image. Second, a bank of directional filters and morphological connected component analysis are applied to detect the pigment networks. Third, features are extracted from the detected image, which can be used in the subsequent stage. Fourth, the classification process is performed by applying feed-forward neural network, in order to classify the region as either normal or abnormal skin. The method was tested on a dataset of 200 dermoscopy images from Hospital Pedro Hispano (Matosinhos), and better results were produced compared to previous studies.

1. Introduction

Malignant melanoma is the most frequent type of skin cancer and the detection of the early stage is a challenging and fundamental task for dermatologists, with the early diagnosis, the patients probability of survival increases greatly. Dermoscopy is a widely common techniqueused to perform an in-vivo observation of pigmented skin lesions.[1].

The pigment network is one of the most relevant dermoscopic structures for its appearance on the body is an indicator of the existence of melanin in deep layers of the skin. The shape of the pigment network can be classified as either Typical or Atypical: typical when the pattern is regularly with a light-to-dark-brown network small uniform space network holes and thin network lines distributed more or less regular throughout the lesion, while atypical is characterized by black, brown, or gray and thickened lines distributed more or less irregularly throughout the lesion. Therefore, an atypical pigment network signals a melanoma[2][3].

Computer Aided Diagnosis (CAD) systems have been proposed by many different groups to identify malignant melanomas in dermoscopy images. These systems use several features, such as color, shape and texture, to classify the images as either normal or abnormal. [2][3][4].In the detection of melanoma dermatologists use the ABCD rule to analyze four parameters (Asymmetry, Border, Colors and Diameter) and distinguish skin lesions or the 7-points checklist which is a scoring method for a set of different classify depending on color, shape and texture [5][4].

An automatic detection of pigment networks is a challenging task, because there is a smooth transition between the color of the pigment network lines and the background. Moreover, presence of dark hair covering the lesions and the existence of specular reflection. In addition, the circular structures such as dots are detected as pigment networks wrongly, or vice-versa [3]. In this paper, four steps have been performed to implement an automatic algorithm for the detection of pigment networks in dermoscopy images. First, the image is pre-processed in order to remove artifacts such as hairs and reflection detection. Second, based on the intensity values and geometrical properties of the pigment networks, a set of tentative pigment network regions are detected. Then, features are extracted from the given lesion and by using a trained classifier, the given lesion is classified in one of two classes (with or without pigment network). The paper is structured as follows. Section 2 provides an overview of the method developed previously. The proposed system is described in details in sections 3-6. The evaluation and the performance of the detection system and the results obtained are discussed in Sections 7. Finally, conclusions are presented in section 8.

2. Previous Work

Sadeghi et al.[6] aim to detect the holes of the network. Three different stages were implemented to detect the holes, first a Laplacian of Gaussian (LoG) filter is used in order to find meshes or cyclic structures. Second, an 8-connected components analysis is applied to transform the filtered image to a set of graphs. Finally, a search for loops or cyclic graphs is carried out. Then the distance between the holes is measured and based on this distance a new graph is created. This graph is used for the detection of pigment network. The same research group presented a new method for classification between Absent, Typical and Atypical [7]. They detected the net structure and extracted structural, geometric, chromatic and texture features, and then they used these features to train a boosting classifier. Two different techniques structural and spectral are proposed by Betta et al.[8] They started to combine these techniques to perform the detection of pigment network. Firstly, they used structural techniques to search for simple shape structures, like lines or dots. Secondly, the spectral technique is based on a Fourier analysis of gray-level image. A sequence of Fast Fourier Transform, high-pass filtering, Inverse Fast Fourier Transform and thresholding were applied. Eventually, the mask from each technique is combined together to provide a network image. This study has been updated and proposed by Di Leo et al.[9] They classified the areas which constitute the network image as either atypical or typical, the existence of typical pigment network or the absence of network are both included in the same class. Chromatic and spatial features such as mean and standard deviation related to the obtained structures were extracted, and used to train a decision tree algorithm.

Grana et al.[10] Proposed an algorithm to undertake the detection of the pigment network. They used Gaussian derivative kernel to detect the net edges and Fisher linear technique to provide the optimal thresholds. Morphological masks are used to complete the line linking process. However, the experiments which has been performed are not focused on the task of distinguish between pigment network and no pigment network.

Barata et al.[3] proposed an algorithm to detect pigment networks using a bank of direction filters and morphological operations. They used two distinctive properties, region pigment network intensity and geometry, features were extracted and an Adaboost algorithm was used to classify the region as either with or without pigment network (normal or abnormal skin).

3. Pre-processing

The purpose of image pre-processing is to improve the image data by removing unwanted distortions from the image or enhancing image features, which can be used as input to other image processing techniques. Pre-processing is an important step in medical image analysis, in order to avoid affecting the extraction of the object by noise, and it facilitates the segmentation process. In dermoscopy images, two main kinds of noise must be removed before doing any step, hairs and reflection artifacts caused by placing gel or oil before capturing the image. Incorrect segmentation of pigment network can be extracted if hairs covering the image are not removed, since hair shapes are similar to pigment network lines. Consequently, the classification of the given region may be incorrect. This process involves two key operations: hairs, reflection artifact detections and removal. The original RGB images were separated to three color components, and the blue component is selected for its best discriminative capacity[11].

3.1. Reflection detection

Refection artifacts and air bubbles appear like noise in dermoscopy images due to placement of oil or gel on the lesion area before capturing the image. To detect this kind of noise, a simple threshold algorithm is applied. Every single pixel *(x,y)* can be detected and classified as a reflection artifact if its intensity value is higher than threshold T_{R1} and if its intensity value minus the average intensity $I_{avg}(x,y)$ of its surrounding neighborhood is higher than threshold T_{R2}, *i:e.*

$$\{I(x,y) > T_{R1}\}and\{(I(x,y) - I_{avq}(x,y)) > T_{R2}\}. \tag{1}$$

where I is the image, $I_{avg}(x,y)$ is the average intensity value in a local neighborhood of the selected pixel which is computed using a local mean filter with dimensions 11x11 and T_{R1}=0.7, T_{R2}=0.098.

3.2. Hair detection

Due to the similarity between hairs and pigment networks in dermoscopy images, it is challenging to detect hairs without any effect of the pigment networks. This similarity may lead to incorrect detection and decrease in classification accuracy since the shape of the hairs similar to the shape of the pigment networks. Then, the same algorithm can be used to detect both. Directional Gabor filters are implemented to extract hair artifacts and pigment networks from dermoscopy images. However, the parameters used in the Gaussian filters at each stage are different [3]. A bank of 64 directional filters has been used to perform hair detection. Let

$$h\theta_i(x,y) = G_1(x,y) - G_2(x,y). \tag{2}$$

be the impulse of a directional filter with angle θ_i where $\theta_i \in [0,\pi]$, i=0.. 64 and G_k is a Gaussian filter:

$$G_K(x,y) = C_K \exp\{-\frac{x'^2}{2\sigma_{x_k}^2} - \frac{y'^2}{2\sigma_{y_k}^2}\}, k = 1,2. \tag{3}$$

The difference between two consecutive filters θ_i and θ_{i+1} is constant and equal to (π/N). Therefore, $N+1$ filter are used. In (3), C_K is a normalization constant and the values of (x',y') are dependent on or related with (x,y) by a rotation of amplitude θ_i.

$$x' = x\cos\theta_i + y\sin\theta_i. \tag{4a}$$

$$y' = y\cos\theta_i - x\sin\theta_i. \tag{4b}$$

Parameters value have been experimentally obtained and set as: σ_{x1} =*20*, σ_{y1}=*6*, σ_{x2}=*20* and σ_{y2}=*0.5*. The size of the mask filter is 41 x 41.

The difference of Gaussians in Equation (2) is used since it allows a better enhancement of directional structures while removing the effect of the background. The image I is filtered by each directional filter. The output of the *ith* directional filter is given by the following convolution:

$$I_i(x,y) = h\theta_i(x,y) * I(x,y). \tag{5}$$

After filtering the image by each directional filter, the combination of the output of $N + 1$ directional filters are estimated and the maximum output at each pixel (x,y) is selected according to the following equation.

$$J(x, y) = max_{i \in \{1,2,..N+1\}}(I_i(x, y)). \tag{6}$$

Where J is the final image obtained from Gabor filter, this image contains both: hairs and pigment network lines. Adjusting the threshold is tricky. Therefore, we used a different approach and produced better results. Image sharpening is applied on the outcome image from Gabor filter, to increase the contrast between different colors, and sharpen the transition from black to white, where all edges become sharp. Then, Sobel filter is applied on the sharpened image with threshold of 0.0145, this process ignores all edges that are not stronger than the threshold value, and detect the rest of the edges as hair. In this case, every single hair is detected and classified as a noise. Therefore, many hairs were not classified in the previous method "Barata's method" are detected and classified. Examples are illustrated in figure 1.

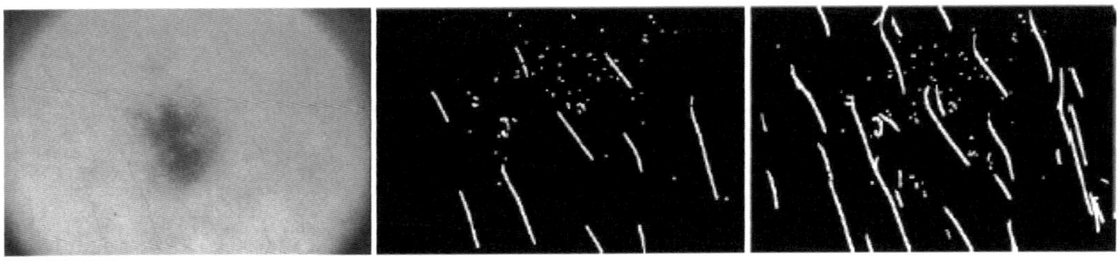

Figure 1. Examples of hair detection process: Original (first column), Barata's method (second column) and our method (third column).

After reflection artifacts and hairs are detected, their binary masks are multiplied by gray scale image $I(x,y)$. This operation leads to appearance of gaps which can be filled by propagating the information from known region which it is the neighborhood into the unknown region [12].

4. Pigment network detection

Pigment network and hair artifacts both have very similar line shape, and they appear with more than one orientation in dermoscopy images. Therefore, the same technique with different parameters can be used to detect the pigment networks from the dermoscopy images. Two main steps have been applied to do so: network enhancement and network detection. First, the output image that was obtained from pre-processing stage is filtered with a bank of directional filters [see (2)-(6)]. Due to the difference of the length and the width of each line stroke of the hairs and pigment network, the parameters used to detect the pigment networks are different from those used to detect the hairs. These parameters were set as: $\sigma_{x1} = 40$, $\sigma_{y1} = 40$, $\sigma_{x2} = 3$ and $\sigma_{y2} = 0.5$ with mask filter dimension of 11 x 11 and $T_H = 0.0185$. Then, to detect the pixels which belong to the net of pigment networks, a threshold value is implemented. The second step is to extracting large linked lines of pigment networks, since it is assumed that the pigment networks consists a set of connected lines. A connected component analysis can be used to identify the connectivity between the pixels. Eight-connectivity criterion is used because it provides the information of vertical, horizontal, and diagonals connections. A threshold value is applied for each connected components, each connected component in the binary image is extracted and classified as a pigment network if its area is greater than threshold value, and it is given by following condition.

$$A(R_c) > A_{min} \tag{7}$$

where R_c is the *cth* connected region, $A(R_c)$ is its area, and A_{min} is a threshold value which was set as 900 in our experiments. By applying the previous condition, all connected components with small area can be classified as noise. An example of pigment network detection is shown in figure 2. The obtained image is a binary "net mask" of all connected regions according to following equation:

$$R = \bigcup_{c:A(R_c)>A_{min}} R_c \tag{8}$$

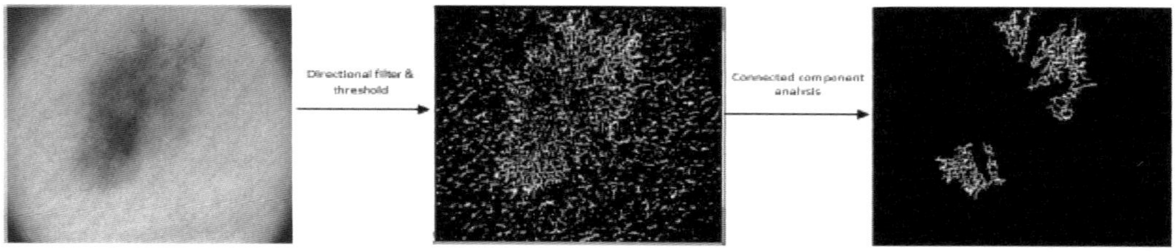

Figure 2. Example of pigment network detection

5. Feature extraction

The third stage of the proposed algorithm aims to extract the relevant features from the final binary images. The density and the regular distribution of the pigment networks are the main properties, and they can be used to identify the images where pigment network is present. By characterizing these two properties of pigment network, five features were defined:

Network/Lesion Ratio: This value computes the area of the network R (see section 4) with the area of the whole lesion

$$Network/Lesion = \frac{A(R)}{A(L)} \tag{9}$$

where $A(R)$ is the area of the detected pigment network and $A(L)$ is the area of the segmented lesion which has been segmented by an experienced dermatologist.

Network/Region Ratio: This value computes the area of the pigment network lines with the area of pigment network regions, because the pigment network consists of holes surrounded by pigmented lines. A simple morphological filling process is applied for all connected regions to get the pigment network region B_c (see figure 3(c)). Then a pigment network regions mask is defined as the union of all the regions.

$$B = \cup_c B_c \tag{10}$$

The network/regions ratio is then obtained as following:

$$Network/Region = \frac{A(R)}{A(B)} \tag{11}$$

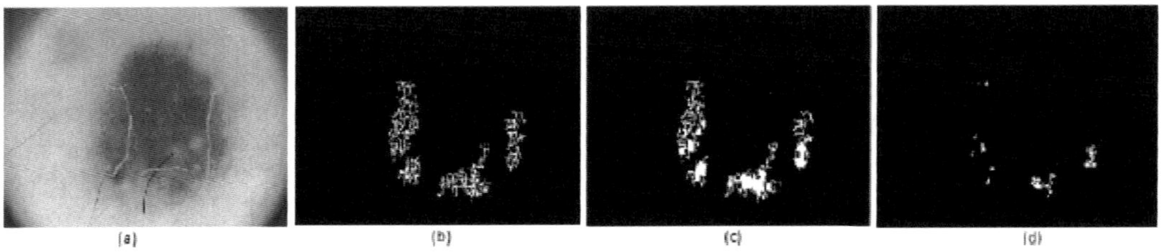

Figure 3. Pigment network masks. (a) original image with pigment network regions highlighted; (b) pigment network detection ;(c) pigment network regions-mask and (d) holes mask.

where $A(R)$ is the area of the detected network R and $A(B)$ is the area of all the pigment network regions present in B.

Number of Holes: As above mentioned, the pigment network consists of holes surrounded by pigment network lines. Therefore, this value computes the total number of the holes of the detected net, it can be simply obtained by subtract R from B. Where R is the detected image and B is the union of all regions. An example can be seen on figure 3(d).

Holes/Lesion Ratio: This value compares the number of the holes with the area of the lesion.

$$Holes/lesion = \frac{H_n}{A(L)} \tag{12}$$

where H_n is the total number of the holes in the detected image and $A(L)$ is the area of the segmented lesion L.

Holes/Region Ratio: This value compares the number of the holes with the area of the detected region.

$$Holes/Region = \frac{H_n}{A(B)} \tag{13}$$

where H_n is the total number of the holes in the detected image and $A(B)$ is the area of all the pigment network regions B.

6. Lesion classification

The final stage of the proposed approach aims to determine the images type, using a machine learning technique. The five features extracted from previous step are used to feed an Artificial Neural Network (ANN) as a binary classifier. As a target: (0) indicates an abnormal skin and (1) indicates a normal skin. A two-layer feed-forward network (hidden layer and output layer) are used with 10 neurons in hidden layer, and scaled conjugate gradient backpropagation with sigmoid function are used to train the network.

7. Experimental results

The proposed approach was tested on a dataset of 200 dermoscopy images (84 with pigment network and 116 without pigment network) from the database of Hospital Pedro Hispano, Matosinhos. The images are labelled with two classes: melanoma and non-melanoma. We evaluate the performance using mean square error and confusion matrices. Quantitative comparison between various methods is difficult because different datasets and criteria have been used. We have evaluated our method against Barata et al [3] as they have the same objectives and they are based on the same dataset. Table 1 shows the comparison result. It indicates that our method by using Sobel filter after applying Gabor filter on the image to extract almost all hairs achieved better results and outperforms Barata et al. [3] by three criteria namely: sensitivity, specificity and accuracy. With a note that the results presented in table 1 are obtained by repeating neural network pattern recognition tool 20 times and obtaining the average in all the above mentioned criteria. Our proposed method successfully classified the given image as either normal or abnormal skin, as well as located the pigment network in dermoscopy images which can assist dermatologists in visual diagnosis and making right diagnosis decisions.

Table 1. Results of the lesion classification SE-sensitivity, SP-specificity and AC-accuracy.

Work	SE	SP	AC
Barata et al	91.1%	82.1%	86.2%
Proposed method	92.3%	95.0%	90.0%

8. Conclusion

An approach of detection of pigment networks from dermoscopy images has been described. Gabor filter is combined with Sobel filter to perform hairs detection, and Barata's method has been used to extract pigment networks from dermoscopy images. The features extracted from the detected images are used to train a feed-forward neural network classifier. The proposed algorithm successfully achieved 92.3% sensitivity, 95.0% specificity and 90.0% classification accuracy.

Acknowledgment

The database used in this study was acquired by Teresa Mendonc, Pedro M. Ferreira, Jorge S. Marques, Andre R. S. Marcal, and Jorge Rozeira from Pedro Hispano Hospital. The authors wish to thank Catarina Barata for sharing.

References

[1] dermoscopy.org/atlas/base.htm
[2] Arroyo J L G and Zapirain B G 2014 Computers in biology and medicine 44 144-157
[3] Barata C, Marques J S and Rozeira J 2012 Biomedical Engineering, IEEE Transactions on 59 2744-2754
[4] Messadi M, Cherifi H and Bessaid A 2014 Journal of Convergence Information Technology 9 21
[5] Walter F M, Prevost A T, Vasconcelos J, Hall P N, Burrows N P, Morris H C, Kinmonth A L and Emery J D 2013 Br J Gen Pract 63 e345-e353
[6] Sadeghi M, Razmara M, Lee T K and Atkins M S 2011 Computerized Medical Imaging and Graphics 35 137-143
[7] Sadeghi M, Razmara M, Wighton P, Lee T K and Atkins M S 2010 Modeling the dermoscopic structure pigment network using a clinically inspired feature set Medical Imaging and Augmented Reality (Springer) pp 467-474
[8] Betta G, Di Leo G, Fabbrocini G, Paolillo A and Sommella P 2006 Dermoscopic image-analysis system: estimation of atypical pigment network and atypical vascular pattern Medical Measurement and Applications, 2006. MeMea 2006. IEEE International Workshop on (IEEE) pp 63-67
[9] Di Leo G, Liguori C, Paolillo A and Sommella P 2008 An improved procedure for the automatic detection of dermoscopic structures in digital elm images of skin lesions Virtual Environments, Human-Computer Interfaces and Measurement Systems, 2008. VECIMS 2008. IEEE Conference on (IEEE) pp 190-194
[10] Grana C, Cucchiara R, Pellacani G and Seidenari S 2006
[11] Silveira M, Nascimento J C, Marques J S, Marcal A R, Mendonca T, Yamauchi S, Maeda J and Rozeira J 2009 Selected Topics in Signal Processing, IEEE Journal of 3 35-45
[12] Criminisi A, Perez P and Toyama K 2003 Object removal by exemplar-based inpainting Computer Vision and Pattern Recognition, 2003. Proceedings. 2003 IEEE Computer Society Conference on vol 2 (IEEE) pp II-721

Detection of Melanoma Skin Cancer in Dermoscopy Images

Khalid Eltayef, Yongmin Li and Xiaohui Liu

Computer Science Department, Brunel University London, Uxbridge, UB8 3PH, UK

E-mail: khalid.eltayef@Brunel.ac.uk, Yongmin.Li@Brunel.ac.uk,
XiaoHui.Liu@brunel.ac.uk

Abstract. Malignant melanoma is the most hazardous type of human skin cancer and its incidence has been rapidly increasing. Early detection of malignant melanoma in dermoscopy images is very important and critical, since its detection in the early stage can be helpful to cure it. Computer Aided Diagnosis systems can be very helpful to facilitate the early detection of cancers for dermatologists. In this paper, we present a novel method for the detection of melanoma skin cancer. To detect the hair and several noises from images, pre-processing step is carried out by applying a bank of directional filters. And therefore, Image inpainting method is implemented to fill in the unknown regions. Fuzzy C-Means and Markov Random Field methods are used to delineate the border of the lesion area in the images. The method was evaluated on a dataset of 200 dermoscopic images, and superior results were produced compared to alternative methods.

1. Introduction

Skin cancers can be classified into melanoma and non-melanoma. Melanoma is a malignancy of the cells which gives the skin its color (melanocytes) and it can invade nearby tissues. Moreover, it spreads through the whole human body and it might cause to patient death and non-melanoma which is rarely spread to other parts of the human body. Malignant melanoma is the most aggressive type of human skin cancers and its incidence has been rapidly increasing [1] [2] [3] [4]. Nevertheless, it is also the most treatable type of skin cancer if detected or diagnosed at an early stage [5]. The diagnosis of melanoma in early stage is a challenging and fundamental task for dermatologists since some other skin lesions may have similar physical characteristics. Dermoscopy is considered as the widely common technique used to perform an in-vivo observation of pigmented skin lesions [6]. In early detection of malignant melanoma, dermoscopic images have great potential, but their interpretation is time consuming and subjective, even for trained dermatologists. Therefore, the need to build a system which can assist dermatologists to get right decision for their diagnosis has become very important.

Computer Aided Diagnosis (CAD) systems have been proposed by many different researchers to identify malignant melanomas in dermoscopy images. CAD systems based on medical knowledge try to mimic the performance of dermatologists for the detection of pigmented region [3] [6] [7]. Image segmentation stage is the most important one since it affects the accuracy of the subsequent steps. However, dermoscopic image segmentation and detect the lesion area is a challenging task since there is a smooth transition between the color of the lesion area and the background, it means there is low contrast between the pigmented lesion and background. Moreover, presence of dark hair covering the lesions and existence a few artifacts such as air bubbles and lighting reflection. This paper presents an automatic method for skin cancer detection on dermoscopy images. First, the image is pre-processed in order to remove the noise and enhance its quality. Second, Fuzzy C-Means (FCM) and Markov Random Field (MRF) are used to segment the pigmented lesion from images. Upon comparison, the proposed method provided good performance in achieving automatic image segmentation over dermoscopy images. The paper is organized as follows. Section 2 provides an overview of the methods developed previously. An overview of the MRF method is described in Section 3. The proposed method is described in details in Sections 4. The experiments and result analysis are discussed in Sections 5. Finally, conclusions are presented in Section 6.

2. Previous Work

Yukse et al. [8] proposed a method which used type-2 fuzzy logic technique for automatic threshold determination to detect the border of the pigmented skin lesion. Celebi et al. [9] presented a method to segment skin lesion in dermoscopy images through statistical region merging method. The method is a technique developed to segment images based on region growing and merging. A method to segment pigmented lesion images through iterative stochastic region merging has been implemented by Wong et al. [10]. K-means clustering method has been applied by Castillejos et al. [11]. The authors presented a novel approach to segment and detect the border of the skin lesion based on the wavelet transform for K-Means, Fuzzy C-Means and Cluster Preselection Fuzzy C-Means methods. Zhou et al. [12] proposed a new type of dynamic energy for the segmentation of pigmented skin lesions in dermoscopy images. The authors combined the classical gradient vector ow (GVF) model with the mean shift method. Threshold method is used by Pirnog et al [13] to extract the lesion area from health skin in dermoscopy images. The RGB image is converted into HSV representation, the image histogram of the saturation (S) image component is computed. Therefore, the authors selected the threshold value and determined the binary mask. A new mean shift approach based on FCM method is proposed by Zhou et al [14] to segment dermoscopy images. It is more effective than the FCM method and less computational time than the mean shift method. Abbas et al [15] proposed a novel perceptually oriented approach for melanoma border detection by combing region and edge-based segmentation techniques. A hill-climbing method is performed to detect the region-of-interest (ROI) and an adaptive thresholding is applied to determine the optimal lesion border. A new approach for lesion segmentation is proposed by Pennisi et at [16]. Closing operation is used to remove the hair and few outlier pixels. And therefore, two segmentation processes are implemented in parallel, providing two different images. One image is built by detecting the skin region and the other one is created by applying edge detected with Delaunay Triangulation. Finally, the authors combine these tow images to extract the final lesion area. Bi et al [17] proposed a new automatic melanoma detection method for dermoscopic images via multi-scale lesion-biased representation and joint reverse classification. They tried to represent the skin lesions using multiple of closely related histograms derived from different rotations and scales of the image.

3. Overview of the MRF Method

The MRF method is a statistical model, but can be used for segmentation methods, it was introduced in image analysis by Geman and Geman [18]. MRF theory provides a tool for modelling a vision problem within the Bayes framework using spatial continuity. The image pixels are indexed by a rectangular patch S and each image pixel s is characterized by the gray level y_s from the set $y = y_s : s \in S$. The labelling process consists of accurately labelling each image pixel $s \in S$ with a class label representing the pattern class in the image. A label set is defined as $\Lambda = 1, 2,..C$ where C is the number of classes. A labelling is indicated by $x = x_s : x_s \in \Lambda, s \in S$ where $x_s = l$ denotes that the class label l is assigned to the pixel s. The goal is to find the labelling $\hat{x}$ of the image, which is the estimation of the true but unknown labelling x^*. According to the MAP estimate [18] we have

$$\hat{x} = argmaxP(x|y) \quad (1)$$

According to Bayes theorem we have

$$P(x|y) = \frac{P(y|x)P(x)}{P(y)} \tag{2}$$

Where *P(x)* is the prior density of the labelling *x* and *P(y|x)* is the conditional probability density of the image *y*.

The prior probability of the image *P(y)* is independent of the labeling *x*, therefore, it is rewritten as follows:

$$\hat{x} = argmax\{P(y|x).P(x)\} \tag{3}$$

The Gaussian distribution is used with the assumption of the existence of Gaussian noise in images. Therefore, the possibility of pixel *s* with the assumption of belonging to class x_s is equal to y_s and can be calculated as follows:

$$P(y_s|x_s) = \frac{1}{\sqrt{2\pi\sigma_{xs}^2}}\exp\{-\frac{(y_s - \mu_{xs})^2}{2\sigma_{xs}^2}\} \tag{4}$$

Based on the conditional independence assumption of *y*, the conditional density *P(y|x)* takes the form of

$$P(y|x) = \prod_{s \in S} P(y_s|x_s) \tag{5}$$

Therefore, *P(y|x)* can be written as follows.

$$P(y|x) = \prod_{s \in S}[\frac{1}{\sqrt{2\pi}}\exp(-\frac{(y_s - \mu_{xs})^2}{2\sigma_{xs}^2} - \log(\sigma_{xs}))] \tag{6}$$

This can be written as

$$P(y|x) = \frac{1}{(2\pi)^{\frac{s}{2}}}\exp[-\sum_{s \in S}(\frac{(y_s - \mu_{xs})^2}{2\sigma_{xs}^2} + \log(\sigma_{xs}))] \tag{7}$$

where μ_{xs} and σ_{xs} indicate the mean and variance of the class *x*, respectively.

The prior model is based on 2D MRF and assumes that the adjacent pixels are probably having the same class label. The Hammersley-Clifford theorem [19] establishes a relation between the MRF and Gibbs distribution. According to this theorem, the prior model *P(x)* is given by a Gibbs distribution with respect to the neighbourhood system N and it takes the form of

$$P(x) = \frac{1}{Z}\exp[-\sum_{c \in C} V_c(x)] \tag{8}$$

where *Z* is the normalization constant or partition function, $V_c(x)$ is the potential function for clique *c* and *C* is the set of all cliques in the image. According to the equations (7) and (8), we can rewrite the posterior probability as:

$$P(x|y) = \propto exp[-U(x)] \tag{9}$$

Where the energy function *U(x)* has the form

$$U(x) = [\sum_{s \in S} \frac{(y_s - \mu_{xs})^2}{2\sigma_{xs}^2} + \sum_{s \in S} \log(\sigma_{xs}) + \sum_{c \in C} V_c(x)] \tag{10}$$

4. Proposed Method

The proposed approach is divided into two steps, image pre-processing and image segmentation. First, the pre-processing stage includes reflection artifact, hairs detection and removal. Second, the segmentation step includes applying the FCM method on images in order to get two benefits: estimate the initial parameters and segment the lesion area and implementing the MRF method based on the previous parameters. Full details will be described in the following section.

4.1. Dermoscopic Image Pre-processing

Pre-processing stage is responsible for detecting and reducing amount of artifacts from the image. In dermoscopy images, this step is mandatory since many dermoscopy images contain a lot of noise such as reflection artifacts and hairs which are covering the lesion area. Incorrect segmentation of pigmented lesion can be obtained if hairs covering the images and reflection artifacts are not removed. This process involves two key operations: hairs, reflection artifact detection and removal. The original RGB images were separated to three color components and the blue component is selected for its best discriminative capacity [20].

4.1.1. Reflection Detection. Reflection artifacts and air bubbles appear like noise in dermoscopy images. To detect this kind of noise, a simple threshold method is applied. Every single pixel (x,y) can be detected and classified as a reflection artifact if its intensity value is higher than threshold T_{R1} and if its intensity value minus the average intensity $I_{avg}(x,y)$ of its surrounding neighborhood is higher than threshold T_{R2} ,*i.e.*

$$\{I(x,y) > T_{R1}\} and \{(I(x,y) - I_{avq}(x,y)) > T_{R2}\}. \tag{11}$$

where I is the image, $I_{avg}(x,y)$ is the average intensity value in a local neighborhood of the selected pixel which is computed using a local mean filter with dimensions 11x11 and $T_{R1}=0.7$, $T_{R2}=0.098$.

4.1.2. Hair Detection and Inpainting. For effective segmentation in dermoscopy images, hairs should be extracted and removed. Many methods have been applied for hair removal and they were mostly based on adaptive thresholding and morphological operations or used median filter method [9]. The directional Gabor filters are implemented to extract the hairs from dermoscopy images. However, the parameters used in the Gaussian filters at each stage are different. A bank of 64 directional filters has been used to perform the hair detection. The image is filtered by each directional filter with different parameters and the difference of Gaussians is used followed by finding the local maximum at each pixel location. Therefore, the threshold method is applied to classify each pixel as either hair or background. A detailed explanation of the method can be obtained from [6].

After reflection artifacts and hairs are detected, their binary masks are multiplied by gray scale image. This operation leads to appearance of gaps which can be filled by propagating the information from known region which it is the neighborhood into the unknown region. Image inpainting method is implemented to fill in the unknown regions and remove the hairs with reflection artifacts. Patch priorities (data term and confidence term) are computed in the borders of the unknown regions based on their neighbours. And therefore, the patch with the highest priority is filled out with data extracted from the source region, and then the patch priorities are updated. This continues until no more gaps exist. Details of the method can be found in [21]. By this way all reflection artifacts and hairs are removed from dermoscopy images. Examples are illustrated in figure 1.

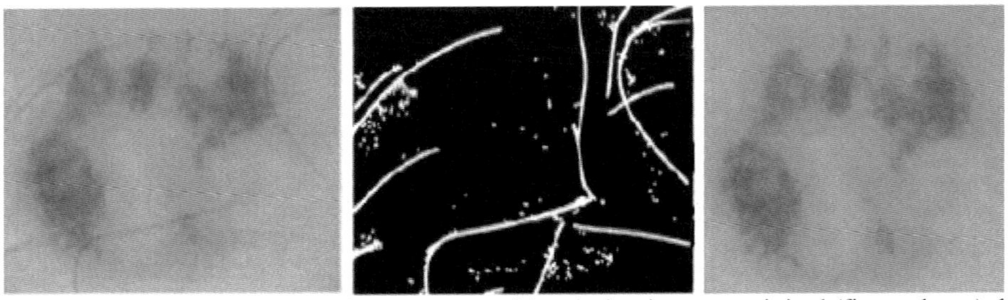

Figure 1. Examples of hair detection process and inpainting images: original (first column), hair detection (second column) and inpainted image (third column).

4.2. Skin Lesion Segmentation

The segmentation stage is one of the most important and challenging steps in dermoscopic image processing. It must be fast and accurate because the subsequent steps such as feature extraction, classification and final decision are dramatically dependent on the performance of the segmentation step. As we mentioned previously, segmentation process for dermoscopy images is extremely difficult due to existence of several factors such as: the low contrast between the lesion and the healthy skin, variance of colors inside the pigmented region and other artifacts. Useful information can be obtained by applying suitable segmentation method. FCM and MRF were incorporated to perform the final segmentation of the images.

The FCM method is used to initiate the segmentation process. The pixels of an input image are divided into two clusters: cancerous pixels as the foreground (ROI) and normal skin pixels as the background. The major drawback with the FCM method is that it deals with pixels individually by their intensity values only, so lacks the capacity to model the overall appearance of a local neighbourhood region. In order to address this issue, The MRF method is implemented to refine the previous segmentation of the image. The main goal of the image segmentation using the MRF method is to minimize the energy function or maximize the probability of pixel allocation to a cluster by using Maximum A Post Priority (MAP) equation (3). The iterated conditional modes (ICM) method is performed in order to minimize the energy function. In this paper we assume that one pixel has 8-neighbors. Then the clique potential is defined on pairs of neighbouring pixels:

$$V_c(xi, xj) = (1 - i_{xi,xj}) \tag{12}$$

where $i_{xi,xj} = 0$ if $xi \neq xj$ and 1 if $xi = xj$

To refine the image segmentation, we obtained the mean and the variance which can be used as initial parameters for the MRF method. The MRF method is then iterated as follows. Each pixel is assigned to a cluster based on the highest probability $P(y|x)$ using equation (7). Then calculate the prior probability using equation (8). All pixels will be assigned again to different classes by getting the minimum cost, equation (10). This process continues until there is no further change between clusters. Finally, Morphological operations such as filling were used to fill in the small holes inside the segmented lesion. As well as, several parts that related to the background were removed, by comparing its area with the lesion area.

Table 1: Segmentation performance on the complete dataset SE-sensitivity, SP-specificity and AC-accuracy.

Method	SE	SP	AC
FCM	88.0%	95.1%	92.0%
Proposed method	93.2%	98.0%	94.0%

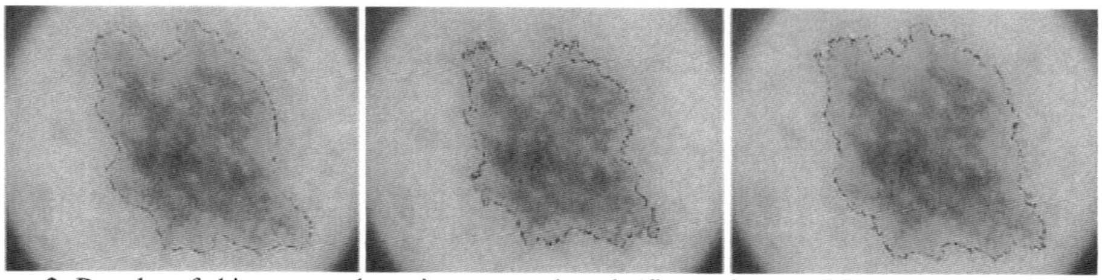

Figure 2: Results of skin cancer detection: ground truth (first column), FCM (second column) and proposed method (third column).

5. Results and Discussions

The proposed approach was tested on the PH2 dataset [22] which provides 200 dermoscopic images, including 80 common nevi, 80 atypical nevi, and 40 melanomas. Manually segmentations were also available and used as ground truth. Quantitative comparison between various methods is difficult because different datasets and criteria have been used. Three criteria namely: sensitivity, specificity and accuracy are used to evaluate the performance of our proposed method. Segmentation of skin cancer was carried out using FCM and proposed method. And therefore, the comparison of these methods was performed with the lesions obtained by expert dermatologist in order to evaluate the performance of our method. Figure 2 illustrates the results of the manual segmentation by expert radiologist together with the results by two methods. In general, the proposed approach has the best performance in terms of the accuracy. For instance, in the second column, the FCM method did not detect the whole lesion area which leads to misclassification. As well as, the edges of the test images are not close to the real boundaries of the lesions, which mean part of the lesion was classified as background. Images in the third column show the results obtained by our method. As it can be seen, the obtained edges are more close to the actual boundaries of the skin lesions and its shape is very similar to that of the ground truth compared to the previous method. The experimental results of the total dermoscopy images are described in table 1. All results confirm that the proposed approach outperforms the FCM method. Eventually, the obtained experimental results indicated that the proposed method is suitable to apply for detection of skin cancer.

6. Conclusions

An approach to the detection of melanoma skin cancer from dermoscopy images has been presented in this paper. To facilitate the process of image enhancement and segmentation, hairs and several artifacts were detected and removed by applying the Gabor filter and image inpainting respectively. Regions of melanoma cancer are extracted by combining the Fuzzy C-Means and Markov Random Field methods. The particles obtained by applying FCM method are preserved, then the binary image is created and incorporated with the MRF method. The results of the proposed method were compared with the ground truth lesions. Our experimental results indicated that the proposed method provided high accuracy of skin lesion segmentation; it successfully achieved 93.2% sensitivity, 98.0% specificity and 94.0% accuracy.

Acknowledgement

The database used in this study was acquired by Teresa Mendonc, Pedro M. Ferreira, Jorge S. Marques, Andre R. S. Marcal, and Jorge Rozeira from Pedro Hispano Hospital [22]. The authors wish to thank Catarina Barata for sharing.

References

[1] Xu L, Jackowski M, Goshtasby A, Roseman D, Bines S, Yu C, Dhawan A and Huntley A 1999 Image and Vision Computing 17 65-74

[2] Smaoui N and Bessassi S 2013 International Journal of Computer Vision and Signal Processing 3 10-17

[3] Arroyo J L G and Zapirain B G 2014 Computers in biology and medicine 44 144-157

[4] Trotter S C, Sroa N, Winkelmann R R, Olencki T and Bechtel M 2013 Journal of Clinical & Aesthetic Dermatology 6

[5] Sadeghi M, Razmara M, Lee T K and Atkins M S 2011 Computerized Medical Imaging and Graphics 35 137-143

[6] Barata C, Marques J S and Rozeira J 2012 IEEE transactions on biomedical engineering 59 2744-2754

[7] Messadi M, Cherifi H and Bessaid A 2014 Journal of Convergence Information Technology 9 21

[8] Yuksel M E and Borlu M 2009 IEEE Transactions on Fuzzy Systems 17 976-982

[9] Emre Celebi M, Kingravi H A, Iyatomi H, Alp Aslandogan Y, Stoecker W V, Moss R H, Malters J M, Grichnik J M, Marghoob A A, Rabinovitz H S et al. 2008 Skin Research and Technology 14 347-353

[10] Wong A, Scharcanski J and Fieguth P 2011 IEEE Transactions on Information Technology in Biomedicine 15 929-936

[11] Castillejos H, Ponomaryov V, Nino-de Rivera L and Golikov V 2012 Computational and mathematical methods in medicine 2012

[12] Zhou H, Schaefer G, Celebi M E, Iyatomi H, Norton K A, Liu T and Lin F 2010 Skin lesion segmentation using an improved snake model 2010 Annual International Conference of the IEEE Engineering in Medicine and Biology (IEEE) pp 1974-1977

[13] Pirnog I, Preda R O, Oprea C and Paleologu C 2015 Automatic lesion segmentation for melanoma diagnostics in macroscopic images Signal Processing Conference (EUSIPCO), 2015 23rd European (IEEE) pp 659-663

[14] Zhou H, Schaefer G, Sadka A H and Celebi M E 2009 IEEE Journal of Selected Topics in Signal Processing 3 26-34

[15] Abbas Q, Garcia I F, Emre Celebi M, Ahmad W and Mushtaq Q 2013 Skin Research and Technology 19 e490-e497

[16] Pennisi A, Bloisi D D, Nardi D, Giampetruzzi A R, Mondino C and Facchiano A 2016 Computerized Medical Imaging and Graphics

[17] Bi L, Kim J, Ahn E, Feng D and Fulham M 2016 Automatic melanoma detection via multi-scale lesion-biased representation and joint reverse classification 2016 IEEE 13th International Symposium on Biomedical Imaging (ISBI) (IEEE) pp 1055-1058

[18] Geman S and Geman D 1984 IEEE Transactions on pattern analysis and machine intelligence 721-741

[19] Besag J 1986 Journal of the Royal Statistical Society. Series B (Methodological) 259-302

[20] Silveira M, Nascimento J C, Marques J S, Marcal A R, Mendonca T, Yamauchi S, Maeda J and Rozeira J 2009 IEEE Journal of Selected Topics in Signal Processing 3 35- 45

[21] Criminisi A, Perez P and Toyama K 2003 Object removal by exemplar-based inpainting Computer Vision and Pattern Recognition, 2003. Proceedings. 2003 IEEE Computer Society Conference on vol 2 (IEEE) pp II-721

[22] Mendonca T, Ferreira P M, Marques J S, Marcal A R and Rozeira J 2013 Ph 2-a dermoscopic image database for research and benchmarking 2013 35th Annual International Conference of the IEEE Engineering in Medicine and Biology Society (EMBC) (IEEE) pp 5437-5440